Invading Nature - Springer Series in Invasion Ecology

Volume 8

More information about this series at http://www.springer.com/series/7228

Matej David • Stephan Gollasch
Editors

Global Maritime Transport and Ballast Water Management

Issues and Solutions

Springer

Editors
Matej David
Dr. Matej David Consult
Korte, Izola, Slovenia

Stephan Gollasch
Gollasch Consulting (GoConsult)
Hamburg, Germany

Additional material can be downloaded from http://extras.springer.com

ISBN 978-94-017-9366-7 ISBN 978-94-017-9367-4 (eBook)
DOI 10.1007/978-94-017-9367-4
Springer Dordrecht Heidelberg New York London

Library of Congress Control Number: 2014955396

Printed on acid-free paper

Springer is part of Springer Science+Business Media (www.springer.com)

Foreword

It is widely accepted that more than 90 % of cargoes in international trade are safely transported by ships throughout the world, and the carriage of ballast water plays an essential role in guaranteeing the safe navigation and operation of such ships. At the same time, though, ballast water poses an environmental threat by serving as a vehicle to transport live unwanted species across the oceans. According to different estimates, up to 10 billion tonnes of ballast water is transported around the world by ships annually, and several thousands of microbial, plant and animal species may be carried globally in ballast water. When these species are discharged into new environments, they may become established and can also turn invasive, thus severely disrupting the receiving environments with the potential to harm human health and the local economy. The global economic impacts of invasive marine species are difficult to quantify in monetary terms, but are likely to be of the order of tens of billions of US dollars per year. Consequently, the introduction of harmful aquatic organisms and pathogens to new environments, including via ships' ballast water, has been identified as one of the four greatest anthropogenic threats to the world's oceans.

The International Maritime Organization (IMO), the United Nations' specialized agency responsible for the safety and security of shipping and the prevention of marine pollution by ships, first responded to this issue by developing guidelines and recommendations aimed at minimizing the transfer of live organisms and pathogens by exchanging ballast water at sea, since experience had shown that ballast water exchange in deep waters reduces the risk of species transfers. At the same time, it was recognized that higher levels of protection could be reached with other protective measures, e.g. through ballast water treatment.

It also became clear at the time that a self-standing international legal instrument for the regulation of ballast water management would be necessary to avoid regulatory action by authorities at national, provincial and even local levels. This could have resulted a fragmented, patchwork-like ballast water management approach which had to be avoided by all possible means in an eminently cross-border industry like shipping. Consequently, IMO developed the globally applicable International Convention for the Control and Management of Ships' Ballast Water and Sediments (BWM Convention), which was adopted in February 2004 at a diplomatic conference in London. This instrument will enter into force 12 months after the date on which more than 30 states, with combined merchant fleets not less than 35 % of the gross tonnage of the world's merchant shipping, have ratified it. As of December 2013, 38 states representing 30.38 % of the world merchant shipping gross tonnage had ratified the BWM Convention.

IMO has also joined forces with the Global Environment Facility (GEF) and the United Nations Development Programme (UNDP) to implement the Global Ballast Water Management Programme (GloBallast), which was followed by the GloBallast Partnerships Programme. A key objective of these programmes is to provide assistance, mainly to developing countries, for the implementation of the BWM Convention.

The BWM Convention introduces new requirements for port States and ships all around the world, although its implementation is a complex process. Despite the global efforts of industry, member states and IMO over many years, efficient, economically feasible, environmentally friendly and safe methods of preventing the translocation of harmful organisms via ballast water are still being developed. The implementation of some of the ballast water management methods becomes even more complicated due to the difficulties encountered in their applicability because of the differences in shipping patterns and geographical specifics. The shipping industry on one side and coastal states on the other are confronted with serious obstacles when trying to find simple solutions to the extent that turnkey solutions may need to be developed on a case-by-case or port-by-port basis, this without causing an excessive burden to the shipping industry and, consequently, to the global trade.

With great interest and appreciation, I note that this book summarizes comprehensively the current knowledge regarding the multifaceted ballast water issue. It provides an overview of the possible solutions to the complex issue of ballast water management and also outlines consequences and implications to address the ballast water "problem" following the provisions of the BWM Convention. It delivers an excellent overview regarding ships' ballast operations; environmental and other aspects of the issue; and international requirements as well as an in-depth analysis of possible ways to approach or manage the challenge in the most effective way. The editors and main authors are scientists from different disciplines, including university professors with maritime and biological expertise, who have been involved or are leading researchers in this field and have participated in the policymaking processes at IMO, at national and regional levels.

I am convinced that this book will be an invaluable tool for university students interested in marine environment protection and, most of all, will provide much-needed assistance to maritime administrations when trying to ratify and implement the BWM Convention.

Motril, Spain Miguel Palomares
December 2013
Former Director of the IMO Marine Environment Division

Foreword

The rapid growth of global economic trade and the seemingly unlimited human mobility around the world, commencing in the mid-1800s, opened many windows of opportunity for trading goods not only between population centers but also into remote places of the world. In the twenty-first century, transportation by trucks, trains and planes is surpassed by far in volume and distance travelled by the shipping and boating industries – trans- and inter-oceanically via container ships, bulk carriers, and tankers and coastally by both cargo vessels and a vast fleet of recreational and fishing vessels. It thus does not come as a surprise that the issue of unintentional transmission of organisms (including pathogens) across oceans and continents has reached a new dimension that is of serious concern to maintain and sustain ecosystem integrity and ecosystem services.

Aquaculturists in coastal and marine waters have been aware of the problems of transfers of exotic species since the end of World War II, being especially affected by the unintentional introductions of fouling organisms and disease agents. While the aquaculture industry was often blamed for self-contamination (which was certainly a valid point and partially true with disastrous examples), we know today that many of the problems with exotic fouling organisms affecting aquaculture and other stakeholders also originated from the shipping industry through the long-term uncontrolled release of ballast water and transfer via hull, sea chest, and other fouling.

Aquatic biodiversity and environmental health have been on the agenda of ecologists for decades. Most concern has been expressed for the potential of "loss of biodiversity" in light of increasing anthropogenic pressures. This concern has been expressed by many organizations, while national and international regulatory authorities try to include biodiversity issues into environmental management schemes. However, early on in the biodiversity debate, fewer scientists pointed to the fact that we are not only dealing with the "loss of biodiversity" but also with a "change" or "increase" of species diversity due to human intervention and that these changes may also be considered as threats to ecosystem stability and services. Thus, some recent literature has argued that *adding* species to natural communities is beneficial, but these arguments typically do not address the fundamental changes that accompany such additions, such as the often vast decrease in the abundance of native species (even if these still remain, somewhere) and the concomitant cascades in altering energy flow, competition, and predator–prey relationships.

Australia, New Zealand, the United States and Canada provided pioneering research work in the area of marine bioinvasions and ballast water by delineating the dimensions of the problem commencing in the 1970s and 1980s. In Europe and the rest of the world, studies on the dimension of the problem started at least a decade later. Commencing in the 1990s, international conventions and organizations (such as the United Nations' International Maritime Organization (IMO), responsible for the safety and security of global shipping and the prevention of marine pollution by ships) began to be concerned about and involved in the promulgation of regulatory frameworks to minimize the risks associated with the increasingly huge volumes of ballast water transfer and biofouling on commercial and recreational vessels. Similarly, over the past two decades, national regulatory frameworks have been developed in a number of countries. All of these management scenarios, however, depend on sound and solid research results to properly and effectively reduce the risk of transfer of (potentially) harmful organisms.

The authors of this book are among the pioneers who intensively studied the role of shipping and have been at the forefront (in cooperation with others worldwide) to promote the development of methods on how to (a) monitor the fate of non-indigenous species transferred by ballast water, (b) standardize mitigation and control procedures for practical application by industry and regulatory authorities, and (c) develop the much-needed risk assessment and "hotspots" identification where protective action is needed most. Their work, together with many other scientists and organizations, contributed to the preparation of the International Convention for the Control and Management of Ships' Ballast Water and Sediments, adopted by IMO in 2004.

This book is very timely, providing a comprehensive state-of-the-art synthesis: during the past two decades, tremendous progress had been made in research to understand both the importance of these transmission vectors and the environmental risks associated with them. The authors have contributed greatly both through original research and practical testing and extensive review work to our present knowledge on mitigation strategies and treatment procedures. The present volume builds and expands on previous overviews where the authors have been instrumental in providing concepts and guidance to help developing solutions to the problem.

The undersigned, having been involved in cooperative work with the authors over many years, are pleased to see this progress reported and summarized in a format that will not only be of great value to experts in the field but also provide both the background and the current state of knowledge to a much broader audience interested in issues related to the unintentional global transfer of species. The engagement of a wide audience via this book's modern and practical summary of global ballast water management will assist greatly in encouraging all stakeholders to more vigorously implement the required management schemes that will reduce invasions and thus their impact on our environment and economy.

Neu Wulmstorf, Germany Harald Rosenthal
Williams College, Mystic Seaport James T. Carlton
Marine Studies Program, Mystic, CT, USA
January 2014

Contents

Contributors

Matej David Dr. Matej David Consult, Korte, Izola, Slovenia
E-mail: matej.david@siol.net

Brian Elliott European Maritime Safety Agency, Lisbon, Portugal
E-mail: brian.elliott@emsa.europa.eu

Stephan Gollasch Gollasch Consulting (GoConsult), Hamburg, Germany
E-mail: sgollasch@aol.com

Chad Hewitt Faculty of Science and Engineering, University of Waikato, Hamilton, New Zealand
E-mail: chewitt@waikato.ac.nz

Karina Keast Australian Government Department of Agriculture, Fisheries and Forestry, Canberra, Australia
E-mail: karina.keast@daff.gov.au

Erkki Leppäkoski Department of Biosciences, Åbo Akademi University, Turku, Finland
E-mail: erkki.leppakoski@gmail.com

Dan Minchin Marine Organism Investigations, Ballina, Killaloe, Co Clare, Ireland
E-mail: moiireland@yahoo.ie

Naomi Parker Ministry for Primary Industries; Strategy, Systems and Science Policy Directorate; Policy Branch, Wellington, New Zealand
E-mail: naomi.parker@mpi.govt.nz

Chris Wiley Transport Canada Marine Safety, Sarnia, Canada
E-mail: chris.wiley@dfo-mpo.gc.ca

Abbreviations

BWDA	Ballast water discharge assessment
BWE	Ballast water exchange
BWEA	Ballast water exchange area
BWM	Ballast water management
BWM Convention	International Convention for the Control and Management of Ships' Ballast Water and Sediments
BWMS	Ballast water management systems
BWRA	Ballast water risk assessment
BWRF	Ballast water reporting form
BWS	Ballast water sampling
cfu	Colony forming units
CME	Compliance monitoring and enforcement
D-1 standard	Ballast Water Exchange Standard (BWM Convention)
D-2 standard	Ballast Water Performance Standard (BWM Convention)
DSS	Decision support systems
HAOP	Harmful aquatic organisms and pathogens
IMO	International Maritime Organization
LME	Large marine ecosystem
MARPOL	International convention for the prevention of pollution of ships
MEPC	Marine Environment Protection Committee
NM	Nautical miles
PRC	Pump rate capacity
PSA	Port State authority
PSC	Port State control
psu	Practical salinity units
RA	Risk assessment

Introduction

Matej David and Stephan Gollasch

Abstract Today global shipping transports over 90 % of the world's overseas trade and trends anticipate that it will continue to play an increasing role world-wide. Shipping operations inevitably include also pressures on natural environments. The most recent waterborne threat is the transfer of harmful aquatic organisms and pathogens with ballast water and sediments releases, which may result in harmful effects on the natural environment, human health, property and resources globally. The significance of the ballast water issue was already addressed in 1973 by the International Maritime Organization (IMO) as the United Nations specialised agency for the regulation of international maritime transport at the global scale. Committed work by many experts, scientists, politicians, IGOs and NGOs at IMO resulted in the adoption if the *International Convention for the Control and Management of Ships' Ballast Water and Sediments* (BWM Convention) in February 2004, which is now to be ratified and implemented. Work on ballast water management issues has also shown to be very complex, hence there are no simple solutions. Nevertheless, the BWM Convention represents a globally uniform framework for the implementation of ballast water management measures, and different supporting tools like risk assessment and decision support systems have been developed to support its efficiency. In this chapter the reader is introduced to various ballast water issues and responses to it. The intention of this book and the overview of its content is also presented.

Keywords Vessels • Ballast water • Ballast water management • Harmful aquatic organisms and pathogens • International maritime organization • Ballast water management convention • Risk assessment • Decision support system

---

M. David (✉)
Dr. Matej David Consult, Korte, Izola, Slovenia
e-mail: matej.david@siol.net

S. Gollasch
Gollasch Consulting (GoConsult), Hamburg, Germany
e-mail: sgollasch@aol.com

M. David, S. Gollasch (eds.), *Global Maritime Transport and Ballast Water Management*, Invading Nature - Springer Series in Invasion Ecology 8,
DOI 10.1007/978-94-017-9367-4_1

General Introduction

The continuous intensification of the globalization of trade and production increased the demand for new, faster and more frequent linkages among trading and commodity production areas. These transport demands can only be met by maritime shipping because of its inherent technical and technological advantages and properties. The shipping industry has reacted to these needs with new and more frequent connections, increased vessels cargo and passenger capacity, and new vessel types and technologies.

Today global shipping transports over 90 % of the world's overseas trade (IMO 2013). Future trends anticipate that global and local shipping play an increasing role world-wide. Intensified shipping and related developments has also resulted in disasters of unprecedented dimensions. Widely known examples include the *Titanic* in 1912, *Torrey Canyon* in 1967, *Amoco Cadiz* in 1978, *Exxon Valdez* in 1989, *Estonia* 1994, *Sea Empress* in 1996, *Erika* in 1999, and *Prestige* in 2003 (David 2007). Such disasters resulted in the loss of human lives, property and/or caused significant damage to coastal ecosystems. In addition another inevitable consequence of shipping disasters is the pollution of the environment caused by a variety of pollutants.

Apart from harmful effects as consequences of shipping disasters, regular shipping activities cause other negative environment effects, e.g., sea pollution through the discharges of oily water and sewage from vessels, air pollution from exhaust gases emitted by the vessel's machinery, pollution of water and marine organisms by toxic protective underwater hull coatings (anti-fouling paints), and one of the most recent waterborne threats – the transfer of harmful aquatic organisms and pathogens (HAOP) with ballast water and sediments releases (e.g., Carlton et al. 1990, 1995; Gollasch 1996; Ruiz et al. 1997, 1999, 2000; Carlton 1999; Hewitt 2002; Hewitt et al. 1999; David et al. 2007; Nellemann et al. 2008). Given its 'mysterious' nature in combination with severely harmful effects on the natural environment, human health and the global economy, the problem has attracted attention of scientists and the public worldwide, which was particularly advanced in the 1980s and 1990s due to severe impacts of only a few introduced species.

What is the problem? Vessels need additional weight as a precondition for safe navigation in cases when they are not carrying cargo or are not fully or equally laden. The weight adding material is referred to as ballast. Historically, ballast was solid (e.g., sand, rocks, cobble, iron). With the introduction of iron, replacing wood, as basic vessel building material in the middle of the nineteenth century, the doors were opened to new ballasting technologies. Loading of water (i.e., ballast water) in cargo holds or ballast water tanks has shown to be easier and more time efficient compared to solid ballast. Therefore, water as ballast was adopted as a new practice of increasing importance. Many different types of vessels have different structures of ballast tanks, as well as different ballast system capacities. Vessels ballast water operations are related to vessel type, vessel construction, cargo operations and weather conditions. However, there are no clear limits among all these factors, but the decision on ballast water operations is under the discretion of the chief officer and direct control of the captain, who is responsible for the vessels stability and safety. Nowadays vessels fundamentally rely on ballast water for safe operations. A model for the assessment of ballast water discharges has been developed and tested. It is estimated that global ballast water discharges from vessels engaged in the international seaborne trade in 2013 would be approximately 3.1 billion tonnes (see chapter "Vessels and Ballast Water").

Water loaded as ballast from a vessel's surrounding environment contains suspended matter and organisms. Ballast water sampling studies have shown that various bacteria, plant and animal species can survive in the ballast water and ballast tank sediment (e.g., Medcof 1975; Carlton 1985; Williams et al. 1988; Locke et al. 1991; Hallegraeff and Bolch 1991; Carlton and Geller 1993; Gollasch 1996; Gollasch et al 2000, 2002; Hamer et al. 2001; Murphy et al. 2002; David et al. 2007; McCollin et al. 2008; Briski et al. 2010, 2011). Some organisms stay viable in ballast tanks for several months duration (e.g., Gollasch 1996; Gollasch et al. 2000) or longer (Hallegraeff and Bolch 1991). Estimates indicate that 3,000–4,000 (Carlton and Geller 1993; Gollasch 1996) and possibly even 7,000 (Carlton 2001) different species are transferred daily via ballast water. Species types found range from unicellular algae to fish (e.g., Gollasch et al. 2002; David et al. 2007). Of those, more than 850 are known as successfully introduced and established into new regions (Hayes and Sliwa 2003). It was concluded that each vessel has the potential to introduce a species and that any single introduced species has the potential to cause a significant negative impact to the recipient environment (e.g., Gollasch 1996). Therefore, loading ballast water and sediment in one port and discharging in another represents a potential risk to transfer HAOP into new environments (see chapter "The Transfer of Harmful Aquatic Organisms and Pathogens with Ballast Water and Their Impacts").

The United Nations also recognised the transfer of HAOP as one of the four greatest anthropogenic pressures to the world's oceans and seas, causing global environmental changes, and posing a threat also to human health, property and resources. Ballast water is one of the prime vectors of this global issue (e.g., Carlton 1985, 1989, 1992, 1996a, b; Wiley 1997; Gollasch et al. 2002; Bax et al. 2003; Bailey et al. 2005; Davidson and Simkanin 2012). The unwanted impacts caused by introduced species are manifold and include changes of species biogeography, biodiversity modifications, introduction of predators, bloom-forming harmful algae, ecosystem engineers, parasites and disease agents resulting in economic problems of marine resource users, such as loss in fisheries, fouling of industrial water pipes and on fishing or aquaculture gear. Even negative impacts on human health are reported because, e.g., harmful algae causing amnesic, diarrhetic or paralytic shellfish poisoning and *Vibrio cholerae* as well as other disease agents were found in ballast water (e.g., Hallegraeff 1993, 1998; Rigby and Hallegraeff 1994; Carlton 1996a, b; Ruiz et al. 2000; van den Bergh et al. 2002; Hayes and Sliwa 2003; Bauer 2006; Gollasch et al. 2009; Romero et al. 2011). In total more than 1,000 aquatic non-indigenous and cryptogenic[1] species are known from Europe (Gollasch 2006; Vila et al. 2010), and Hewitt and Campbell (2010), Hayes and Gollasch

[1] Cryptogenic species are species which cannot reliable be assigned as being non-indigenous or native because their origin is uncertain (Carlton 1996a, b).

(both unpublished), suggest >2,000 aquatic non-indigenous species have been introduced world-wide. The monetary impact caused by these species is difficult to quantify (van den Bergh et al. 2002). However, comprehensive studies concluded that the estimated yearly damage or control costs of introduced aquatic non-indigenous species is $14.2 billion in the USA (Pimentel et al. 2005) and costs for repair, management and mitigation measures of such species in Europe was estimated to more than 1.2 billion Euro annually (Shine et al. 2010) (see chapter "The Transfer of Harmful Aquatic Organisms and Pathogens with Ballast Water and Their Impacts").

Following the primary species introduction from, e.g., the coasts of one continent to another, secondary spread within the recipient continents coastal waters may occur because introduced species may be further transferred by, e.g., coastal or local shipping, pleasure craft, fisheries etc., or may also spread by natural means (e.g., Minchin et al. 2005; Simkanin et al. 2009; Rup et al. 2010; Bailey et al. 2011; Darling et al. 2012; David et al. 2013) thereby increasing their impact (see chapter "The Transfer of Harmful Aquatic Organisms and Pathogens with Ballast Water and Their Impacts").

The significance of the ballast water issue was already addressed in a 1973 International Maritime Organization (IMO) Resolution (IMO 1973). IMO as the United Nations specialised agency for the regulation of international maritime transport at the global scale, was tasked to deal with this issue further. After more than one decade of intensive and committed work by many experts, scientists, politicians, IGOs and NGOs at IMO, the final text of the *International Convention for the Control and Management of Ships' Ballast Water and Sediments* (BWM Convention) was completed and adopted in February 2004 at a diplomatic conference in London (IMO 2004; Gollasch et al. 2007). The BWM Convention introduced new BWM related requirements for port States and vessels all around the world. However, the implementation of this Convention is far from being simple. After the adoption of the BWM Convention several countries and regions have implemented (voluntary) ballast water management approaches (Gollasch et al. 2007; David 2007; David and Gollasch 2008) (see chapters "Policy and Legal Framework and the Current Status of Ballast Water Management Requirements" and "Ballast Water Management Under the Ballast Water Management Convention").

Due to global efforts of industry, Member states and IMO, efficient, financially feasible, environmentally friendly and safe methods of preventing the translocation of HAOP via ballast water were developed. More than 30 ballast water management systems (BWMS) have already been certified (type approved) so that most vessels can today be equipped with such systems. We are aware that this is a very fast developing area and market, at least 20 more systems are currently in the certification process (see chapter "Ballast Water Management Systems for Vessels").

The BWM Convention is at the moment of this writing not yet in force, but does today represent a solid and uniform framework for preventive measures to avoid HAOP introductions and it needs to be implemented by individual countries or joint approaches. The BWM Convention enters into force 12 months after the date on which more than 30 states, with combined merchant fleets not less than 35 % of the gross tonnage of the world's merchant shipping, have signed this Convention. As of December 2013, 38 states ratified the BWM Convention, representing 30.38 % of the world merchant shipping gross tonnage (for an update visit Status of Conventions at www.imo.org).

Nonetheless it must be emphasized that efficient ballast water management (BWM) does not imply the prevention of HAOP introductions at any cost, thereby laying an additional burden on and generating higher costs for the shipping industry. Undoubtedly, the cost of prevention should not be higher than the benefits it generates.

Conditioned by the lack of on board installed BWMS on existing vessels, ballast water exchange (BWE) is today the most widespread available BWM method also approved by the BWM Convention. Nevertheless, ballast water exchange has drawbacks which make it inefficient or even impracticable under certain conditions (e.g., on shorter voyages where "intended routes" are too close to the shore, attain insufficient water depths, a lack of knowledge of the presence of HAOP in the water exchange area). Further, other issues related to an efficient BWM system arise which are outside of the vessels' responsibility, e.g., targeting of vessels for ballast water sampling as part of port State compliance control procedures.

As a result, countries wishing to protect their seas, human health, property and resource from the introduction of HAOP with ballast water are confronted with a significant challenge. Given that BWM requirements may result in inefficiencies, lower safety margins and higher costs in the shipping industry, the reasons described above make the 'blanket approach' (i.e., mandatory BWM for all ships) unjustifiable in a range of different local conditions. An alternative to the blanket approach is the 'selective approach' where BWM is required for selected vessels. This selection should be based on a suite of information needs and procedural decisions to aid transparent and robust BWM decisions. Such systems have been developed in a variety of applications where a large number of complex decisions must be made in a consistent, transparent and defensible manner. These systems are typically referred to as decision support systems (DSS). Such a DSS as applied to BWM implies adjusting the intensity level of BWM measures to each voyage based on risk assessment (RA), and recommends also compliance monitoring and enforcement (CME) actions (see chapters "Ballast Water Management Under the Ballast Water Management Convention", "Ballast Water Management Systems for Vessels", "Risk Assessment in Ballast Water Management", "Ballast Water Sampling and Sample Analysis for Compliance Control" and "Ballast Water Management Decision Support System").

A BWM DSS provides essentially needed support to responsible agencies for the implementation of effective BWM measures. The introduction of BWM practices adds burden and costs mostly to the shipping industry, on the other side, their efficiency is critical. In light of these, the BWM DSS needs to provide for (David 2007):

- an effective protection against the introduction of HAOP;
- proper RA as one of the key elements of the BWM DSS;
- local specifics are addressed in direct relation with the effectiveness of the BWM (e.g., geographical, hydrological, meteorological, important resources, shipping patterns, regulatory regime);

- a selection of most effective and safe BWM methods according to the RA;
- the consideration of impacts to the shipping industry (including safety);
- the consideration of impacts on international trade;
- timely decision making;
- the reduction of subjectiveness in the decision process; and
- a consistent and transparent decision making process.

A uniform DSS methodology and RA concerning HAOP introductions via ballast water has not yet been developed. Several foundations have already been laid, e.g., Australian DSS (Hayes and Hewitt 1998, 2000), GloBallast[2] Ballast Water Risk Assessment (GloBallast 2003), Det Norske Veritas (DNV) Environmental Ballast Water Management Assessment – EMBLA (Behrens et al. 2002), and BWM RA and DSS for Slovenia (David 2007). More recently BWRA according to the BWM Convention requirements was developed for HELCOM (David et al. 2013) and OSPAR. Currently BWRA and BWM DSS for European Seas is being developed under the EU-funded VECTORS[3] project, and for the Adriatic Sea under the IPA Adriatic strategic project BALMAS.[4] Yet the complexity and intrinsically modern character of the problem leaves several questions, as revealed by the inefficiency of these applied systems, unanswered. The need for answers bears vital significance for the international environment, the goal being the future implementation of an efficient BWM system in tandem with considerations for a sustainable shipping industry (see chapters "Risk Assessment in Ballast Water Management", "Ballast Water Management Decision Support System" and "Ballast Water Management Decision Support System Model Application").

Intention of This Book

According to our knowledge this is the first comprehensive book on BWM worldwide. This book provides an overview of the possible solutions to the complex issue of BWM and will further outline consequences and implications to address the ballast water "problem" following the provisions of the BWM Convention. There is a need for good insights to the ship ballast operations, environmental and other aspects of the issue as well as international requirements. Further in-depth knowledge is needed on options how to approach and manage it in a most effective way, especially considering specifics on a case-by-case basis. The editors and authors of this book are scientists of different disciplines including professors of universities in the maritime sphere and biological arena who have been involved in or are

[2] GEF/UNDP/IMO, Global Ballast Water Management Program.

[3] Vectors of Change in Oceans and Seas Marine Life, Impact on Economic Sectors (VECTORS), European Community's Seventh Framework Programme (FP7/2007–2013) under Grant Agreement No. [266445].

[4] Ballast Water Management System for Adriatic Sea Protection (BALMAS), IPA Adriatic Cross-Border Cooperation Programme strategic project.

leading researchers in this field. This includes the involvement in the policy making processes at the highest international (IMO), national and regional levels. Experience of this group has been gained through years of committed work in this field, which gave an unique opportunity to gain specific knowledge and experience to offer an in-depth insight and some possible solutions to the related issues. Complimentary, the book contributions reflect the industry, administrations and academic views regarding BWM. Therefore, the book is expected to be of primary interest to students and scientists in various fields, including maritime transport, naval architecture, biology, decision and policy making at national and international levels, especially related to the shipping industry and environmental protection. The book is also written to be of interest to the wider public to broaden the scope of audience and to raise awareness to the topic.

Book Content

After this general introductory chapter, the book continues to describe vessels' ballast water systems, considering stability, structural and safety aspects as well as ballast water volumes being carried by ships and how its discharge (in ports) can be calculated. Next, the types and dimensions of organisms transported with ballast water and their impact is described followed by a chapter which comprehensively summarizes worldwide ballast water policies and regulations implemented to avoid species introductions. The BWM Convention as overarching instrument and its supporting guidelines are introduced by also mentioning the port and flag State requirements. Exemptions from and additional BWM measures as well as BWM exceptions are explained. In continuation, a comprehensive overview of BWMS is given. Recommendations and options for compliance control measurements with the BWM Convention's standard are provided, separated in indicative and in detailed ballast water sampling and sample processing methods. This is followed by a description of the integration of RA, BWM and CME in a DSS. The RA exemptions process is shown in detail highlighting the RA principles and the need for a precautionary approach. Flow charts guide the reader through a RA model for granting exemptions from BWM requirements. While the RA result is a simple risk quantifying answer (high, medium, low), an approach is needed when a decision on "what to do" is to be taken. This DSS considers the RA results and forms the core part of this book. Theoretical and practical profiles of the ballast water RA and DSSs are presented and analysed as BWM tools. These provide a solid framework for the DSS model. The DSS model is presented in the form of flow charts as a step by step approach from the highest level to the details. The generic DSS model is further analysed decision by decision and element by element, also considering their interactions. This BWM DSS approach provides a mechanism to aid transparency and consistency in the decision process regarding BWM needs. The BWM DSS model is then validated in a case study, by using real ballast water discharge data of the Port of Koper, Slovenia as well as data on vessel voyages, including vessel movements, main routes, navigational constraints and ballast water patterns, i.e., amount of ballast water to be managed per vessel and type, ballast water exchange (BWE) capacity rates per vessel type and source ports. The book ends with BWM related conclusions also identifying knowledge gaps and highlighting further research needs.

References

Bailey SA, Duggan IC, Jenkins PT, MacIsaac HJ (2005) Invertebrate resting stages in residual ballast sediment of transoceanic ships. Can J Fish Aquat Sci 62:1090–1103

Bailey SA, Deneau MG, Jean L, Wiley CJ, Leung B, MacIsaac HJ (2011) Evaluating efficacy of an environmental policy to prevent biological invasions. Environ Sci Technol 45:2554–2561

Bauer M (ed) (2006) Harmful algal research and response: a human dimensions strategy. National Office for Marine Biotoxins and Harmful Algal Blooms, Woods Hole Oceanographic Institution, Woods Hole

Bax N, Williamson A, Aguero M, Gonzalez E, Geeves W (2003) Marine invasive alien species: a threat to global biodiversity. Mar Policy 27:313–323

Behrens HL, Haugom GP, Mordal Bakke S (2002) EMBLA Concept: risk assessment methodology and impact and consequence methodology scheme. DNV Report, Det Norske Veritas, Hövik

Briski E, Bailey S, Cristescu ME, MacIsaac HJ (2010) Efficacy of 'saltwater flushing' in protecting the Great Lakes from biological invasions by invertebrate eggs in ships' ballast sediment. Freshw Biol 55:2414–2424

Briski E, Bailey S, MacIsaac HJ (2011) Invertebrates and their dormant eggs transported in ballast sediments of ships arriving to the Canadian coasts and the Laurentian Great Lakes. Limnol Oceanogr 56(5):1929–1939

Carlton JT (1985) Transoceanic and interoceanic dispersal of coastal marine organisms: the biology of ballast water. Oceanogr Mar Biol Annu Rev 23:313–371

Carlton JT (1989) Man's role in changing the face of the ocean: biological invasions and implications for conservation of near-shore environments. Conserv Biol 3:265–273

Carlton JT (1992) Introduced marine and estuarine molluscs of North America: an end-of-the-20th-century perspective. J Shellfish R 11:489–505

Carlton JT (1996a) Marine bioinvasions: the alteration of marine ecosystems by nonindigenous species. Oceanography 9:1–7

Carlton JT (1996b) Biological invasions and cryptogenic species. Ecology 77(6):1653–1655

Carlton JT (1999) Scale and ecological consequences of biological invasions in the world's oceans. In: Sandlund OT, Schei PJ, Viken A (eds) Invasive species and biodiversity management. Kluwer Academic Publishers, Dordrecht, pp 195–212

Carlton JT (2001) Introduced species in U.S. coastal waters: environmental impacts and management priorities. Pew Oceans Commission, Arlington

Carlton JT, Geller JB (1993) Ecological roulette: the global transport of nonindigenous marine organisms. Science 261:78–82

Carlton JT, Thompson JK, Schemel LE, Nichols FH (1990) Remarkable invasion of San Francisco Bay (California, USA) by the Asian clam Potamocorbula amurensis 1. Introduction and dispersal. Mar Ecol Prog Ser 66:81–94

Carlton JT, Reid DM, van Leeuwen H (1995) Shipping study: the role of shipping in the introduction of non-indigenous aquatic organisms to the coastal waters of the United States (others than the Great Lakes) and an analysis of control options. National Sea Grant College Programme/ CT Sea Grant Project R/ES 6

Darling JA, Herborg L-M, Davidson IC (2012) Intracoastal shipping drives patterns of regional population expansion by an invasive marine invertebrate. Ecol Evol 2(10):2552–2561

David M (2007) A decision support system model for ballast water management of vessels: doctoral dissertation. University of Ljubljana, Portorož
David M, Gollasch S (2008) EU shipping in the dawn of managing the ballast water issue. Mar Pollut Bull 56:1966–1972
David M, Gollasch S, Cabrini M, Perkovič M, Bošnjak D, Virgilio D (2007) Results from the first ballast water sampling study in the Mediterranean Sea – the Port of Koper study. Mar Pollut Bull 54:53–65
David M, Gollasch S, Leppäkoski E (2013) Risk assessment for exemptions from ballast water management – the Baltic Sea case study. Mar Pollut Bull 75:205–217
Davidson IC, Simkanin C (2012) The biology of ballast water 25 years later. Biol Invasion 14:9–13
GloBallast (2003) Ballast water risk assessment, user guide for the BWRA database/GIS System. GEF/UNDP/IMO, Global Ballast Water Management Program. URS Meridian-GIS
Gollasch S (1996) Untersuchungen des Arteintrages durch den internationalen Schiffsverkehr unter besonderer Berücksichtigung nichtheimischer Arten, Doctoral Dissertation (in German), Univ. Hamburg, Verlag Dr. Kovac, Hamburg
Gollasch S (2006) Overview on introduced aquatic species in European navigational and adjacent waters. Helgol Mar Res 60:84–89
Gollasch S, Lenz J, Dammer M, Andres HG (2000) Survival of tropical ballast water organisms during a cruise from the Indian Ocean to the North Sea. J Plankton Res 22(5):923–937
Gollasch S, Macdonald E, Belson S, Botnen H, Christensen J, Hamer J, Houvenaghel G, Jelmert A, Lucas I, Masson D, McCollin T, Olenin S, Persson A, Wallentinus I, Wetsteyn B, Wittling T (2002) Life in ballast tanks. In: Leppäkoski E, Gollasch S, Olenin S (eds) Invasive aquatic species of Europe: distribution, impacts and management. Kluwer Academic Publishers, Dordrecht, pp 217–231
Gollasch S, David M, Voigt M, Dragsund E, Hewitt CL, Fukuyo Y (2007) Critical review of the IMO International Convention on the Management of Ships' Ballast Water and Sediments. Harmful Algae 6:585–600
Gollasch S, Haydar D, Minchin D, Wolff WJ, Reise K (2009) Introduced aquatic species of the North Sea coasts and adjacent brackish waters. In: Rilov G, Crooks J (eds) Biological invasions in marine ecosystems: ecological, management and geographic perspectives. Ecological studies, vol 204. Springer, Heidelberg, pp 507–525
Hallegraeff GM (1993) A review of harmful algal blooms and their apparent global increase. Phycologia 32:79–99
Hallegraeff GM (1998) Transport of toxic dinoflagellates via ships' ballast water: bioeconomic risk assessment and efficacy of possible ballast water management strategies. Mar Ecol Prog Ser 168:297–309
Hallegraeff GM, Bolch CJ (1991) Transport of toxic dinoflagellate cysts via ship's ballast water. Mar Pollut Bull 22:27–30
Hamer JP, Lucas IAN, McCollin TA (2001) Harmful dinoflagellate resting cysts in ships' ballast tank sediments; potential for introduction into English and Welsh waters. Phycologia 40:246–255
Hayes KR, Hewitt CL (1998) A risk assessment framework for ballast water introductions. CRIMP Technical Report 14, Division of Marine Research, CSIRO, Hobart
Hayes KR, Hewitt CL (2000) A risk assessment framework for ballast water introductions – volume II. CRIMP Technical Report 21, Division of Marine Research, CSIRO, Hobart
Hayes KR, Sliwa C (2003) Identifying potential marine pests – a deductive approach applied to Australia. Mar Pollut Bull 46:91–98
Hewitt CL (2002) The distribution and diversity of tropical Australian marine bioinvasions. Pac Sci 56:213–222
Hewitt CL, Campbell ML (2010) The relative contribution of vectors to the introduction and translocation of invasive marine species. Department of Agriculture, Fisheries and Forestry (DAFF), Canberra
Hewitt CL, Campbell ML, Thresher RE, Martin RB (eds) (1999) Marine biological invasions of Port Phillip Bay, Victoria. CSIRO Center for Research on Introduced Marine Pests (CRIMP), Technical Report 20, Hobart

IMO (1973) Final act of the international conference on marine pollution 1973, Resolution 18. International Maritime Organization, London
IMO (2004) International convention for the control and management of ships' ballast water and sediments, 2004. International Maritime Organization, 13 Feb 2004, London
IMO (2013) Introduction to IMO. http://www.imo.org/About/Pages/Default.aspx. Last accessed in Oct 2013
Locke A, Reid DM, Sprules WG, Carlton JT, van Leeuwen HC (1991) Effectiveness of mid-ocean exchange in controlling freshwater and coastal zooplankton in ballast water. Canadian Technical Report. Fish Aquat Sci 1822:1–93
McCollin TA, Shanks AM, Dunn J (2008) Changes in zooplankton abundance and diversity after ballast water exchange in regional seas. Mar Pollut Bull 56:834–844
Medcof JC (1975) Living marine animals in a ships' ballast water. Proc Natl Shellfish Assoc 65:54–55
Minchin D, Gollasch S, Wallentinus I (2005) Vector pathways and the spread of exotic species in the sea. ICES Cooperative Research Report 271. International Council for the Exploration of the Sea, Copenhagen
Murphy KR, Ritz D, Hewitt CL (2002) Heterogeneous zooplankton distribution in a ship's ballast tank. J Plankton Res 24(7):729–734
Nellemann C, Hain S, Alder J (eds) (2008) In dead water – merging of climate change with pollution, over-harvest, and infestations in the world's fishing grounds. United Nations Environment Programme, GRID-Arendal
Pimentel D, Zuniga R, Morrison D (2005) Update on the environmental and economic costs associated with alien-invasive species in the United States. Ecol Econ 52:273–288
Rigby GR, Hallegraeff GM (1994) The transfer and control of harmful marine organisms in shipping ballast water: behaviour of marine plankton and ballast water exchange on the MV "Iron Whyalla". J Mar Environ Eng 1:91–110
Romero MLJ, Kotaki Y, Lundholm N, Thoha H, Ogawa H, Relox JR, Terada R, Takeda S, Takata Y, Haraguchi K, Endo T, Lim PT, Kodama M, Fukuyo Y (2011) Unique amnesic shellfish toxin composition found in the South East Asian diatom *Nitzschia navis-varingica*. Harmful Algae 10:456–462
Ruiz GM, Carlton JT, Grosholtz ED, Hines AH (1997) Global invasions of marine and estuarine habitats by non-indigenous species: mechanisms, extent and consequences. Am Zool 37:621–632
Ruiz GM, Fofonoff P, Hines AH, Grosholz ED (1999) Non-indigenous species as stressors in estuarine and marine communities: assessing invasion impacts and interactions. Limnol Oceanogr 44:950–972
Ruiz GM, Fofonoff PW, Carlton JT, Wonham MJ, Hines AH (2000) Invasion of coastal marine communities in North America: apparent patterns, processes, and biases. Annu Rev Ecol Syst 31:481–531
Rup MP, Bailey SA, Wiley CJ, Minton MS, Miller AW, Ruiz GM, MacIsaac HJ (2010) Domestic ballast operations on the Great Lakes: potential importance of Lakers as a vector for introduction and spread of nonindigenous species. Can J Fish Aquat Sci 67:256–268
Shine C, Kettunen M, Genovesi P, Essl F, Gollasch S, Rabitsch W, Scalera R, Starfinger U, ten Brink P (2010) Assessment to support continued development of the EU Strategy to combat invasive alien species. Final Report for the European Commission. Institute for European Environmental Policy (IEEP), Brussels
Simkanin C, Davidson I, Falkner M, Sytsma M, Ruiz G (2009) Intra-coastal ballast water flux and the potential for secondary spread of non-native species on the US west coast. Mar Pollut Bull 58:366–374
van den Bergh JCJM, Nunes PALD, Dotinga HM, Kooistra WHCF, Vrieling EG, Peperzak L (2002) Exotic harmful algae in marine ecosystems: an integrated biological–economic–legal analysis of impacts and policies. Mar Policy 26:59–74

Vila M, Basnou C, Pyšek P, Josefsson M, Genovesi P, Gollasch S, Nentwig W, Olenin S, Roques A, Roy D, Hulme PE, DAISIE Partners (2010) How well do we understand the impacts of alien species on ecosystem services? A pan-European cross-taxa assessment. Front Ecol Environ 8:135–144
Wiley C (1997) Aquatic nuisance species: nature, transport, and regulation. In: D'Itri FM (ed) Zebra mussels and aquatic nuisance species. Ann Arbor Press, Chelsea
Williams RJ, Griffiths FB, Van der Wal EJ, Kelly J (1988) Cargo vessel ballast water as a vector for the transport of nonindigenous marine species. Estuar Coast Shelf Sci 26:409–420

Vessels and Ballast Water

Matej David

Abstract Commercial vessels are built for the transport of various cargoes or passengers. When a vessel is not fully laden, additional weight is required to provide for the vessel's seaworthiness, e.g. to compensate for the increased buoyancy which can result in the lack of propeller immersion, inadequate transversal and longitudinal inclination, and other stresses on the vessel's hull. The material used for adding weight to the vessel is referred to as ballast. Historically, ballast material was solid, but after the introduction of iron as basic vessel building material in the middle of the nineteenth century, loading of water (i.e., ballast water) in cargo holds or tanks had shown to be easier and more efficient. Even when a vessel is fully laden it can require ballast water operations due to a non-equal distribution of weights on the vessel, weather and sea conditions, an approach to shallow waters, and the consumption of fuel during the voyage. As a result of these factors, vessels fundamentally rely on ballast water for safe operations as a function of their design and construction. This chapter describes vessel's ballast water systems, ballast tank designs, ballasting and deballasting processes as well as safety and legislative aspects of ballast water operations. In addition a detailed ballast water discharge assessment model is provided. Using concepts of this model an estimation of global ballast water discharges from vessels engaged in the international seaborne trade was estimated as 3.1 billion tonnes in 2013.

Keywords Ballast water • Vessels design • Ballast water system • Ballast water tank design • Ballasting and deballasting processes • Ballast water safety and legislative aspects • Ballast water discharge assessment

---

M. David (✉)
Dr. Matej David Consult, Korte, Izola, Slovenia
e-mail: matej.david@siol.net

M. David, S. Gollasch (eds.), *Global Maritime Transport and Ballast Water Management*, Invading Nature - Springer Series in Invasion Ecology 8,
DOI 10.1007/978-94-017-9367-4_2

The Importance of Ballast for Vessels

Commercial vessels are built for the transport of various commodities or people by the sea or inland waterways. When a vessel is not fully laden, additional weight is required to compensate for the increased buoyancy that can result in:

- the lack of propeller immersion,
- inadequate transversal inclination, i.e., heeling,
- inadequate longitudinal inclination, i.e., trim,
- static and dynamic stresses on the vessel's hull including shear and torsion forces, bending moments and slamming, and
- static and dynamic transversal and longitudinal instability,

in order to provide for the vessel's seaworthiness. This implies that not only commercial vessels, but also some other vessels (e.g., navy vessels, bigger pleasure boats) use ballast water to provide for adequate seaworthiness (David 2007).

The material used for adding weight to the vessel is referred to as ballast. Historically, ballast material was solid (e.g., sand, rocks, cobble, iron). After the introduction of iron, replacing wood, as basic vessel building material in the middle of the nineteenth century, the doors were opened to new technologies. Loading of water (i.e., ballast water) in cargo holds or tanks (i.e., ballast water tanks) was shown to be easier and more efficient, and hence was adopted as a new practice of increasing importance.

A vessel deemed to be "not fully laden" is a situation when she is not at her maximum allowed draught; i.e., when her carrying capacity in terms of weight, i.e., deadweight (DWT), is not fully exploited. This is typically a dynamic situation during cargo operations in a port; i.e., a vessel will experience changes in loading as it loads and/or unloads cargo. This condition may also result from either the lack of cargo available for transport, or occurs when cargo is light and the total volume of a vessel's cargo spaces becomes a limiting factor (David 2007). However, even when a vessel is fully loaded it can require ballast water operations due to a non-equal distribution of weights on the vessel; i.e., loading of non-homogeneous cargoes, e.g., general cargoes, very heavy cargoes or heavy containers on top of light containers.

Other dynamic factors may also require ballast water operations, such as weather and sea conditions on the route, the approach to shallow waters, and the consumption of fuel and diesel oil during the voyage. According to expected weather conditions, a vessel would sail in a heavy ballast condition, i.e., maximum ballast loaded, when expecting bad weather, or a light ballast condition, i.e., partial ballast loaded, when it is ensured that the weather conditions and rough seas will not impair the vessel's stability, e.g., when approaching a port or inland waterways. Vessels would go from heavy ballast to light ballast conditions when safe and weather as well as sea conditions are favourable to consume less fuel, and when in save haven close to a port or at the ports anchorage, to get ready for loading cargo. When approaching shallow waters a vessel may also need to discharge some ballast water to provide for less draught, or when she needs to sail below a bridge she may need to add ballast to provide for lower air draft.[1] In relation to the fuel and diesel oil consumption during a voyage, e.g., a Panamax container vessel consumes approx. 100–180 tonnes of heavy fuel per day, and according to the *International Convention for the Safety of Life at Sea* (SOLAS), 1974, vessels need to be adequately trimmed[2] to provide for optimal hydrodynamics, they need to provide for bridge visibility standards, and for minimum aft draught for adequate propeller immersion.

Some types of vessels, especially Ro-Ro, container and passenger vessels, which load cargo or passengers also very high above the waterline, and cargoes frequently are non-equally distributed, have so called anti-heeling tanks to compensate for transversal unequal distribution of weight and prevent vessel from listing. This is especially important in port during cargo operations. Vessels usually do not load and discharge water in or from the anti-heeling tanks, but have a constant volume of water in these tanks which is than being pumped from one side of the vessel to another.

As a result of these factors, vessels fundamentally rely on ballast water for safe operations as a function of their design and construction.

Vessel's Ballast System

The number, volume and distribution of ballast tanks are vessel type and size related. The ballast tanks can be in the vessel's double bottom (DBT – double bottom tanks), port and starboard along the sides (ST – side tanks or WT – wing tanks), in the bow (FPT – forepeak tank), in the stern (APT – after peak tank), port and starboard underneath the main deck (TST – topside tanks or upper wing tanks), and other (e.g., CT – central tanks). Though FPT and APT tanks are traditional on all types of vessels, some does not have these tanks, e.g., The Hamburg Express class containerships. Some older vessels, mainly tankers, were also using cargo holds (or cargo tanks respectively) to ballast, but today's vessels have tanks that are dedicated only for ballasting, i.e., segregated ballast tanks (see Figs. 1 and 2). The specific case today to ballast in cargo holds may apply to bigger bulk carriers, which may load water in some of the central cargo holds to sail in so called "heavy ballast condition" when exposed to heavy sea conditions.

Ballast tanks are connected with the ballast water pump(s) by a ballast water pipeline. Water from the vessels surrounding area is loaded on the vessel through the vessel sea-chest(s) and strainer(s) (see Fig. 3) via the ballast pipeline to ballast tanks.

Inside the ballast tanks water is loaded and discharged via the ballast water pipeline suction head (see Fig. 4).

Vessels with greater ballast capacity are usually equipped with two ballast pumps (see Fig. 5) in order to ensure ballast water operations are carried out even in case

[1] i.e., the distance from the water to the highest part of the vessel.

[2] i.e., difference between the forward and aft draft, when this exists, means longitudinal list of this vessel; when there is no trim, vessel is on even keel.

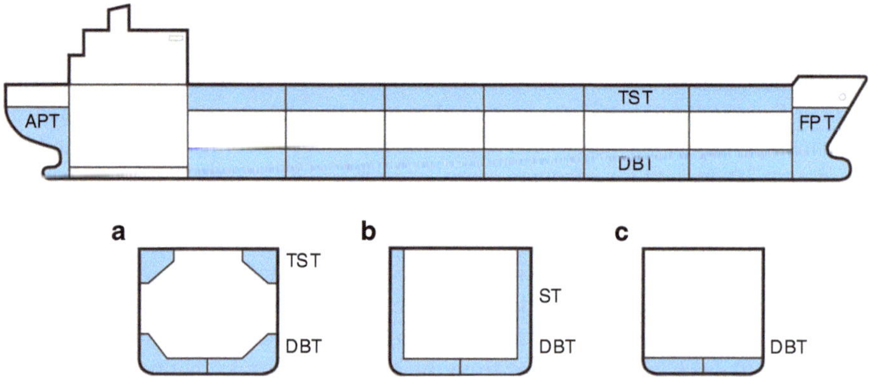

Fig. 1 Ballast tanks on: (**a**) most bulk carriers, (**b**) tankers, container vessels, and some newest bulk carriers, and (**c**) Ro-Ro and general cargo vessels. (*APT* after peak tank, *DBT* double bottom tanks, *FPT* forepeak tank, *ST* side tanks, *TST* topside tanks or upper wing tanks)

Fig. 2 Interior of a DBT (*left*) and ST (*right*) ballast tank on a bulk carrier (Photos: Guy Mali)

of a failure of one pump, while some smaller vessels may use service pumps also for ballast operations.

Ballast tanks may be accessed/entered for maintenance, cleaning and other purposes via manholes or tank hatches. Ballast tanks are equipped with air vents, which allow the air in the ballast tanks to be expelled from the tank to prevent overpressurisation when the ballast tanks are filled, or to let the air in and prevent underpressurisation when ballast tanks are emptied (see Fig. 6).

Fig. 3 Ballast water intake area with the strainer in the front below the walk-on grating connected to the sea-chest

Fig. 4 Ballast water pipeline suction head (Photo: Guy Mali)

Fig. 5 Two ballast pumps of 1,500 m^3 capacity on a container vessel

Fig. 6 An air-vent on the *left*, a sounding pipe in the *center back*, and a TST hatch on the *right* on a bulk carrier

Fig. 7 Discharge of ballast water below the water level from a container vessel

It is absolutely critical to know how much ballast is in each tank to be able to provide for the vessels seaworthiness. On older vessels these measurements are done via sounding pipes (see Fig. 6), and then by means of sounding tables, the quantity of ballast water can be calculated. Most modern ships are equipped with instruments that enable automatic measurements of the quantity of ballast water in ballast tanks, while these still need to be equipped with sounding pipes to allow direct measurements in the case of automatic system failure.

Ballast water is discharged through the overboard discharge, which is on most vessels situated below the water level (see Fig. 7). On some vessels ballast water discharge is situated above the water level, and mainly on bulk-carriers ballast water can be discharged directly from the topside tanks high above the water level (e.g., see chapter "Ballast Water Sampling and Sample Analysis for Compliance Control", Fig. 4).

Vessel Ballast Capacity

The vessel ballast capacity is mainly determined by the vessel cargo capacity in terms of cargo weight, and the speed at which the cargo operations may be conducted. Generally, the more tonnes of cargo a vessel is capable to carry, the more ballast may be needed when sailing without cargo on board, and if the cargo operations on a vessel are very fast, then the ballast uptake or discharge has to be correspondingly fast. The ballast water capacity of a vessel is given in terms of volume of spaces that are available for ballasting expressed in m^3, and in terms of the ballast pumps capacity expressed in m^3/h.

The volumetric ballast water capacity mainly determines the vessels seaworthiness in different static and dynamic conditions. For instance, according to Det Norske Veritas, Rules for Classification of Ships (Part 3, Ch. 1, Sec. 3) (DNV 2000), ships of 20,000 tonnes DWT and above having the class notation Tanker for Oil and ships of 30,000 tonnes DWT and above with the class notation Tanker for Oil Products are required to have segregated ballast tanks. The capacity of segregated ballast tanks is to be at least such that, in any ballast condition at any part of the voyage, including the conditions consisting of lightweight plus segregated ballast only, the ship's draughts and trim can meet each of the following requirements:

The moulded draught amidships (dm) in meters (without taking into account any ship's deformation) is not to be less than:

$$dm = 2.0 + 0.02L \tag{1}$$

where L means length between perpendiculars.

The draughts at the forward and after perpendiculars are to correspond to those determined by the draught amidships (dm) association with the trim (t) by the stern of not greater than

$$t \leq 0.015L \tag{2}$$

In any case the draught at the after perpendicular is not to be less than that which is necessary to obtain full immersion of the propeller(s) (Perkovič and David 2002a).

In general, cargo vessels such as, general cargo, Ro-Ro, e.g., ferries and car carriers, use only small quantities of ballast water, i.e., generally some 20 % of their DWT, with some exceptions even of more than 40 % of DWT for special uses (Capt. Peter Stapleton personal communication). On the other hand, vessels for the transport of liquid and dry bulk cargoes, e.g., tankers, dry-bulk carriers, require significantly larger quantities of ballast water, i.e., mostly between 30 and 50 % of their DWT, what may result to more than 100,000 m^3 of ballast water per vessel. A summary of the ballast water capacities for main ship types identified by different authors is presented in the Table 1 (David et al. 2012).

The ballast water pumps capacity is mainly related to the speed of vessels cargo operations, i.e., how much cargo can be loaded or discharged in a certain period of time, as the ballasting operations are mainly being conducted in the opposite way than the cargo operations. Some vessels may be loading cargo at much higher speeds than the others, hence need much faster ballast pumping rates otherwise the cargo operation may have to be slowed down. Bigger tanker vessels, i.e., crude oil tankers, are the fastest in cargo loading/discharging rates, nowadays conducting cargo operations at 10,000 tonnes/h or even faster, and bigger bulk carriers with up to 6,000 tonnes/h, hence having ballast water pumping capacities in the range of 6,000–15,000 m^3/h.

Container vessels when in most developed ports manage to load or discharge approx. 18–22 containers[3] per crane per hour (Chief Officer Kiril Tereščenko per-

[3] i.e., 40 ft containers or instead of one 40 ft container can be two 20 ft containers loaded or discharged at the same time.

Table 1 Percentage of vessels ballast water capacity in relation to the vessels DWT, based on different vessel types. *BC* bulk carrier, *C* container, *GC* general cargo, *T* tanker, *Pas* passenger, *RR* Ro – Ro (After David et al. (2012) (Reprinted from Decision Support Systems, 53, David M, Perkovič M, Suban V, Gollasch S, A generic ballast water discharge assessment model as a decision supporting tool in ballast water management, 175–185, copyright 2012, with permission from Elsevier)

Vessel type/ DWT	AQIS (1993)	Walters (1996)	Wiley (1997)	Carlton et al. (1995)	Farley (1996)	Cohen (1998)	Dobes (1997)	Hay and Tanis (1998)	Behrens et al. (2003)	Suban (2006)
All vessels			30	38	40	36	33		36	33
BC		41		43				60		33
BC/250,000	30–45									
BC/150,000	30–45									30–45
BC/70,000	36–57									30–45
BC/35,000	30–49									33–57
T		26		38						
T/100,000	40–45									
T/40,000	30/38									43
C		30		32				30–60		35
C/40,000	30–38									28–40
C/15,000	30									30
GC		35						30–60		29
GC/17,000	35									
GC/8,000	38									
Pas/RR	33	38								43

sonal communication) and an experienced crane driver can handle also up to 30 containers per hour (Chief Officer Guy Mali personal communication). The number of gantry-cranes that can be employed at a time depends on the vessel size, port/terminal and priority of vessel. The number of container operations is also very much related to the capacity of containers handling at the terminal. There are usually several, e.g., three to five, cranes in operation at the same time., e.g., in average the container vessel Hamburg Express when in the Port of Rotterdam handles 4,100 containers in ~24 h, what results in approx. 46,000 tonnes of cargo loaded or discharged (Chief Officer Guy Mali personal communication.). In general container vessels manage to be served by ballast water pump capacities in the range of 1.000–3.000 m^3/h, i.e., two pumps, each 500–1,500 m^3/h.

As the port cargo loading and unloading capacities are increasing through time mainly with the use of newer technologies supporting faster cargo operations, newer vessels of similar cargo capacities in general have ballast water systems of higher capacity. An increase in ballast water capacities of new vessels can be expected also in the future.

Ballasting and Deballasting Process

Vessels conduct ballast water operations usually in the port as opposite to the cargo operations, i.e., when a vessel would load cargo, ballast water would be discharged, and when more or heavier cargo is loaded on one side, ballast water would be discharged from that side or loaded/moved to the other side. Ballasting and deballasting may also be conducted during navigation or at the anchorage, depending on the vessel type, weather and sea conditions, and vessel operations.

Ballast water is taken onboard by:

- gravity through opening valves which enables a vessel to take on water into ballast tanks (or cargo holds used for ballast) below the water line;
- pumping water into ballast tanks (or cargo holds used for ballast) above the water line.

Nevertheless, all the water may be taken on board by pumping, instead of using the gravity method.

The tanks are filled according to a predetermined sequence, depending on the type of the vessel and current cargo operation. The ballast tanks are usually filled up to maximum capacity in order to prevent the free surface effects.[4] This "rule",

[4] i.e., movements of water in the tank from side to side and hence changing centres of gravity as well having dynamic side effects, and with this negatively impacting the transversal stability of the vessel; this is especially important for cargo holds and wider ballast tanks; e.g., double bottom, topside.

however, generally does not apply to fore-peak and after-peak tanks since these are frequently filled partially because of trimming the vessel.

Deballasting is conducted in the opposite sequence by:

- gravity through opening the valves that enables a vessel to discharge ballast water into the surrounding environment from ballast tanks (or cargo holds used for ballast) above the water line;
- pumping out the ballast water from ballast tanks (or cargo holds used for ballast) below the water line.

Nevertheless, all ballast water may be discharged into the surrounding environment by pumping, instead of using the gravity method (David 2007).

When tanks are getting close to empty, ballast pumps start loosing suction as they start getting air in the system. The remaining water in tanks after pumping with ballast pumps is in general between 5 and 10 % of ballast water tank volume, what is mainly depending on the vessels trim. The ballast pipes suction heads are usually installed on the back side of the ballast tanks, hence for pumping out most of the remaining ballast water the vessel needs to be trimmed astern, what is also a very general practice. This astern trimming is to compensate the change of trim during the voyage because of fuel consumption from tanks, which are more in the stern part of the vessel, to arrive in the next port of call approximately on even keel. However, when Gollasch and David conducted shipboard tests of different BWM methods we noticed that on vessels which were trimmed ahead about 15 % and more of unpumpable water remained in the tanks during the empty-refill (sequential) BWE. Actually, practice on some newer container vessels has shown that when trimmed ahead the vessel consumes less fuel during navigation probably due to better hydrodynamics, hence these would nowadays usually start the voyage on even keel or even be trimmed ahead (Captain Alok Kumar personal communication, Chief Officer Guy Mali personal communication). When at the start of the voyage a vessel could not be trimmed ahead because of some limitations (e.g., limited maximum draft, required even keel), ballast operations would be conducted at see what is done by internal transfer of ballast water or pumping in some additional ballast water. For almost total deballasting of tanks, i.e., 1–2 % of the ballast water tank volume remaining as unpumpable ballast, a ballast ejector pump is used. This is also so called "stripping" and is done by using the firepump together with the ballast stripping eductor (Chief Officer Guy Mali personal communication).

All ballasting and deballasting activities are usually led by the first (chief) officer, who is responsible for the vessel's stability. Following his instructions, the pumps and valves are operated automatically from a ballast control console or from a computer by an officer (Fig. 8).

Some older vessels do not have an automated control over ballast pumps and valves, then this may be done manually by an engineer, while the bosun (senior deck crewman, ranked below the deck officers) has to monitor the conditions of ballast in the ballast tanks by measuring the water level via sounding pipes at adequate

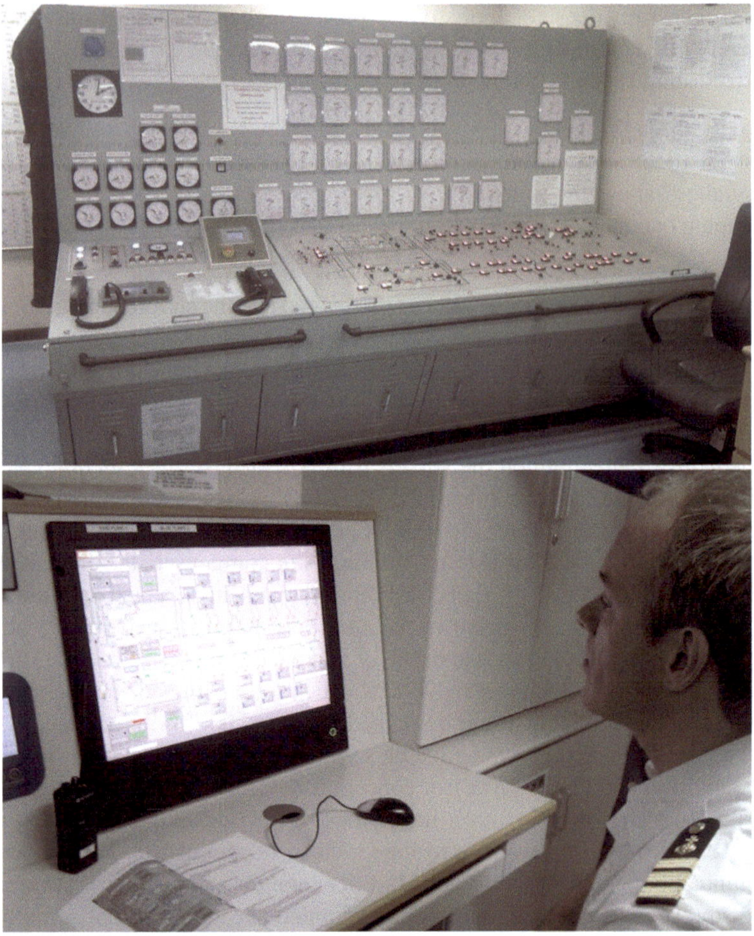

Fig. 8 Ballast control console (*top*) and computer ballast system control (*bottom*)

time intervals, and regularly reporting them to the officer or engineer. The entire ballasting and deballasting process, as well as internal transfers of ballast, has to be recorded in the ship's logbooks (e.g., Ballast Water Handling Log (Chief Officer Guy Mali personal communication). Some states require also Ballast Water Reporting Forms (BWRF).

Safety and Legislative Aspects of Loading and Discharging Ballast

Loading and discharging of cargo and ballast directly affects the transversal and longitudinal stability as well structural integrity of the vessel, and consequently safe navigation and the safeguarding of human lives. Hence, the examination of all

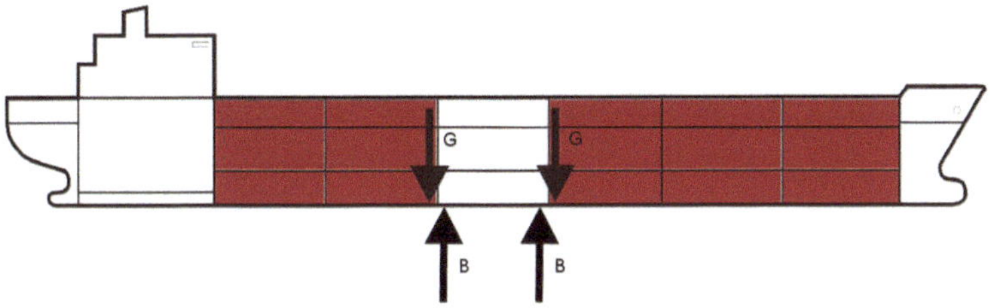

Fig. 9 *Arrows* showing where in this case shear forces act; i.e., where two tank sections next to each other, one being fully ballasted having more gravity (*G*) than the empty tank section, where the buoyancy (*B*) effect is stronger

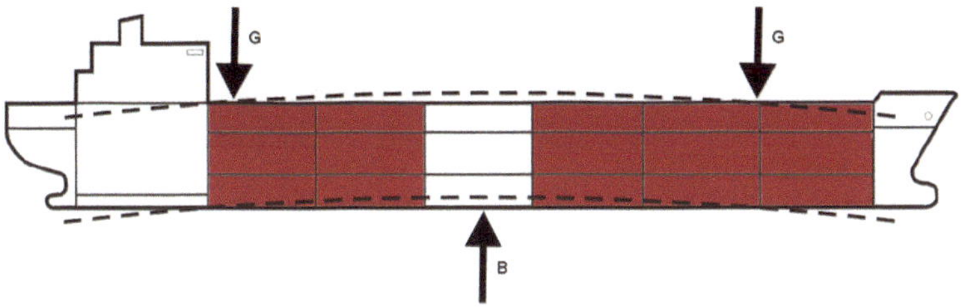

Fig. 10 *Arrows* showing the acting of bending forces with increased buoyancy (*B*) in the amidships and increased gravity (*G*) in fore and aft part, causing longitudinal deflection of the vessel hull, so called hogging

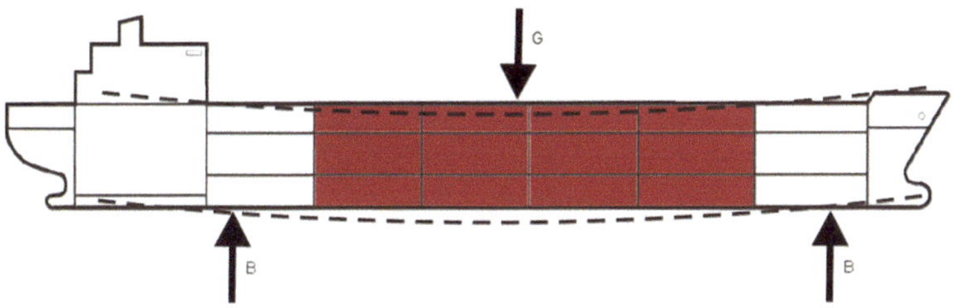

Fig. 11 *Arrows* showing the acting of bending forces with increased buoyancy (*B*) in the fore and aft part and increased gravity (*G*) in the amidships part, causing longitudinal deflection of the vessel hull, so called sagging

ballast-related procedures has to lay special emphasis on safety. The interim phases in loading and discharging ballast water generate changes that usually exert different negative influence on a vessel's stability and induce additional static forces on the vessel hull (see Figs. 9, 10 and 11). Improper management of cargo and ballast operation may result in structural failure of the vessel hull in the port (see Fig. 12) or even results in the vessel to capsize.

When the vessel is sailing, she is exposed to more dynamic conditions as compared to being in a port, influenced from the outside by waves and wind (see Figs. 13 and 14). One of the undesirable effects is that caused by free surfaces affecting vessels stability, where ballast water is able to move inside the tanks if these are not

Fig. 12 Vessel structure that failed because of overstress in hogging (Source: Cornelius de Keyzer, Master Mariner, Port of Rotterdam, the Netherlands)

Fig. 13 Vessel in heavy weather conditions, waves in transversal effect (Source: Cornelius de Keyzer, Master Mariner, Port of Rotterdam, the Netherlands)

Fig. 14 Vessel in heavy weather conditions, waves in longitudinal effect (Source: Cornelius de Keyzer, Master Mariner, Port of Rotterdam, the Netherlands)

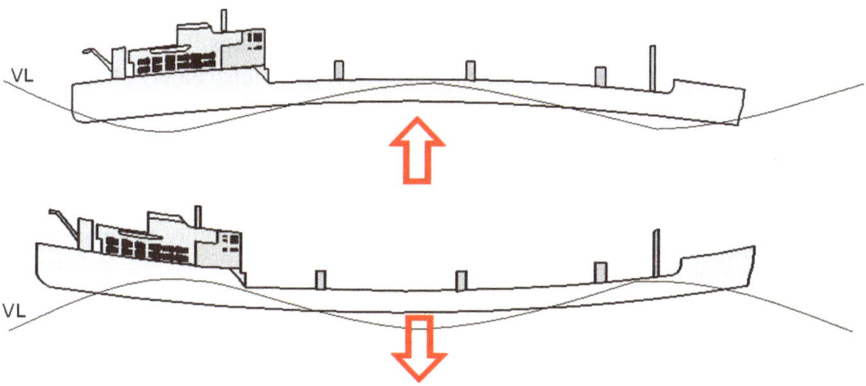

Fig. 15 Longitudinal wave effects on the vessel hull, inducing additional (*top drawing*) or reduced (*bottom drawing*) buoyancy amidships, that may result in vessel structure failure. *VL* water line

completely filled, while waves may exert extensive additional forces on the vessel hull, especially as additional shear forces and bending moments (Perkovič and David 2002a, b) (see Fig. 15).

Inappropriate ballast water operation and with this reduced transversal stability may result in a vessel to flip aside (see Fig. 16).

Safe navigation and the safeguard of human lives at sea are regulated by the SOLAS Convention of IMO. Aiming to increase safety and reduce pollution from

Fig. 16 The Cougar Ace flip (Photo credit: Capt. Kevin Bell, US Fish and Wildlife Service)

ships, on 24th May, 1994, the convention was supplemented with amendments in a new chapter, IX – Management for the Safe Operation of Ships. The main objective of Chapter IX is the mandatory consideration of the International Safety Management Code (ISM Code).

The ISM Code was adopted following the tacit consent procedure. It became mandatory on 1st July, 1998, for all passenger vessels including high speed craft, tankers, chemical tankers, and liquefied gas carriers, including high speed cargo craft of GT 500 or more. On 1st July, 2002, it became mandatory for other cargo ships and mobile offshore drilling units of GT 500 or more.

Based on the ISM Code requirements, all ships listed above are required to conduct ballast water operations in accordance with a previously prepared plan delineated in the Safety Management Manual that has to be available on board the vessel at all times. The responsibility for the preparation of safety plans for specific vessel types is by the shipowner, who also has to provide for regular inspection and the proper functioning of the safety and ballast water system (David 2007).

How Much Ballast Water Vessels Discharge?

As stated above, vessels in general discharge ballast when loading cargo, and the reverse. Logically, it appears that all vessels that load cargo in the port consequently discharge ballast. But in reality the situation is not so simple. Vessels load different types of cargoes, which could be divided into specifically heavy, e.g., metal rolls, steel, iron, ore, carbon, oil; and light cargoes, e.g., grains, timber, paper, vehicles, containers.

In the case of loading a heavy cargo, a vessel will be most probably immersed to her maximum draught, i.e., one of the load lines,[5] hence needs to discharge all ballast to load as much cargo as possible. This means that the vessel will discharge all ballast except the quantity unable to be discharged, and the quantity needed for trimming and heeling where appropriate.

Some vessels in ports usually undertake both cargo loading and discharge operations, e.g., containers, vehicles, general cargoes. In these cases the ballast water situation fully depends on the quantity of discharged and loaded cargo. If the quantity of discharged cargo is greater than that loaded, it is supposed that the ship will not discharge ballast and vice versa.

The quantity of ballast water may also depend on weather conditions. When expecting to sail through bad weather conditions and heavy seas, vessels would be in heavy ballast condition to improve the safety of navigation.

Tanker vessels carrying heavy oil or vessels specialized for the carriage of orange juice, for instance, as a rule return to the port of loading empty and therefore require larger quantities of ballast water for safe navigation. On the other hand, a general cargo and container vessel will when in operation always carry some cargo, i.e., some will be discharged and some loaded at the next port of call. These vessels can thus carry ballast water taken up in different ports. The quantity of ballast water carried, however, primarily depends on the cargo handling operations carried out. Therefore, if a significantly greater quantity of cargo is discharged than loaded, it may be assumed that ballast water will be required on board, and vice versa (David et al. 2012).

However, when a vessel loads a light cargo, her maximum DWT capacity will not be exploited, because the limiting factor becomes the volume available to store the cargo, and not the cargo weight. Some light cargoes are frequently also loaded on the deck as well as in cargo holds. Consequently, the vessel has diminished transversal stability and needs to improve it by adding ballast in her double bottom tanks. A typical example is that of loading timber on deck, and this may also be the case when heavier containers would be loaded on top of lighter containers or on the upper deck of a car carrier.

The above described situations and conditions show that the ballast water operations are related to different vessel types, vessel construction, cargo operations and weather conditions. However, there are no clear limits among all these factors, but the decision on ballast water operations is under the discretion of the chief officer and direct control of the captain, who is responsible for the vessels stability and safety (David et al. 2012).

[5] i.e., appropriate load line according to the IMO Load Line Convention.

Ballast Water Discharge Assessment

Different ballast water studies around the globe involved an assessment of ballast water discharges in a port or wider area (e.g., AQIS 1993; Walters 1996; Wiley 1997; Carlton et al. 1995; Farley 1996; Dobes 1997; Cohen 1998, Hay and Tanis 1998; Behrens et al. 2003; Perkovič et al. 2004). This is clearly one of the information needed for understanding how biological invasions are facilitated (Bailey et al. 2011; Briski et al. 2012; Chan et al. 2013), and this information had impacted the wider public opinion and consequently government administrations, policies etc. It is also very important to understand the ballast water operation patterns to enable provisions of adequate decision support tools for ballast water management (David et al. 2012), i.e., horses for courses. However, having in mind the complexity of ballast water operations, it becomes clear that such assessments are very challenging, and that an accurate ballast water discharge assessment for each vessel call to a port, especially for those that only partially load and unload cargo in the same ports, is almost impossible.

A ballast water discharge assessment model was prepared during two ballast water management studies conducted in Slovenia and the model was applied to the Port of Koper data. For the purpose of a wider application of the model a detailed model verification study was conducted (David et al. 2012) and the model has been applied in different studies to assess ballast water discharges in some ports around Europe (EU FP7 VECTORS project,[6] IPA Adriatic BALMAS project[7]). The model can be used for the assessment of ballast water discharges in past years as well as for a prediction of ballast water operation of a ship calling to a port. In more biological terms, historical data may be helpful when studying vessels and ballast water patterns through time and relating them to known introduced species, to assess biological propagule pressure, as well as background data for RA assumptions. In terms of ballast water management, the ballast water discharge assessment model provides responsible authorities with many tools; e.g., for targeting vessels for adequate BWM measures based on the risk posed, check for false BWM reporting, targeting vessel for compliance monitoring, etc. Furthermore, model calculations may also be used to identify the dimensions of land-based ballast water reception facilities should it be planned to make such facilities available. A ballast water discharge assessment is also helpful to evaluate the environmental acceptability of ballast water treatment systems which use active substances (chemical treatment) to kill organisms. The model may be used to calculate, e.g., the annual amount of ballast water discharges, and in a worst case scenario where all ballast water discharged was assumed to be treated with the same active substance, it could be evaluated if the remaining toxicity of the ballast water at discharge is environmentally acceptable (David et al. 2012).

---

[6] European Community's Seventh Framework Programme under Grant Agreement No. [266445] for the project Vectors of Change in Oceans and Seas Marine Life, Impact on Economic Sectors (VECTORS). http://www.marine-vectors.eu/

[7] IPA Adriatic Cross-Border Cooperation Programme - strategic project Ballast Water Management System for Adriatic Sea Protection (BALMAS), http://www.balmas.eu/

The model is presented in Fig. 17. More details about the model logic, application and accuracy of results are presented in David et al. (2012).

Estimation of Ballast Water Discharges World-Wide

In the past global ballast water discharges were assessed or quoted by, e.g., 10 billion tonnes by Gollasch (1998), and 3.5 billion tonnes by Endresen et al. (2004). At the time these assessments were conducted, the world seaborne trade amounted to around 5 billion tonnes of cargo per year, i.e., in 1995 it was 4.651 billion tones, and in 2000 it was 5.871 billion tonnes (UNCTAD 2006). The Endresen et al. (2004) assessment considered the world seaborne trade to be 8.734 billion tonnes of cargo, 5.434 billion tonnes in international and 3.3 billion tonnes in national seaborne trade.

The ballast water capacity varies as a function of the cargo carrying capacity and ship type, with an average value of 33 % of the vessel's DWT (Suban 2006). However, the ballast capacity is only partially utilized because the vessel's DWT is commonly not fully exploited. First of all it is necessary to consider the fact that the ship is not loading the full DWT capacity. From DWT it is necessary to deduct weight of stores, fuel, fresh water and other weights. This weight usually represents around 5–10 % of a ship's DWT (Suban et al. 2006), hence the ballast water capacity would be about 37 % of the vessels cargo capacity in terms of weight. Secondly, vessels frequently do not exploit also their maximum DWT dedicated to cargo, e.g., different vessels, especially container vessels, car carriers, and general cargo vessels are usually only partially loaded, and bulk carriers when they load light cargoes as grains or wood. The BWDA model (see section "Ballast water discharge assessment") considers all this, hence the estimated discharge would amount to 33 % of the cargo volume in the world seaborne trade, not considering the lightweight cargoes.

The world international seaborne trade in 2011 amounted to 8.748 billion tonnes of cargo (UNCTAD 2012), thus the global ballast water discharges from vessels engaged in the international seaborne trade in 2011 would be about 2.88 billion tonnes. If we want to estimate the global ballast water discharges for 2013, the information needed is not yet available, but needs to be estimated. According to the UNCTAD (2012) data the world wide economic crisis was reflected in the decrease of world seaborne trade especially in 2009, after which it recovered with an annual growth of about 350 million tonnes per year until 2011, while the average annual growth from 2000 to 2011 was about 250 million tonnes per year. Assuming an average annual growth of 300 million tonnes per year as the global economy recovered after the 2009 crisis and continued to grow (UNCTAD 2012), the world international seaborne trade in 2013 would amount to about 9.35 billion tonnes of cargo, thus the global ballast water discharges from vessels engaged in the international seaborne trade in 2013 would be about 3,1 billion tonnes.

The amounts estimated here are much lower than some earlier estimations mentioned above, especially when considering that the global cargo transport today is

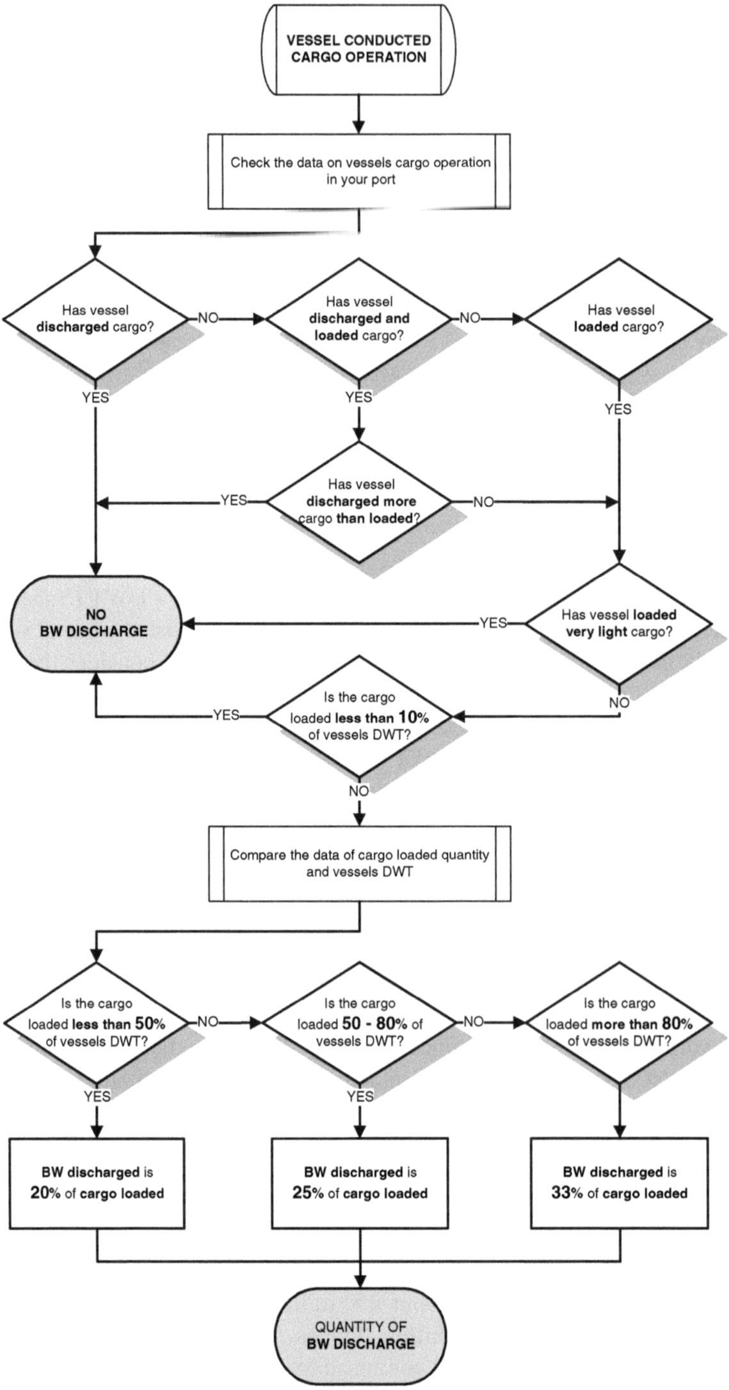

Fig. 17 Ballast water discharge assessment model (David et al. 2012) (Reprinted from Decision Support Systems, 53, David M, Perkovič M, Suban V, Gollasch S, A generic ballast water discharge assessment model as a decision supporting tool in ballast water management, 175–185, copyright 2012, with permission from Elsevier) (This figure can be downloaded from http://extras.springer.com/)

much higher. Nevertheless, it is important to understand, that the volumetric estimation of ballast water discharges is only one very superficial expression when ballast water is seen from a different perspective, the perspective of transfer of harmful aquatic organisms and pathogens. From this perspective, volumes of ballast water being discharged are much less important then what is actually in the ballast water discharged (see chapter "The Transfer of Harmful Aquatic Organisms and Pathogens with Ballast Water and Their Impacts").

Acknowledgements My grateful thanks to Captain P.G.H. Stapleton, Master Mariner, BSc, from UK; Captain Alok Kumar, Master Mariner, BSc, from India; Chief Officer Kiril Tereščenko, MSc, MBA, from Poland; and Chief Officer Guy Mali, BSc, from the Netherlands; for reviewing this chapter and for very valuable comments. I further like to thank my colleague Stephan Gollasch for his valuable editorial suggestions.

The research leading to part of these results has received funding from the European Community's Seventh Framework Programme (FP7/2007–2013) under Grant Agreement No. [266445] for the project Vectors of Change in Oceans and Seas Marine Life, Impact on Economic Sectors (VECTORS). Part of this publication has been produced with the financial assistance of the IPA Adriatic Cross-Border Cooperation Programme – strategic project Ballast Water Management System for Adriatic Sea Protection (BALMAS). The contents of this publication are the sole responsibility of authors and can under no circumstances be regarded as reflecting the position of the IPA Adriatic Cross-Border Cooperation Programme Authorities.

References

AQIS (1993) Ballast water management. Ballast water research series report no 4. Australian Quarantine and Inspection Service, AGPS Canberra

Bailey SA, Chan F, Ellis SM, Bronnenhuber JE, Bradie JN, Simard N (2011) Risk assessment for ship-mediated introductions of aquatic nonindigenous species to the Great Lakes and freshwater St. Lawrence River. DFO Canadian Science Advisory Secretariate. Research Document 2011/104

Behrens HL, Endresen Ø, Mjelde A, Garmann C (2003) Environmental accounting system for Norwegian shipping. EASNoS Phase 1, DNV Rep. no. 2002-1645

Briski E, Bailey SA, Casas-Monroy O, DiBaccio C, Kzaczmarska I, Levings C, MacGillvary ML, McKindsey CW, Nasmith LE, Parenteau M, Piercey GE, Rochon A, Roy S, Simard N, Villac MC, Weise AM, MacIsaac HJ (2012) Relationship between propagule pressure and colonization pressure in invasion ecology: a test with ships' ballast. Proc R Soc B 279:2990–2997

Carlton J, Reid D, van Leeuwen H (1995) The role of shipping in the introduction of nonindigenous aquatic organisms to the coastal waters of the United States (other than the Great Lakes) and an analysis of control options. Shipping study I. USCG report no. CG-D-11-95. National Technical Information Service, Springfield

Chan FT, Bailey SA, Wiley CJ, MacIsaac HJ (2013) Predictive relative risk assessment for ballast-mediated invasions at Canadian arctic ports. Biol Invasion 15(2):295–308

Cohen AN (1998) Ships' ballast water and the introduction of exotic organisms into the San Francisco estuary. Current status of the problem and options for management. A report for the CALFED Category III Steering Committee. San Francisco Estuary Institute, Richmond

David M (2007) A decision support system model for ballast water management of vessels: doctoral dissertation. University of Ljubljana, Portorož

David M, Perkovič M, Suban V, Gollasch S (2012) A generic ballast water discharge assessment model as a decision supporting tool in ballast water management. Decis Support Syst 53:175–185

DNV (2000) Rules for classification of ships. Managing risk – electronic rulebook. Det Norske Veritas, Høvik, Norway

Dobes JC (1997) Ballast water management—a glimpse of the future. ASNE environmental symposium, technical papers

Endresen Ø, Behrens HL, Brynestad S, Andersen AB, Skjong R (2004) Challenges in global ballast water management. Mar Pollut Bull 48:615–623

Farley R (1996) Analysis of overseas vessel transits into the Great Lakes through commercial shipping and resultant distribution of ballast water. University of Michigan College of Engineering Department of Naval Architecture and Marine Engineering paper 331, pp 12–19

Gollasch S (1998) Removal of barriers to the effective implementation of ballast water control and management measures in developing countries (for GEF/IMO/UNDP), London

Hay C, Tanis D (1998) Mid ocean ballast water exchange: procedures, effectiveness and verification. Cawthron report no. 468, Wellington

Perkovič M, David M (2002a) Feasibility of ballast water exchange methods on the open sea. In: David M (ed) 5th International Conference on Traffic Science, ICTS 2001 conference proceedings. Faculty of Maritime Studies and Transport, Portorož, pp 227–240

Perkovič M, David M (2002b) Safety aspects of the ballast water exchange. In: Jenček P, Štrubelj M (eds) 6th International Conference on Traffic Science, ICTS 2002 conference proceedings. Faculty of Maritime Studies and Transport, Portorož, pp 373–384

Perkovič M, Suban V, David M (2004) Ballast water discharges in the Slovenian sea. In: 8th International Conference on Traffic Science – ICTS 2004. Conference proceedings, Nova Gorica

Suban V (2006) Model določanja količine izpuščenega balasta za preteklo obdobje s pomočjo izvedenih podatkov: master thesis. University of Ljubljana, Portorož

Suban V, David M, Perkovič M (2006) Model for the assessment of the quantity of ballast water discharges using ships' traffic data. In: Conference proceedings. 8. International conference maritime transport and infrastructure, Riga, pp 69–75

UNCTAD (2006) Review of maritime transport (2006). Report by the UNCTAD secretariat, United Nations, New York/Geneva

UNCTAD (2012) Review of maritime transport (2012). Report by the UNCTAD secretariat, United Nations, New York/Geneva

Walters S (1996) Ballast water, hull fouling and exotic marine organism introductions via ships – a Victorian study. Publication 494. Environment Protection Authority, Victoria

Wiley C (1997) Aquatic nuisance species: nature, transport, and regulation. In: D'Itri FM (ed) Zebra mussels and aquatic nuisance species. Ann Arbor Press, Chelsea

The Transfer of Harmful Aquatic Organisms and Pathogens with Ballast Water and Their Impacts

Stephan Gollasch, Dan Minchin, and Matej David

Abstract The annual number of new species records world-wide has paralleled shipping and is increasing. For example, in ICES member countries a new introduction forming a new population beyond its natural range occurs approximately every 9 weeks. The introduction of non-indigenous species by ships' ballast water is known since more than 100 years, but it was not until 1970s that the first biological samples from ballast water were taken. Since, more than 1,000 species were identified from ballast tanks, including human pathogens. It was estimated that 3,000–7,000 different species are moved each day around the globe by ships and it was concluded that shipping is the prime species introduction pathway with each vessel having the potential to introduce a species. However, not all species find a suitable situation in the new environment, but it was suggested that >2,000 aquatic non-indigenous species have been introduced world-wide, of which in minimum 850 are likely introduced by ships. Not all introduced species are considered harmful, in some cases this is quite the reverse, as some support important industries. However, a number of introduced species had almost catastrophic and seemingly irreversible impacts and all of the summed impacts amount to considerable costs of billions of Euro annually. Consequently, a precautionary approach suggests that every vessel transporting ballast water should be treated as a potential risk by enabling introductions of harmful species. This chapter summarises key aspects of the current knowledge on species transfers with vessels ballast water.

Keywords Non-indigenous species • Cryptogenic species • Harmful aquatic organisms • Human pathogens • Impact • Invasion rate • Shipping • Ballast water • Biofouling

S. Gollasch (✉)
Gollasch Consulting (GoConsult), Hamburg, Germany
e-mail: sgollasch@aol.com

D. Minchin
Marine Organism Investigations, Ballina, Killaloe, Co Clare, Ireland
e-mail: moiireland@yahoo.ie

M. David
Dr. Matej David Consult, Korte, Izola, Slovenia
e-mail: matej.david@siol.net

M. David, S. Gollasch (eds.), *Global Maritime Transport and Ballast Water Management*, Invading Nature - Springer Series in Invasion Ecology 8,
DOI 10.1007/978-94-017-9367-4_3

Introduction

The first possible occurrence of a non-indigenous species attributed to being introduced in ships' ballast water was by Ostenfeld (1908) who reported the Asian phytoplankton species *Odontella* (*Biddulphia*) *sinensis* following its mass occurrence in the North Sea in 1903. It was not until 70 years later that the first biological ballast water sampling (BWS) study was undertaken by Medcof (1975). This was then followed by several others (e.g., Carlton 1986; Williams et al. 1988; Locke et al. 1991; Hallegraeff and Bolch 1992; Gollasch 1996, 2002; Hamer et al. 2000; Gollasch et al. 2000a, b, 2002; Murphy et al. 2002; David et al. 2007; McCollin et al. 2008; Briski et al 2010, 2011) working in different world regions.

The annual number of new species records world-wide since 1850 has paralleled trade, both by shipping and also aquaculture developments. Improvements to ship design allowing for the construction of faster and bigger vessels has led to shorter voyage durations which almost certainly provide for a higher survival of organisms in ballast tanks, and more frequent discharges of bigger quantities of ballast water.

For example, in ICES member countries these shipping species introduction vectors together with others result in a new introduction forming a new population beyond its natural range approximately every 9 weeks (Minchin et al. 2005).

Biodiversity and environmental health has been on the agenda of aquatic ecologists for several decades and of great concern is the potential of "loss of biodiversity" due to increased anthropogenic pressures. However, already in the early biodiversity debate, few scientists highlighted that we are not only facing a "loss of biodiversity" but also a "change" or "increase" of species diversity due to human intervention. These changes may also be considered as threats to ecosystem health and services (Rosenthal pers. comm.).

Species movements with ships ballast water are in the focus of this chapter, resulting in (a) transport of native species, i.e., movements within their natural region of occurrence, (b) introduction of non-indigenous (also named non-native, alien, exotic, immigrant) species, i.e., species movements to areas where they were previously unknown and (c) movement of cryptogenic species, i.e., those species where it is not known if they are native to a region or whether they have been introduced (Carlton 1996). In each of these three categories some species arrivals are simply an addition to the biological diversity of a region without causing negative impacts, whereas a smaller number of species are considered harmful, e.g., human pathogens, and some can cause drastic changes to the receiving environments with a capability of modifying economies and with consequences for human health (e.g. Gollasch et al. 2009).

Here we describe the extent of species movements with ships ballast water worldwide and also provide some examples of the species that have been transferred and have resulted in different impacts following their arrival.

Definition of Terms

There are many different terms and definitions used around the world describing introduced species and their impacts and there is no common agreement in the scientific community or embedded in regulative/management/policy. The following paragraphs define some of the key terms used in this chapter.

Non-indigenous species are species, or other viable biological substances, that entered an ecosystem beyond its historical known range, including all organisms that have been transferred from one country to another, this includes invasive species, i.e., species causing economic or environmental harm or harm to human health (ANS Task Force 1999). A similar definition refers to non-indigenous species as any individual, group, or population of a species, or other viable biological material, that is intentionally or unintentionally moved by human activities beyond its natural range or natural zone of potential dispersal, including moves from one continent or country into another and moves within a country or region; including all domesticated and feral species, and all hybrids except for naturally occurring crosses between indigenous species. Synonyms: alien, immigrant, introduced, and non-native (EPA 2001). The IMO Guidelines G7 (IMO 2007) defines non-indigenous species as "... any species outside its native range, whether transported intentionally or accidentally by humans or transported through natural processes." This definition goes further compared to the previous ones as it includes natural transport processes while other definitions limit non-indigenous species to human-mediated species movements. It should also be noted that not all non-indigenous species are negatively impacting in the receiving environment.

The negatively impacting species, which are termed invasive species, i.e., are those species which threaten the diversity or abundance of native species; the ecological stability of infested ecosystems; economic (e.g., agricultural, aquacultural, commercial, or recreational) activities dependent on these ecosystems; and/or human health. Synonyms include harmful, injurious, invader, noxious, nuisance, pest, and weed (EPA 2001). As per this definition invasive species could be either native (see outbreak forming species below), cryptogenic or non-indigenous species. A second definition addresses invasive alien species (IAS, based on Olenin et al. 2010) as a subset of established non-indigenous species, which have spread, are spreading or have demonstrated their potential to spread elsewhere and have an adverse effect on one or more of the following: biological diversity, ecosystem functioning, socio-economic values or human health in invaded regions. However, there are also native species which cause concern which becomes in many cases clear when they occur in higher densities, examples include outbreaks of native jellyfish or mass developments of native harmful algae (outbreak forming species).

The term Harmful Aquatic Organisms and Pathogens (HAOP) appears in the IMO Ballast Water Management Convention (BWM Convention) and defines it as being any aquatic organisms or pathogens, which, if introduced into the sea including estuaries, or into fresh water courses, may create hazards to the environment, human health, property or resources, impair biological diversity or interfere with other legitimate uses of such areas (IMO 2004). As a result this term HAOP includes all potentially harmful non-indigenous, cryptogenic and impacting native aquatic species including pathogens.

Natural Species Movements

Many species have the potential to spread by their own means, for example, the migrations over long distances known for, e.g., birds that may carry associates with them that may either attach to them or otherwise infect them. Cladocerans, which are free swimming crustaceans that for part of their life cycle have a relatively smaller resting stage, have been found encrusted on birds feet and so explains how they can be spread between different separated water bodies. Otherwise their spread would not have been possible, as a result birds have been implicated in the spread of many species. Similarly, turtles have been found to spread several species found either entangled or attached to the turtle shell (e.g., Oliverio et al. 1992), such as macroalgae, bryozoans, barnacles, sea squirts, molluscs which were moved over long distances in this way (Pfaller et al. 2008).

Further, ocean currents can move species and under certain rare hydrodynamic conditions, with perturbations in the strength and direction of flow, species can be moved beyond their normal geographic range, perhaps also as a result of climate alterations, for example the increased spread of the sardine, a pilchard, which is occasionally found in the southern North Sea and western Baltic as a result of a rare northeast Atlantic Ocean water inflow and warmer water temperatures (Weber and Frieß 2003).

These natural phenomena result in changes to local species richness and may only appear on a temporary basis within a region, being known as rare guests, or vagrants. Such natural appearances, especially on the fringing ranges of a species where their ability to survive is just possible are a normal part of nature's biodiversity and is often seen as an advantage. In contrast are the human-assisted species movements which can cause irreversible negative impacts.

Human-Assisted Species Movements

In contrast to natural spread, species have been transported since humans started to explore the world. Early movements will have been with solid ballast (and the damp ballast conditions will have allowed for several attaching, sediment dwelling, or otherwise associated, near-shore and intertidal species to survive and become carried) used to stabilise wooden vessels, as attached hull fouling, with boring organisms in hulls, and what might have been carried as cargo (Eldredge and Carlton 2002; Minchin et al. 2005). Many of the movements will have been unintentional and there is little historic record of what might have been transferred several centuries ago.

There will have been many further transmissions during the periods of colonisation and wrecks of vessels may have seeded species in new regions. The more modern forms of transit will have dispensed organisms with solid ballast and utilised water in its place. With this, all ships had, and have, the capacity of spreading species unintentionally (e.g., with ships fouling or ballast, associated with the cargo of vessels or transported on deck) (e.g., Gollasch et al. 2002b; Fofonoff et al. 2003; Minchin et al. 2006, 2009; Carlton and Eldredge 2009).

Ships may transfer organisms over long distances (e.g., across oceans and seas), termed a primary introduction, whereas regional transport is considered as the facilitator of secondary transfer. It should be noted that even short distance transfers are of concern (e.g., Ruiz et al. 2000; David et al. 2007) in order to avoid negative impacts of species when being moved within, e.g., one regional sea or neighbouring waterbodies via inland canals.

In aquatic environments there are seven principal categories as to how species are spread (pathways). Each pathway enables several ways a species may be transferred (vectors). Overall, there are more than 50 recognised vectors (ICES 2005; Minchin et al. 2005). Shipping is considered to be the principal pathway worldwide, by which species are spread. The prime vectors involving shipping are ballast water and sediments accumulated at the tank bottom as well as hull fouling, where also free-living (non-fouling) species were found (Faubel and Gollasch 1996; Gollasch 2002). Species have also been transported as fouling and free living stages in sea chests (Coutts and Dodgshun 2007), as fouling inside ballast tanks, with anchor chains and as fouling in the engine cooling water pipework as well as with cargo. In some cases several vectors may be responsible for the transmission of a single species (Minchin et al. 2007a, b, c).

Ballast water also contains sediments, usually obtained in estuarine areas and shallow turbid bays. These sediment accumulations, that can range from silt to sands, settle on the bottom ballast tanks, providing a niche for infaunal organisms. As a consequence ballast tanks offer three different habitats to species (1) the water itself, (2) sediments at the tank bottom and (3) the tank walls for fouling organisms.

Species in Ballast Water Tanks

According to expert estimates, 3,000–4,000 different species are moved each day around the globe by ships (Carlton and Geller 1993; Gollasch 1996). More recent estimates indicate that the number of species in transit with ships is most probably in the range of 7,000 every day (Carlton 2001) and this does not take into account the transfer of microorganisms such as bacteria and pathogens. While even the general estimates vary greatly, the dimension of species transmission must be regarded as being exceptionally high (often referred to as colonization pressure) and it was concluded that each vessel has the potential to introduce a species (Gollasch 1996).

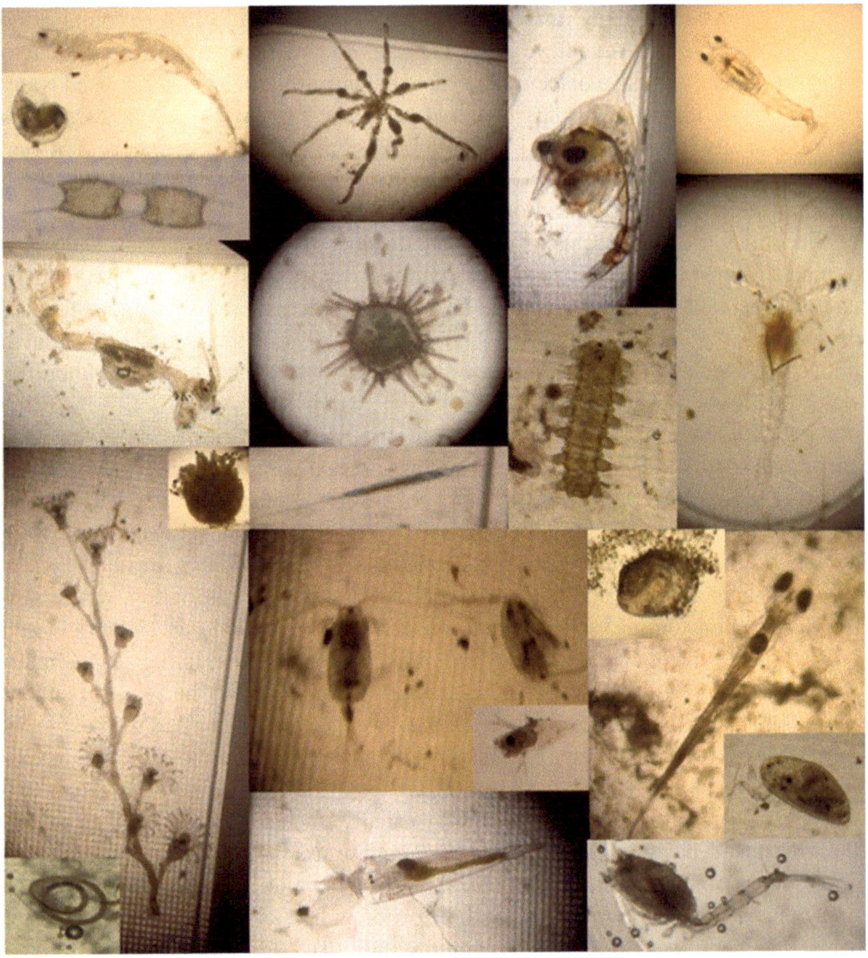

Fig. 1 Variety of species found in ballast water samples documenting that also fragile organisms survive the ballasting processes

It is the free-living, often larval, stages in the species life-cycle that are most likely to be transported with ballast water (Hewitt and Campbell 2010). The Fig. 1 shows some examples of organisms which were found in ballast water samples. Very often such stages may be taken-up during the night, since many planktonic organisms undergo vertical migrations to appear higher in the water column during darkness. These vertical migration patterns are widely recognised within marine and freshwater environments. Adult stages of bottom living organisms may also become entrained in ballast water uptake once they occur in the water column. This may be due to strong currents, storm activity or nearby dredging operations which stir up bottom sediment and organisms.

Ballast water studies conducted since the 1980s in different parts of the world have shown that ships, to an enormous extent, facilitate the transfer of aquatic

Table 1 Summary of European BWS studies indicating each study source, when it was conducted, number of vessels sampled, number of samples and number of taxa identified

Source	When	Vessels sampled	Samples taken	Taxa identified
Belgium	1995–1998	5	32	28
Denmark	2000–2001	1	8	4
England & Wales	1996–1999	132	265	320
Germany	1992–1999	198	215	521
Lithuania	1999–2000	11	22	90
Netherlands	1999–2000	17	23	88
Norway	1996–1999	51	12	184
Scotland	1994–1997	127	226	327
Slovenia	2003	15	90	134
Sweden	1996	3	>3	41
EU-CA	1998–1999	5	705	67
Total	1994–2003	565	1598	[a]More than 1,000

Gollasch (1996), Macdonald and Davidson (1998), Gollasch et al. (2000a, b), Olenin et al. (2000), Gollasch et al. (2002) and David et al. (2007)

[a]An approximation was made because several taxa were identified in more than one study

organisms across natural barriers (e.g., Williams et al. 1988; Hallegraeff and Bolch 1992; Carlton and Geller 1993; Hay 1990; Gollasch 1996; Macdonald and Davidson 1998; Ruiz et al. 2000; Gollasch et al. 2000a, b, 2002; Olenin et al. 2000; Murphy et al. 2002; David et al. 2007; Briski et al. 2010, 2011). A summary of European shipping studies revealed that 1,598 ballast water samples were collected between 1992 and 2003 on 565 vessels of different origin (see Table 1).

The diversity of living organisms (including native, cryptogenic and non-indigenous species) found during the European BWS studies included viruses, bacteria including human pathogens, fungi, protozoa, algae (unicellular phytoplankton algae and macroalgae), invertebrates and fish. Crustaceans, molluscs and polychaetes, as well as algae, were the dominant groups found in samples and consisted of more than 1,000 identified species. The majority that occurred within ballast were small in body dimensions and better able to survive the physical forces generated by the vessel pumps during the ballasting process. Nevertheless, fishes of up to 15 cm have been found within tanks (Gollasch et al. 2002) which was also documented during BWS events when testing the performance of ballast water treatment systems (Gollasch and David, own observation). A list of all animals, plants and bacteria groups found in the European BWS studies undertaken until 2002 is available in Leppäkoski et al. (2002a, b). Since this study was completed, further studies were conducted (e.g., David et al. 2007; Drake et al. 2007; Dobbs 2008; McCollin et al. 2008; Briski et al. 2010, 2011) and altogether they provided sufficient information to support the need for ballast water management actions.

The majority of organisms taken-up in ballast water expire at an exponential rate during the first 3–5 days in a ballast tank due to a wide range of conditions that occur within them (e.g., McCollin et al. 2008; Gollasch et al. 2000a, b; Olenin et al. 2000). Ballast tanks are, for most organisms, unfavourable habitats, there is an

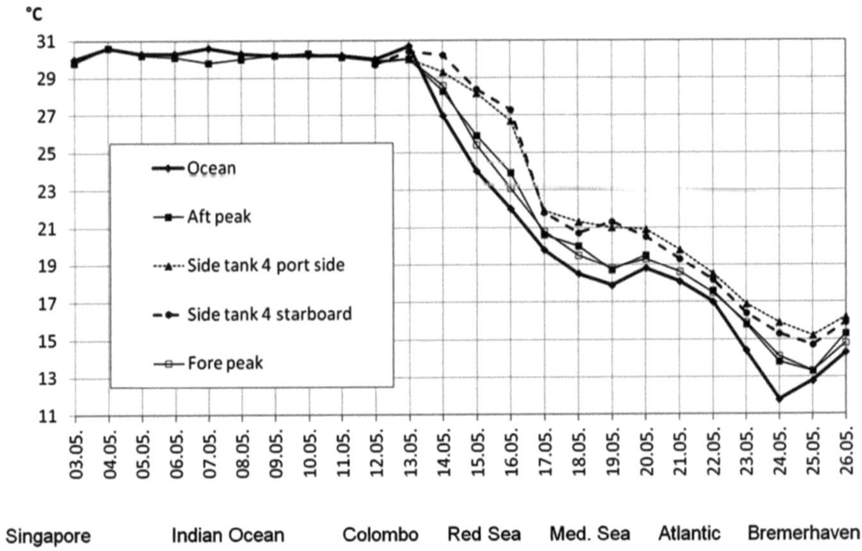

Fig. 2 Ballast water tank temperatures and sea surface temperatures during a voyage from Singapore to Bremerhaven (Reprinted from Gollasch et al. 2000a, copyright 2000, with permission from RightsLink Service, Oxford University Press)

absence of light and there can be limited resources such as oxygen, food, lack of shelter and the varying temperatures that may take place during a voyage that may considerably differ from the ballast water uptake area. Some inter-oceanic ship voyages can expose ships ballast water to wide ranges of temperature between tropical and temperate regions, this is because the ballast water gradually assumes a similar temperature to that of the ambient sea surface a ship passes through, as can happen with passages between the Pacific and Atlantic (see Fig. 2). In winter these changes can be extreme.

Although voyage duration affects the survival of organisms in ballast water, it is not a ballast water management option to retain the water onboard for long periods in order to ensure that all organisms die. Some organisms can survive 3 or more months between ballast uptake and discharge. In one case, where daily samples of ballast were taken during a voyage, a crustacean (a harpacticoid copepod), greatly increased their numbers, and had most probably reproduced inside the ballast tank during the voyage (Gollasch et al. 2000a). Some organisms can survive in ballast sediments for long periods when they develop resting stages. Some phytoplankton species, in particular dinoflagellates, several of which can generate toxins, form such resting stages (cysts) which may settle to the bottom sediment in a ballast tank. These may remain viable (in a dormant state) despite unfavourable conditions from months to years. This poses a risk since as viable cyst-forming species may be discharged during deballasting with disturbed sediments after several voyages from their uptake or with sediments when removed during tank cleaning (Hallegraeff and Bolch 1992).

Table 2 Summary of numbers of different organisms per litre

	No. tanks sampled	Numbers and range per litre	Mean SD per litre
Zooplankton	429	0–172	4.64 ± 0.71
Phytoplankton	273	$1–49.7 \times 10^6$	$299 \times 10^3 \pm 183 \times 10^3$
Bacteria	11	$2.4 \times 10^8–1.9 \times 10^9$	$8.3 \times 10^8 \pm 1.7 \times 10^7$
Viruses	7	$0.6 \times 10^9–14.9 \times 10^9$	$7.4 \times 10^9 \pm 2.9 \times 10^9$

From Gollasch and McCollin (2003) and IMO (2003)

There is not only a high diversity of species in transit with ships, but also large numbers of individuals that may be transported and that might survive to a destination port region. The overall numbers of organisms recorded from ballast water have been reported by the ICES/IOC/IMO Study Group on Ballast and Other Ships Vectors (SGBOSV) under the four headings: virus-like particles, bacteria, phytoplankton and zooplankton (see Table 2). The purpose was to provide guidance for the development of ballast water discharge standards for the BWM Convention. Any estimates of the numbers of the different groups to be in transit are likely to be underestimates because species that reside within sediments, and those planktonic species that pass through the plankton nets, using the standard mesh sizes 55 and 80 μm, during BWS, do not get considered (Gollasch and McCollin 2003). During the performance test of ballast water treatment systems more than 29,000 zooplankton organisms greater than or equal to 50 μm in minimum dimension per m^3 and more than 47,000 phytoplankton cells greater than 4 μm in minimum dimension per milliliter have been found in pumped ballast (Gollasch and David, unpublished). Such great numbers of living organisms taken up during ballasting indicates a high probability of a viable population evolving following discharge in new environments, often referred to as propagule pressure.

Chain of Events for a Species Introduction

The previous section of this chapter has shown that an enormous number of species in high organism concentrations are being transferred with ballast water. However, only the transport of a species does not result in a colonization of a new region, there is a chain of events that a species must endure in order to become established within a new environment (Carlton 1986; Hayes 1998). This starts with the uptake of ballast water. As many species have seasonal planktonic stages it is during those periods of abundance that sufficient surviving numbers may go on to later form a viable inoculum at discharge. Suspended sediments can also result in cysts and benthic biota becoming transmitted with the same ultimate capability. Having survived the uptake process the voyage(s) must be endured followed by the trauma during discharge. On arrival sufficient numbers will be needed to establish a population. The numbers required to develop new populations is generally unknown but theoretically some species can generate new populations with low numbers (Bailey et al. 2009). Survival depends on their tolerance to the conditions in the new environment and the degree of dispersal following discharge. Very often these windows of opportunity for establishment may depend on the precise location of ballast release. The colonization success may depend on the season during which arrival takes place, in cold climate regions warm water species may only survive discharge in summer, and might not subsequently survive any winter. Unless a species can reproduce a colonisation cannot evolve. Once a founder population is formed a species can then be spread by a wider range of human activity processes but also by natural processes.

Transfer and Impacts of Non-indigenous Species

Hewitt and Campbell (2010), Hayes and Gollasch (both unpublished) suggest >2,000 aquatic non-indigenous species have been introduced world-wide, of which in minimum 850 are thought to have been introduced by ships (Hayes and Sliwa 2003). There are some world regions that have greater numbers of recorded aquatic non-indigenous species present, these have often been in port regions, in sheltered bays and estuaries in regional seas (see Fig. 3).

In Europe >1,000 non-indigenous species are recorded from coastal and adjacent waters. The numbers of non-indigenous species in European seas have different patterns to all other world regions, this is because more than 50 % of the introductions occur in the Mediterranean Sea with more than 650 species records of which, at least

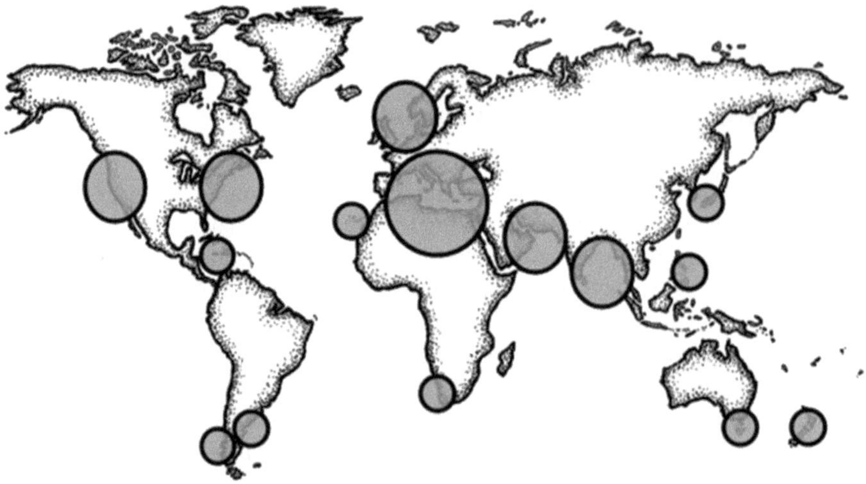

Fig. 3 Hot spots of invasive marine species. *Small circles*: <150 species, *medium circles*: 150–250 species and *large circle*: >250 species (see text for references)

325 are established. The North Sea makes up 16.2 % of the known non-indigenous species. The lowest numbers occur in Arctic waters where 18 non-indigenous species were recorded, making up only 1.3 % of the European component (Vermeij 1991; O'Mahony 1993; Boudouresque et al. 1994; Leppäkoski 1994; Eno and Clark 1994; Gollasch 1996; Olenin and Leppäkoski 1999; Reise et al. 1999; Leppäkoski and Olenin 2000; Ricciardi and MacIsaac 2000; Zaitsev and Ozturk 2001; Aladin et al. 2002; Berger and Naumov 2002; Eldredge and Carlton 2002; Golani et al. 2002; Gomoiu et al. 2002; Goulletquer et al. 2002; Hopkins 2002; Carlton and Eldredge 2009; Leppäkoski et al. 2002a, b, 2009; Minchin and Eno 2002; Occhipinti-Ambrogi 2002; Ozturk 2002; Grigorovich et al. 2003; Hewitt et al. 2004, 2007, 2009; Zenetos et al. 2004; CIESM 2005; Jensen and Knudsen 2005; Pancucci-Papadopoulou et al. 2005; Reise et al. 2005; Streftaris et al. 2005; Wolff 2005; Olenin 2005; Cardigos et al. 2006; Gollasch and Nehring 2006; Gollasch 2006; Gollasch et al. 2009; Alexandrov et al. 2007; Gittenberger 2007; Kerckhof et al. 2007; Cook et al. 2008; Olenina et al. 2010; Verlaque et al. 2010; AquaNIS 2013.[1] Katsanevakis et al. 2013).

Most species introductions almost certainly go unnoticed. Some species, either gradually or rapidly expand their populations to become invasive, a time when they become easily recognised, usually some years after an arrival. However, the great majority of non-indigenous species that are introduced are not perceived to cause harm, but it is those, that result in some form of impact, that are of concern.

The impacts of introduced species vary greatly and can cause considerable harm by modifying natural environments with consequent long-term impacts (see Box 1). While there are a comparatively small number of invasive species among all non-indigenous species that arrive, those that have impacts may have serious consequences that may endure for a considerable time. In the extreme cases these negative consequences are almost catastrophic and seemingly irreversible (e.g. Hayes and Sliwa 2003).

Box 1: Nature of Impacts

Impacts on biodiversity: predation on native communities, alteration of habitat structure and re-organisation of the trophic web, importation of diseases and disease agents, alterations of the genome (Olenin et al. 2007).

Economic losses: impacts on aquaculture production, impacts on fisheries resources, fouling of abstraction piping, impact on recreational resources.

Human health concerns: infectious cholera strains, other diseases, toxins generated by algae that contaminate foods, outbreaks of stinging jellyfish affecting swimmers, bathers cut feet on bivalve shells.

[1] AquaNIS is the information system on aquatic non-indigenous and cryptogenic species currently being developed in the European Community's Seventh Framework Programme (FP7/2007-2013) under Grant Agreement No. [266445] for the project Vectors of Change in Oceans and Seas Marine Life, Impact on Economic Sectors (VECTORS).

Not all introduced species are considered harmful, in some cases this is quite the reverse, as some provide for important industries providing employment and the sustained production of valued products. Examples include the many clam species, oysters and shrimp that have been cultivated. However, some may become so prolific to create some unwanted effects, such as the recent expansion and fouling of the Pacific oyster in the North Sea following increased recruitment, arising from changes in climate (Reid and Valdés 2011). However a latent threat to the environment, human health, property or resources remains as a non-impacting species may turn into an invasive species at a later stage.

Great harm can be caused by the introduction of one harmful species. For instance, the Chinese mitten crab *Eriocheir sinensis* has resulted in economic damage to pond fisheries and damage to river banks from burrowing with the resultant accumulations leading to increased dredging costs (e.g. Gollasch 1999). The zebra mussel *Dreissena polymorpha*, originally from the Black Sea region, has expanded its range in Europe and now is extensively distributed in North America. It has resulted in environmental changes to lakes and rivers; but, on account of its ability to attach to surfaces with byssal threads, has fouled abstraction piping and thrash racks of power stations and municipal water supplies, and continues to do so (Hebert et al. 1989; Carlton and Geller 1993; Johnson and Padilla 1996; van der Velde et al. 2010). The predatory sea star *Asterias amurensis* arrived to Australia from the north-west Pacific and has caused significant changes to bottom dwelling communities, some of economic importance (Buttermore et al. 1994; Byrne and Morrice 1997; Rossa et al. 2003). A further predator, the comb-jelly *Mnemiopsis leidyi*, was inadvertently introduced to the Black Sea from the eastern coast of the Americas. Its vast numbers resulted in heavy predation on zooplankton, including the larval stages of commercially important fishes (GESAMP 1997; Vinogradov et al. 2005). Although as a result of a further comb-jelly introduction that fed on *M. leidyi* its abundance declined in the Black Sea it appeared in the Caspain Sea carried by shipping using the interconnecting Volga-Don-Canal (Ivanov et al. 2000). It has since appeared in the Kiel Bight and has spread to several Baltic countries and to the southern North Sea (Javidpour et al. 2006) and it also expanded southwards to the Eastern Mediterranean. The North Sea invasion was overlooked for some time as the species was initially misidentified as a native comb-jelly (Faasse and Bayha 2006). Using taxonomic identification with microsatellites it was possible, for the first time for comb-jellies, to show that there have been two separate invasions of *M. leidyi* colonizing European waters from two North American source areas. The results show one originating from or near the Gulf of Mexico having arrived to the Black Sea and the North and Baltic Seas population was traced to New England populations (Reusch et al. 2010).

All of the summed impacts amount to a considerable economic cost which is difficult to quantify. In the USA alone a comprehensive study concluded that the estimated annual damage and/or control costs addressing introduced aquatic non-indigenous species is $14.2 billion (Pimentel et al. 2005). A recent summary for Europe that includes costs for repair, management and the mitigation of impacts results at more than 12 billion Euros annually (Shine et al. 2010). However, this cost includes terrestrial impacts on habitats and on services. For the aquatic sector the costs are thought to be 10–15 % of this amount.

Transfer and Impacts of Potentially Harmful Phytoplankton Species

Ballasted seawater may contain 30–>100 phytoplankton species including those being potentially toxic or others forming harmful algal blooms. These are unicellular microalgae, most usually these are diatoms and dinoflagellates, and may occur at levels of a thousand to a million or more cells per litre (Hallegraeff 1993, 1995, 1998). As a result these have great potential for global transfer and 'successfully' introduction of these species.

Over 100 years ago it was claimed that the centric diatom *Odontella sinensis*, known from tropical and subtropical coasts of the Indo-Pacific, had arrived in the ballast water of a merchant vessel, and had spread to become sufficiently abundant in the North Sea to result in plankton blooms in 1903 (Ostenfeld 1908). These blooms had no known harmful effects. It was not until the 1970s, the introduction of further centric diatom, *Coscinodiscus wailesii*, to the North Sea which clogged fishing nets due to extensive mucilaginous accretions (Boalch and Harbour 1977; Laing and Gollasch 2002). For many other phytoplankton species their origin is unknown which is also due to the taxonomic uncertainties with many phytoplankton species (see cryptogenic species below).

In Australia, an investigation found that 80 % of vessels contained approximately 30 culturable diatom species, including the potentially toxic *Pseudonitzschia* species that can cause Amnesic Shellfish Poisoning (ASP) (Forbes and Hallegraeff 1998) which can debilitate humans following consumption of contaminated shellfish. Further, cultures of viable dinoflagellates *Alexandrium catenella*, *A. tamarense* and *Gymnodinium catenatum*, all known for the toxins they can produce and consequent impacts on human health, were extracted from the ballast water of 5 % of the vessels arriving from Japan and Korea (Hallegraeff and Bolch 1992). Studies of vessels entering British ports confirmed the presence of *A. minutum*, *A. catenella and A. tamarense* in 17 % of ballast water samples (Hamer et al. 2001). In one case, a single ballast tank contained as many as 300 million viable *Alexandrium tamarense* cysts (Hallegraeff and Bolch 1992). The occurrence of such numbers in ballast water discharges may well have contributed to the widespread distribution of this species. In addition, the potentially ichthyotoxic dinoflagellate *Pfiesteria piscicida* has been confirmed using molecular probes in ballast water entering US ports (Doblin et al. 2002). In conclusion, the presence of potentially harmful marine microalgae in ballast water has been firmly established.

Dinoflagellates do not always need to form blooms in order to result in toxic events. They can occur at comparatively low densities sufficient to render cultured filter-feeding molluscs toxic. However, there are national programmes that regularly monitor for these toxins to ensure that both cultured and wild molluscs are safe for human consumption. Sudden outbreaks of toxic dinoflagellate species have taken place world-wide and have been attributed to ballast water releases (Hallegraeff 1993; David et al. 2007). On occasion their occurrence and also collapses of blooms of non-toxic species can cause de-oxygenation events to result in losses to aquaculture production, fishery landings and high mortalities of bottom living species. Although monitoring programmes exist, human casualties are also reported each year due to consumption of toxin contaminated seafood and it was found that ballast water and the sediment contained in the tanks, are one of the main (if not *the* main) transfer vectors of potentially toxic dinoflagellates (Hallegraeff 1993; David et al. 2007).

Transfer and Impacts of Cryptogenic Species

There are many species whose status is unclear because they may be native species that have recently been recognised or undergone an outbreak and their native range is not clearly known. Those species not demonstrably native or introduced are termed cryptogenic species (Carlton 1996). There are several examples that include the fouling brackish water barnacle *Balanus improvisus*, the bivalve *Mya arenaria* and the ship-worm *Teredo navalis*. Due to the taxonomic uncertainties many phytoplankton species (i.e., dinoflagellates and diatoms) are seen as cryptogenic species as many are now known from many different world regions and their identification is often a highly specialised skill, improved in recent decades using new technologies (Gómez 2008). This group of species is of special concern as many are potentially toxin producers which affects many resource users (see above).

Mya Arenaria

Already the Vikings sailed the seas and their activities may have resulted in the introduction of the North American bivalve *Mya arenaria* to Europe (Petersen et al. 1992). It was suggested that Vikings when returning from North America may have kept live *Mya arenaria* onboard either intentionally as fresh food, or unintentionally may have imported them with the solid ballast on their vessels. Excavations at Haithabu, Germany, a Viking trade hub in the Baltic Sea, revealed enormous numbers of ballast stones at and near the landing pier (see Fig. 4), supporting the probability of a species introduction with this solid ballast.

Viking ships are likely to have explored sheltered estuaries in North America, and these environments would likely have had large numbers of *Mya arenaria*.

Fig. 4 Ballast stones excavated near the vessel landing pier of the Viking trade hub Haithabu, Germany

Very likely, on account of the great importance of each vessel when not in use they were likely to have been carried on muddy shore, a habitat the soft clam also occupies, and so before any journey a supply of food may have been readily available. However, Wolff (2005) stated that the transfer of *Mya* to Europe by the Vikings poses a problem. Except for an occasional event when a vessel may have been driven off course by gales, there was no direct Viking shipping activity between North America and Europe (Marcus 1980). Greenlanders sailed to North America more frequently and also travelled between Greenland and Norway. However these voyages were not undertaken by the same vessels. It may therefore be possible that *Mya* was first introduced from North America to Greenland and subsequently from Greenland to Europe (Ockelmann 1958; Petersen 1978, 1999). In contrast, a different scenario describes that there was a gradual re-expansion of this mussel into Europe following the last glaciation period from a southern locality. So in this case it remains uncertain whether this mussel was introduced or it naturally recolonised Europe.

Teredo Navalis

The shipworm, which is not actually a worm, but a 'worm-like' mollusc, which bores into wood, is one of the oldest invaders known. They naturally spread with wooden material, within the hulls of wooden vessels from early times and is so widely spread that its native origin has become obscured. It was first recorded in Europe in 1731 (Sellius 1733), when it destroyed wooden dyke gates in the Netherlands, causing a terrible flood. At this time the Dutch believed it was introduced from Asia, possibly sent as a punishment from God (www.waddensea.org). Many naval engagements at sea may have been lost on account of the weakening effects of the boreholes on the hull and many vessels will have been disabled and wrecked on account of this damage. It was also proposed that the vessels of the Spanish armada, while waiting in French and Portuguese harbours to prepare for the invasion of England in 1588, may have been weakened in stability by the ship-worm so that the fight was lost. It was proposed that the 'shipworm' originated in the North Atlantic area (Schütz 1961) on account of its tolerance to low temperatures. This could support a possible origin from northern or southern Atlantic waters. Nowadays *T. navalis* is known to occur in Northern Europe, Indonesia, Japan, Australia, Brazil, the Atlantic and Pacific US and Canadian coasts. Many attempts were made to deter their colonisation of hulls over the centuries. Today they continue to have impacts, but on account of the usage of steel as vessel hull material, these impacts are to harbour pilings, as has happened in recent years on the Kiel Canal, Germany.

Transfer and Impacts of Human Pathogens

Human pathogens and microorganisms are also transferred with ship's ballast water (Ruiz et al. 2000; Drake et al. 2001, 2007; Casale 2002; Dobbs and Rogerson 2005; Dobbs 2008).

Particular strains of cholera, have and continue to, threaten human health worldwide. There is evidence that ships spread the pathogenic strain of this bacterium, *Vibrio cholera* O1. In 1991 the virulent form was found in Mobile Bay, Alabama, in the Gulf of Mexico (McCarthy and Khambaty 1994). It appeared in oysters that had filtered the virus arising from the discharges of ballast water (Motes et al. 1994). During a standard inspection, the US Food and Drug Administration isolated *Vibrio cholerae* O1 from the stomach content of a fish caught in Mobile Bay. The strain was similar to that found in Latin America where many humans died (Casale 2002). Indeed, the epidemic occurring in Peru was directly related to ships ballast discharges and spread to many regions in South America. In 1991 more than a million people had become infected; and by 1994 there were >10,000 victims although it is believed that their number was underestimated due to inappropriate coverage. This particular form of cholera had previously been known only in Bangladesh (Casale 2002) thereby highlighting the likelihood that ballast water transport contributed to this disease outbreak.

A study of ships in Hamilton and Toronto (Canada), at the entrance to the Great Lakes of North America, conducted in 1995 found within 71 ballast water samples a frequency of 45 % of the faecal coliform bacteria, *Escherichia coli*, and 80 % of the samples contained enterococcal bacteria (Whitby et al. 1998). Furthermore, streptococcal bacteria were found in four ballast water samples taken during this Canadian study.

Future Issues and Concerns

Monitoring of ballast water receiving habitats to document newly introduced species is rarely undertaken. Coastal monitoring programmes exist, but in many cases they lack sampling stations in ports where the ballast water is discharged or taken up. Only when new introduced species are recognized soon after introduction an eradication programme is advisable, should the newly found species cause concern. Regular monitoring using a rapid assessment approach is more likely to encounter targeted species at an earlier time (Minchin 2007b) so that those found at an early stage might be eradicated (Bax 1999). The longer a species occurs unnoticed the more unlikely an eradication programme will be successful, as during the intervening time, the species is likely to have spread over a wider area.

The identification of species is often dependent upon taxonomic skills which are not easily acquired. It may well be possible that such services, due to the reducing number of specialists, will become less available creating a consequent confusion in the area of biogeography and biological invasion science. Taxonomic skills are needed as the invasion status of an organism can normally only be assessed when the species level is identified. The lack of taxonomic expertise may also lead to overlooking introduced species, as had happened with *Mnemiopsis leidyi* in the North Sea. The presence of this species was possibly overlooked for almost a decade as the comb jellies found were confused with native species (Van Ginderdeuren et al. 2012).

Although there is an impressively high number of new non-indigenous and/or harmful species being introduced every day all around the world, relatively few cases of 'successful' invasions have been recognised. Despite relatively few invasions, a number of cases have had significant (almost catastrophic) and seemingly irreversible impacts. Consequently, a precautionary approach suggests that every vessel transporting ballast water should be treated as a potential risk by enabling introductions of harmful species.

Acknowledgements The research leading to these results has received funding from the European Community's Seventh Framework Programme (FP7/2007–2013) under Grant Agreement No. [266445] for the project Vectors of Change in Oceans and Seas Marine Life, Impact on Economic Sectors (VECTORS).

References

Aladin NV, Plotnikov IS, Filippov AA (2002) Invaders in the Caspian Sea. In: Leppäkoski E, Gollasch S, Olenin S (eds) Invasive aquatic species of Europe, distribution, impacts and management. Kluwer Academic Publishers, Dordrecht, pp 351–359

Alexandrov B, Boltachev A, Kharchenko T, Lyashenko A, Son M, Tsarenko P, Zukinsky V (2007) Trends of aquatic alien species invasions in Ukraine. Aquat Invasion 2(3):215–242

ANS Task Force (1999) Aquatic Nuisance Species (ANS) Task Force. http://anstaskforce.gov/default.php. Accessed 12 Feb 1999

AquaNIS (2013) Information system on aquatic non-indigenous species. World Wide Web electronic publication. www.corpi.ku.lt/databases/aquanis, version 2.36+. Accessed 13 Nov 2013

Bailey SA, Vélez-Espino LA, Johannsson OE, Koops MA, Wiley CJ (2009) Estimating establishment probabilities of Cladocera introduced at low density: an evaluation of the proposed ballast water discharge standards. Can J Fish Aqua Sci 66(2):261–276

Bax NJ (1999) Eradicating a dreissenid from Australia. Dreissena! 10:1–5

Berger VJA, Naumov AD (2002) Biological invasions in the White Sea. In: Leppäkoski E, Gollasch S, Olenin S (eds) Invasive aquatic species of Europe, distribution, impacts and management. Kluwer, Dordrecht, pp 235–239

Boalch GT, Harbour DS (1977) Unusual diatom off the coast of south west England and its effect on fishing. Nature 269:687–688

Boudouresque CF, Briand F, Nolan C (1994) Introduced species in European coastal waters. European Commission. In: Boudouresque CF, Briand F, Nolan C (eds) Ecosystem research report 8

Briski E, Bailey S, Cristescu ME, MacIsaac HJ (2010) Efficacy of 'saltwater flushing' in protecting the Great Lakes from biological invasions by invertebrate eggs in ships' ballast sediment. Freshw Biol 55:2414–2424

Briski E, Bailey S, MacIsaac HJ (2011) Invertebrates and their dormant eggs transported in ballast sediments of ships arriving to the Canadian coasts and the Laurentian Great Lakes. Limnol Oceanogr 56(5):1929–1939

Buttermore RE, Turner E, Morrice MG (1994) The introduced northern Pacific seastar *Asterias amurensis* in Tasmania. Mem Qld Mus 36(1):21–25

Byrne M, Morrice MG (1997) Introduction of the northern Pacific asteroid *Asterias amurensis* to Tasmania: reproduction and current distribution. Mar Biol 127(4):673–685

Cardigos F, Tempera F, Avila S, Goncalves J, Colac A¸ Santos RS (2006) Non-indigenous marine species of the Azores. Helgol Mar Res 60:160–169

Carlton JT (1986) Transoceanic and interoceanic dispersal of coastal marine organisms. The biology of ballast water. Oceanogr Mar Biol Ann Rev 23:313–371

Carlton JT (1996) Biological invasions and cryptogenic species. Ecology 77(6):1653–1655

Carlton JT (2001) Introduced species in U.S. coastal waters: environmental impacts and management priorities. Pew Oceans Commission, Arlington

Carlton JT, Eldredge LG (2009) The introduced and cryptogenic marine and estuarine animals and plants of the Hawaiian archipelago, vol 4, Bishop museum bulletin in cultural and environmental studies. Bishop Museum Press, Honolulu

Carlton JT, Geller JB (1993) Ecological roulette: the global transport of non-indigenous marine organisms. Science 261:78–82

Casale GA (2002) Ballast water – a public health issue? GloBallast Programme, IMO London, Ballast Water News. Issue 8

CIESM (2005) Alien marine organisms introduced by ships in the Mediterranean and Black seas. CIESM workshop monographs 20

Cook EJ, Ashton G, Campbell M, Coutts A, Gollasch S, Hewitt C, Liu H, Minchin D, Ruiz G, Shucksmith R (2008) Non-native aquaculture species releases: implications for aquatic ecosystems. In: Holmer M, Black K, Duarte CM, Marbà N, Karakassis I (eds) Aquaculture in the ecosystem. Springer, Heidelberg, pp 155–184

Coutts ADM, Dodgshun TJ (2007) The nature and extent of organisms in vessel sea-chests: a protected mechanism for bioinvasions. Mar Pollut Bull 54(7):875–886
David M, Gollasch S, Cabrini M, Perkovič M, Bošnjak D, Virgilio D (2007) Results from the first ballast water sampling study in the Mediterranean Sea – the Port of Koper study. Mar Pollut Bull 54:53–65
Dobbs FC (2008) Ship ballast tanks: how microbes travel the world. Microbiol Today 35:78–81
Dobbs FC, Rogerson A (2005) Ridding ships' ballast water of microorganisms. Is it even possible to remove, kill, or "inactivate" all of them—and if so, should we try? Environ Sci Tech 39:259A–264A
Doblin MA, Drake LA, Coyne KJ, Rublee PA, Dobbs FC (2002) Pfiesteria species identified in ships' ballast water and residuals: a possible vector for introductions to coastal areas. In: Steidinger KA, Landsberg JL, Tomas CR, Vargo GA (ed) Harmful Algae 2002. Florida Fish and Wildlife Conservation Commission, Florida Institute of Oceanography and Intergovernmental Oceanographic Commission of UNESCO. St. Petersburg, FL, pp 317–319
Drake LA, Choi K-H, Ruiz GM, Dobbs FC (2001) Global re-distribution of bacterioplankton and virioplankton communities. Biol Invasion 3:193–199
Drake LA, Doblin MA, Dobbs FC (2007) Potential microbial bioinvasions via ships' ballast water, sediment, and biofilm. Mar Pollut Bull 55:333–341
Eldredge LG, Carlton JT (2002) Hawaiian marine bioinvasions: a preliminary assessment. Pac Sci 56:211–212
Eno N C, Clark RA (1994) A review of non-native marine species in British waters. Appendix 1: information sheets for non-native marine species. Joint Nature Conservation Committee
EPA (2001) United States Environmental Protection Agency http://www.epa.gov/gmpo/2001annualreportfinal.pdf. Accessed 14 July 2001
Faasse MA, Bayha KM (2006) The ctenophore *Mnemiopsis leidyi* A. Agassiz 1865 in coastal waters of the Netherlands: an unrecognized invasion? Aquat Invasion 1(4):270–277
Faubel A, Gollasch S (1996) *Crptostylochus hullensis* sp. nov. (Polycladida, Acotylea, Platyhelminthes): a possible case of transoceanic dispersal on a ship's hull. Helgol Mar Res 50(4):533–537
Fofonoff PW, Ruiz GM, Steves B, Carlton JT (2003) In ships or on ships? Mechanisms of transfer and invasion for non-native species to the coasts of North America. In: Ruiz GM, Carlton JT (eds) Invasive species: vectors and management strategies. Island Press, Washington, pp 152–182
Forbes E, Hallegraeff GM (1998) Transport of potentially toxic *Pseudo-nitzschia* diatom species via ballast water. In: Proceedings of the 15th international diatom symposium, Perth, pp 509–520
GESAMP (1997) Opportunistic settlers and the problem of the ctenophore *Mnemiopsis leidyi* invasion in the Black Sea. GESAMP Rep Stud 58:1–84
Gittenberger A (2007) Recent population expansions of non-native ascidians in the Netherlands. J Exp Mar Biol Ecol 342:122–126
Golani D, Orsi-Relini L, Massutí E, Quignard JP (2002) CIESM atlas of exotic species in the Mediterranean, vol 1 Fishes. Briand F (ed) CIESM Publishers, Monaco
Gollasch S (1996) Untersuchungen des Arteintrages durch den internationalen Schiffsverkehr unter besonderer Berücksichtigung nichtheimischer Arten. Diss., Univ. Hamburg; Verlag Dr. Kovac, Hamburg
Gollasch S (1999) Current status on the increasing abundance of the Chinese Mitten Crab *Eriocheir sinensis* H. Milne Edwards, 1854 in German rivers. United States Fish and Wildlife Service
Gollasch S (2002) The importance of ship hull fouling as a vector of species introductions into the North Sea. Biofouling 18(2):105–121
Gollasch S (2006) Overview on introduced aquatic species in European navigational and adjacent waters. Helgol Mar Res 60:84–89
Gollasch S, McCollin T (2003) Report of the ICES/IOC/IMO Study Group on Ballast and Other Ship Vectors (SGBOSV). ICES, Copenhagen

Gollasch S, Nehring S (2006) National checklist for aquatic alien species in Germany. Aqua Invasion 1(4):245–269
Gollasch S, Lenz J, Dammer M, Andres HG (2000a) Survival of tropical ballast water organisms during a cruise from the Indian Ocean to the North Sea. J Plankton Res 22:923–937
Gollasch S, Rosenthal H, Botnen H, Hamer J, Laing I, Leppäkoski E, MacDonald E, Minchin D, Nauke M, Olenin S, Utting S, Voight M, Wallentinus I (2000b) Fluctuations of zooplankton taxa in ballast water during short-term and long-term ocean-going voyages. Int Rev Hydrobiol 85(5–6):597–608
Gollasch S, MacDonald E, Belson S, Botnen H, Christensen JT, Hamer JP, Houvenaghel G, Jelmert A, Lucas I, Masson D, McCollin T, Olenin S, Persson A, Wallentinus I, Wetsteyn LPMJ, Wittling T (2002) Life in ballast tanks. In: Leppäkoski E, Gollasch S, Olenin S (eds) Invasive aquatic species of Europe distribution, impact and management. Kluwer Academic Publishers, Dordrecht, pp 217–231
Gollasch S, Haydar D, Minchin D, Wolff WJ, Reise K (2009) Introduced aquatic species of the North Sea coasts and adjacent brackish waters. In: Rilov G, Crooks J (eds) Biological invasions in marine ecosystems: ecological, management and geographic perspectives, vol 204, Ecological studies. Springer, Heidelberg, pp 507–525
Gómez F (2008) Phytoplankton invasions: comments on the validity of categorising the non-indigenous dinoflagellates and diatoms in European seas. Mar Pollut Bull 56(4):620–628
Gomoiu M-T, Alexandrov B, Shadrin N, Zaitsev Y (2002) The Black Sea – a recipient, donor and transit area for alien species. In: Leppäkoski E, Gollasch S, Olenin S (eds) Invasive aquatic species of Europe, distribution, impacts and management. Kluwer, Dordrecht, pp 341–350
Goulletquer P, Bachelet G, Sauriau PG, Noel P (2002) Open Atlantic coast of Europe – a century of introduced species into French waters. In: Leppäkoski E, Gollasch S, Olenin S (eds) Invasive aquatic species of Europe. Distribution, impacts and management. Kluwer Academic Publishers, Dordrecht, pp 276–290
Grigorovich IA, Therriault TW, MacIsaac HJ (2003) History of aquatic invertebrate invasions in the Caspian Sea. Biol Invasion 5(1–2):103–115
Hallegraeff GM (1993) Human activities and the apparent global increase of harmful algal blooms. In: Sixth international conference on toxic marine phytoplankton 92
Hallegraeff GM (1995) Transport of toxic dinoflagellates via ship's ballast water: an interim review. Bd. O:15, ICES Annual Science Conference, Aalborg
Hallegraeff GM (1998) Transport of toxic dinoflagellates via ships' ballast water: bioeconomic risk assessment and efficacy of possible ballast water management strategies. Mar Ecol Progr Ser 168:297–309
Hallegraeff GM, Bolch CJ (1992) Transport of diatom and dinoflagellate resting spores in ships' ballast water: implications for plankton biogeography and aquaculture. J Plankton Res 14(8):1067–1084
Hamer JP, McCollin TA, Lucas IAN (2000) Dinofagellate cysts in ballast tank sediments: between tank variability. Mar Pollut Bull 40(9):731–733
Hamer JP, Lucas IAN, McCollin TA (2001) Harmful dinoflagellate resting cysts in ships' ballast tank sediments; potential for introduction into English and Welsh waters. Phycologia 40:246–255
Hay CH (1990) The dispersal of sporophytes of *Undaria pinnatifida* by coastal shipping in New Zealand, and implications for further dispersal of Undaria in France. Br Phycol J 25:301–313
Hayes KR (1998) Ecological risk assessment for ballast water introductions: a suggested approach. ICES J Mar Sci 55:201–212
Hayes KR, Sliwa C (2003) Identifying potential marine pests—a deductive approach applied to Australia. Mar Pollut Bull 46:91–98
Hebert PDN, Muncaster BW, Mackie GL (1989) Ecological and genetic studies on *Dreissena polymorpha* (Pallas): a new mollusc in the Great Lakes. Can J Fish Aquat Sci 46:1587–1591
Hewitt CL, Campbell ML (2010) The relative contribution of vectors to the introduction and translocation of invasive marine species. Commissioned by The Department of Agriculture, Fisheries and Forestry (DAFF), Canberra

Hewitt CL, Campbell ML, Thresher RE, Martin RB, Boyd S, Cohen BF, Currie DR, Gomon MF, Keough MJ, Lewis JA, Lockett MM, Mays N, McArthur MA, O'Hara TD, Poore GCB, Ross DJ, Storey MJ, Watson JE, Wilson RS (2004) Introduced and cryptogenic species in Port Phillip Bay, Victoria, Australia. Mar Biol 144:183–202

Hewitt CL, Campbell ML, Schaffelke B (2007) Introductions of marine macroalgae – accidental transfer pathways and mechanisms. Bot Mar 50:326–337

Hewitt C, Gollasch S, Minchin D (2009) The vessel as a vector – biofouling, ballast water and sediments. In: Rilov G, Crooks J (eds) Biological invasions in marine ecosystems: ecological, management and geographic perspectives, vol 204, Ecological studies. Springer, Heidelberg, pp 117–129

Hopkins CCE (2002) Introduced marine organisms in Norwegian waters, including Svalbard. In: Leppäkoski E, Gollasch S, Olenin S (eds) Invasive aquatic species of Europe, distribution, impacts and management. Kluwer, Dordrecht, pp 116–119

ICES (2005) International Council for the Exploration of the Sea (ICES) Code of practice on the introduction and transfers of marine organisms. Copenhagen

IMO (2003) Comments on draft Regulation E-2. Concentrations of organisms delivered in ships' ballast water in the absence of any treatment: establishing a baseline for consideration of treatment efficacy. Submitted by the International Council for the Exploration of the Sea (ICES) to the Marine Environment Protection Committee as document MEPC 49/2/21

IMO (2004) International convention on the control and management of ship's ballast water and sediments. International Maritime Organization, London. www.imo.org

IMO (2007) Guidelines for risk assessment under regulation A-4 of the BWM convention (G7). IMO, Marine Environment Protection Committee, Resolution MEPC.162(56), 13 July 2007. International Maritime Organization, London

Ivanov VP, Kamakin AM, Ushivtzev VB, Shiganova T, Zhukova O, Aladin N, Wilson SI, Harbison GR, Dumont HJ (2000) Invasion of the Caspian Sea by the comb-jellyfish *Mnemiopsis leidyi* (Ctenophora). Biol Invasion 2(3):255–258

Javidpour J, Sommer U, Shiganova T (2006) First record of *Mnemiopsis leidyi* A. Agassiz 1865 in the Baltic Sea. Aquat Invasion 1(4):299–302

Jensen KR, Knudsen J (2005) A summary of alien marine benthic invertebrates in Danish waters. Oceanol Hydrobiol Stud 34(S1):137–162

Johnson LE, Padilla DK (1996) Geographic spread of exotic species: ecological lessons and opportunities from the invasion of the zebra mussel *Dreissena polymorpha*. Biol Conserv 78:23–33

Katsanevakis S, Gatto F, Zenetos A, Cardoso AC (2013) How many marine aliens in Europe? Manag Biol Invasion 4(1):37–42

Kerckhof F, Haelters J, Gollasch S (2007) Alien species in the marine and brackish ecosystem: the situation in Belgian waters. Aquat Invasion 2(3):243–257

Laing I, Gollasch S (2002) *Coscinodiscus wailesii* – a nuisance diatom in European waters. In: Leppäkoski E, Gollasch S, Olenin S (eds) Invasive aquatic species of Europe: distribution, impacts and management. Kluwer Academic Publishers, Dordrecht, pp 53–55

Leppäkoski E (1994) Non-indigenous species in the Baltic Sea. In: Boudouresque CF, Briand F, Nolan C (eds) Introduced species in European Coastal Waters. Report of an international workshop. Bd. 8, European Commission, Luxembourg, pp 67–75

Leppäkoski E, Olenin S (2000) Non-native species and rates of spread: lessons from the brackish Baltic Sea. Biol Invasion 2:151–163

Leppäkoski E, Gollasch S, Gruszka P, Ojaveer H, Olenin S, Panov V (2002a) The Baltic – a sea of invaders. Can J Fish Aquat Sci 59:1175–1188

Leppäkoski E, Gollasch S, Olenin S (eds) (2002b) Invasive aquatic species of Europe distribution, impact and management. Kluwer Academic Publishers, Dordrecht

Leppäkoski E, Shiganova T, Alexandrov B (2009) European enclosed and semi-enclosed seas. In: Rilov G, Crooks J (eds) Biological invasions in marine ecosystems: ecological, management and geographic perspectives, vol 204, Ecological studies. Springer, Heidelberg, pp 529–547

Locke A, Reid DM, Sprules WG, Carlton JT, van Leeuwen HC (1991) Effectiveness of mid-ocean exchange in controlling freshwater and coastal zooplankton in ballast water. Can Tech Rep Fish Aquat Sci 1822:1–93
Macdonald EM, Davidson RD (1998) Ballast water project. Fisheries Research Services report 3/97. FRS Marine Laboratory, Aberdeen
Marcus GJ (1980) The conquest of the North Atlantic. The Boydell Press, Woodbridge
McCarthy SA, Khambaty FM (1994) International dissemination of epidemic *Vibrio cholera* by cargo ship ballast and other non-potable waters. Appl Environ Microbiol 60(7):2597–2601
McCollin T, Shanks AM, Dunn J (2008) Changes in zooplankton abundance and diversity after ballast water exchange in regional seas. Mar Pollut Bull 56:834–844
Medcof JC (1975) Living marine animals in a ships' ballast water. Proc Natl Shellfish Assoc 65:54–55
Minchin D (2007a) A checklist of alien and cryptogenic aquatic species in Ireland. Aquat Invasion 2(4):341–366
Minchin D (2007b) Rapid coastal survey for targeted alien species associated with floating pontoons in Ireland. Aquat Invasion 2(1):63–70
Minchin D (2007c) Aquaculture and transport in a changing environment: overlap and links in the spread of alien biota. Mar Pollut Bull 55:302–313
Minchin D, Eno C (2002) Exotics of coastal and inland waters of Ireland and Britain. In: Leppäkoski E, Gollasch S, Olenin S (eds) Invasive aquatic species of Europe: distribution, impact and management. Kluwer, Dordrecht, pp 267–275
Minchin D, Gollasch S, Wallentinus I (2005) Vector pathways and the spread of exotic species in the sea. ICES cooperative research report 271
Minchin D, Floerl O, Savini D, Occhipinti-Ambrogi A (2006) Small craft and the spread of exotic species. In: Davenport J, Davenport JL (eds) The ecology of transportation: managing mobility for the environment. Springer, Berlin, pp 77–97
Minchin D, Gollasch S, Cohen AN, Hewitt C, Olenin S (2009) Characterizing vectors of marine invasions. In: Rilov G, Crooks J (eds) Biological invasions in marine ecosystems: ecological, management and geographic perspectives, vol 204, Ecological studies. Springer, Heidelberg, pp 109–115
Motes M, DePaola A, Zywno-Van GS, McPhearson M (1994) Occurrence of toxigenic *Vibrio cholera* O1 in oysters in Mobile Bay, Alabama: an ecological investigation. J Food Protect 57(11):975–980
Murphy KR, Ritz D, Hewitt CL (2002) Heterogeneous zooplankton distribution in a ship's ballast tanks. J Plankton Res 24(7):729–734
Occhipinti-Ambrogi A (2002) Current status of aquatic introductions in Italy. In: Leppäkoski E, Gollasch S, Olenin S (eds) Invasive aquatic species of Europe, distribution, impacts and management. Kluwer, Dordrecht, pp 311–324
Ockelmann WK (1958) The zoology of East Greenland. Marine lamellibranchiata. Medd Grønland 122(4):1–256
Olenin S (2005) Invasive aquatic species in the Baltic States. Monograph. University of Klaipeda, Coastal Research and Planning Institute
Olenin S, Leppäkoski E (1999) Non-native animals in the Baltic Sea: alteration of benthic habitats in coastal inlets and lagoons. Hydrobiologia 393:233–243
Olenin S, Gollasch S, Jonushas S, Rimkute I (2000) En-route investigations of plankton in ballast water on a ships' voyage from the Baltic Sea to the open Atlantic coast of Europe. Int Rev Hydrobiol 85:577–596
Olenin S, Minchin D, Daunys D (2007) Assessment of biopollution in aquatic ecosystems. Mar Pollut Bull 55(7–9):379–394
Olenin S, Alemany FA, Cardoso C, Gollasch S, Goulletquer P, Lehtiniemi M, McCollin T, Minchin D, Miossec L, Occhipinti Ambrogi A, Ojaveer H, Jensen KR, Stankiewicz M, Wallentinus I, Aleksandrov B (2010) Marine strategy framework directive – Task Group 2 report. Non-indigenous species. Luxembourg: Office for Official Publications of the European Communities

Olenina I, Wasmund N, Hajdu S, Jurgensone I, Gromisz S, Kownacka J, Toming K, Vaiciūtė OS (2010) Assessing impacts of invasive phytoplankton: the Baltic Sea case. Mar Pollut Bull 60:1691–1700

Oliverio M, Gerosa G, Cocco M (1992) First record of *Pinctada radiata* (Bivalvia:Pteriidae) epibiont on the loggerhead sea turtle *Caretta caretta* (Chelonia:Cheloniidae). Boll Malacologico 28(5–12):149–152

O'Mahony JHT (1993) Phytoplankton species associated with imports of the Pacific oyster *Crassostrea gigas* from France to Ireland. ICES CM 1993/F:26. 8

Ostenfeld CH (1908) On the immigration of Biddulphia sinensis Grev. and its occurrence in the North Sea during 1903–1907. Medd Komm Havunders Ser Plankton 1(6):1–46

Ozturk B (2002) The Marmara Sea, a link between the Mediterranean and the Black Sea. In: Leppäkoski E, Gollasch S, Olenin S (eds) Invasive aquatic species of Europe, distribution, impacts and management. Kluwer, Dordrecht, pp 337–340

Pancucci-Papadopoulou MA, Zenetos A, Corsini-Foka M, Politou CY (2005) Update of marine aliens in Hellenic waters. Medit Mar Sci 6:147–158

Petersen GH (1978) Life cycles and population dynamics of marine benthic bivalves from the Disko Bugt area of West Greenland. Ophelia 17:95–120

Petersen GH (1999) Five Recent *Mya* species, including three new species and their fossil connections. Polar Biol 22:322–328

Petersen KS, Rasmussen KL, Heinemeier J, Rud N (1992) Clams before Columbus? Nature 359:679

Pfaller JB, Frick MG, Reich KJ, Williams KL, Bjorndal KA (2008) Carapace epibionts of loggerhead turtles (*Caretta caretta*) nesting at Canaveral National Seashore, Florida. J Nat Hist 42(13–14):1095–1102

Pimentel D, Zuniga R, Morrison D (2005) Update on the environmental and economic costs associated with alien-invasive species in the United States. Ecol Econ 52:273–288

Reid PC, Valdés L (2011) ICES status report on climate change in the North Atlantic. ICES cooperative research report no. 310

Reise K, Gollasch S, Wolff WJ (1999) Introduced marine species of the North Sea coasts. Helgol Meeresunters 52:219–234

Reise K, Dankers N, Essink K (2005) Introduced species. In: Essink K, Dettmann C, Farke H, Laursen K, Lüerßen G, Marencic H, Wiersing W (eds) Wadden Sea water quality status report 2004. Wadden Sea ecosystem no. 19. Common Wadden Sea Secretariat, Wilhelmshaven, pp 155–161

Reusch TBH, Bolte S, Sparwek M, Moss AG, Javidbour J (2010) Microsatellites reveal origin and genetic diversity of Eurasian invasions by one of the world's most notorious marine invader, *Mnemiopsis leidyi* (Ctenophora). Mol Ecol 19:2690–2699

Ricciardi A, MacIsaac HJ (2000) Recent mass invasion of the American Great Lakes by Ponto-Caspian species. Trends Ecol Evol 15(2):62–65

Rossa DJ, Johnsona CR, Hewitt CL (2003) Variability in the impact of an introduced predator (*Asterias amurensis*: Asteroidea) on soft-sediment assemblages. J Exp Mar Biol Ecol 288:257–278

Ruiz GM, Fofonoff PW, Carlton JT, Wonham MJ, Hines AH (2000) Invasion of coastal marine communities in North America: apparent patterns, processes, and biases. Ann Rev Ecol Syst 31:481–531

Schütz L (1961) Distribution and dispersal of the borer *Teredo navalis* L. and its entering of the Kiel Canal near Kiel. Kieler Meeresforsch 17:228–236

Sellius G. (1733) Historia naturalis teredinis seu xylophagi marini tubulo conchoidis speciatim Belgici. Trajecti ad Rehenum, [xxxiv]+356

Shine C, Kettunen M, Genovesi P, Essl F, Gollasch S, Rabitsch W, Scalera R, Starfinger U, ten Brink P (2010) Assessment to support continued development of the EU Strategy to combat invasive alien species. Final report for the European Commission. Institute for European Environmental Policy (IEEP), Brussels

Streftaris N, Zenetos A, Papathanassiou E (2005) Globalisation in marine ecosystems: the story of non-indigenous marine species across European seas. Oceanogr Mar Biol Ann Rev 43:419–453

Van der Velde G, Rajagopal S, bij de Vaate A (2010) The zebra mussel in Europe. Backhuys Publishers, Leiden

Van Ginderdeuren K, Hostens K, Hoffman S, Vansteenbrugge L, Soenen K, De Blauwe H, Robbens J, Vincx M (2012) Distribution of the invasive ctenophore *Mnemiopsis leidyi* in the Belgian part of the North Sea. Aquat Invasion 7(2):163–169

Verlaque M, Ruitton S, Mineur F, Boudouresque C-F (2010) CIESM atlas of exotic macrophytes in the Mediterranean Sea. http://www.ciesm.org/atlas/appendix4.html visited 22 Oct 2010)

Vermeij GJ (1991) Anatomy of an invasion: the trans-Arctic interchange. Paleobiology 17(3):281–307

Vinogradov ME, Shushkina EA, Lukasheva TA (2005) Population dynamics of the ctenophores *Mnemiopsis leidyi* and *Beroe ovata* as a predator-prey system in the near-shore communities of the Black Sea. Oceanology 45(1):S161–S167

Weber W, Frieß CC (2003) Are southern species expanding into the North Sea and Baltic? Inf Fischwirtsch Fischereiforsch 50(4):163–165

Whitby GE, Lewis DP, Shafer M, Wiley CJ (1998) A microbiological, chemical and physical survey of ballast water on ships in the Great Lakes. In: Proceedings eighth international zebra mussel & aquatic nuisance species conference, Sacramento, pp 102–109

Williams RJ, Griffiths FB, Van der Wal EJ, Kelly J (1988) Cargo vessel ballast water as a vector for the transport of non-indigenous marine species. Estuar Coast Shelf Sci 26:409–420

Wolff W (2005) Non-indigenous marine and estuarine species in the Netherlands. Zool Med 79:1–116

Zaitsev Y, Ozturk B (eds) (2001) Exotic species in the Aegean, Marmara, Black Azov and Caspian seas Turkish Marine Research Foundation, Istanbul

Zenetos A, Gofas S, Russo G, Templado J (2004) CIESM atlas of exotic species in the Mediterranean, vol 3 Molluscs. CIESM Publishers, Monaco

Policy and Legal Framework and the Current Status of Ballast Water Management Requirements

Stephan Gollasch, Matej David, Karina Keast, Naomi Parker, and Chris Wiley

Abstract There is a wealth of policy and management options addressing species introductions including conventions, treaties, multilateral agreements and codes of practices. Together these instruments support an internationally consistent management of specific transport vectors, quarantine or other biosecurity measures. This chapter lists selected global legal frameworks addressing species introductions. Chronologically, the first international instrument on unintentional introductions may have been the International Health Regulations issued in 1969 by the World Health Organization (WHO). These regulations were prepared to support public health care operations and to ensure the prevention of the spread of epidemics (e.g. plague, cholera). This chapter addresses legal frameworks addressing species introductions with the focus on ballast water related policy and legal frameworks. It gives an update on the current status of ballast water management requirements world-wide. A number of countries have taken the approach to nationally implement ballast water management requirements. We describe that most of these national requirements are based upon the IMO Ballast Water Exchange Standard,

S. Gollasch (✉)
Gollasch Consulting (GoConsult), Hamburg, Germany
e-mail: sgollasch@aol.com

M. David
Dr. Matej David Consult, Korte, Izola, Slovenia
e-mail: matej.david@siol.net

K. Keast
Australian Government Department of Agriculture, Fisheries and Forestry, Canberra, Australia
e-mail: karina.keast@daff.gov.au

N. Parker
Ministry for Primary Industries; Strategy, Systems and Science Policy Directorate; Policy Branch, Wellington, New Zealand
e-mail: naomi.parker@mpi.govt.nz

C. Wiley
Transport Canada Marine Safety, Sarnia, Canada
e-mail: chris.wiley@dfo-mpo.gc.ca

M. David, S. Gollasch (eds.), *Global Maritime Transport and Ballast Water Management*, Invading Nature - Springer Series in Invasion Ecology 8,
DOI 10.1007/978-94-017-9367-4_4

some countries refer to the Ballast Water Performance Standard and a minority addresses land-based ballast water reception facilities.

Keywords Non-indigenous species policy • Harmful aquatic organisms policy • Ballast water management policy • Ballast water management legal frameworks • Ballast water management requirements

Policy and Legal Framework for Ballast Water Management

There is a wealth of policy and management options to combat the introduction of species including conventions, treaties, multilateral agreements and codes of practices which aim to support an internationally consistent management of specific transport vectors, quarantine or other biosecurity measures (Campbell et al. 2009; Hewitt et al. 2009). These instruments regulate species transfers, control their release or address mitigation measures for introduced species populations by, e.g., eradication programmes (see Fig. 1). This chapter focusses on ballast water related policy and legal frameworks and gives an update on the current status of ballast water management (BWM) requirements world-wide.

The *International Convention for the Control and Management of Ships' Ballast Water and Sediments, London 2004* (BWM Convention) is considered as the basic global framework for BWM measures. International and national legislation provide for the prevention of harmful impacts caused by discharges of Harmful Aquatic Organisms and Pathogens (HAOP) via ballast water. Port States need to prevent unlawful acts of vessels flying their flag (i.e., Flag state obligations), as well as those occurring in their jurisdictional waters (i.e., Port State obligations) (IMO 2004). The BWM Convention and related BWM measures are addressed more in detail in chapter "Ballast Water Management Under the Ballast Water Management Convention".

Global Legal Frameworks Addressing Species Introductions

Chronologically, the first international instrument to address unintentional introductions may have been the *International Health Regulations* issued in 1969 by the World Health Organization (WHO).[1] These regulations were prepared to provide support to public health care operations and to ensure the prevention of the spread of epidemics (e.g., plague, cholera).

The first international instrument to include marine species introductions may be the *RAMSAR Convention or Convention on Wetlands of International Importance*

[1] http://whqlibdoc.who.int/publications/1983/9241580070.pdf, last accessed 02.10.2012.

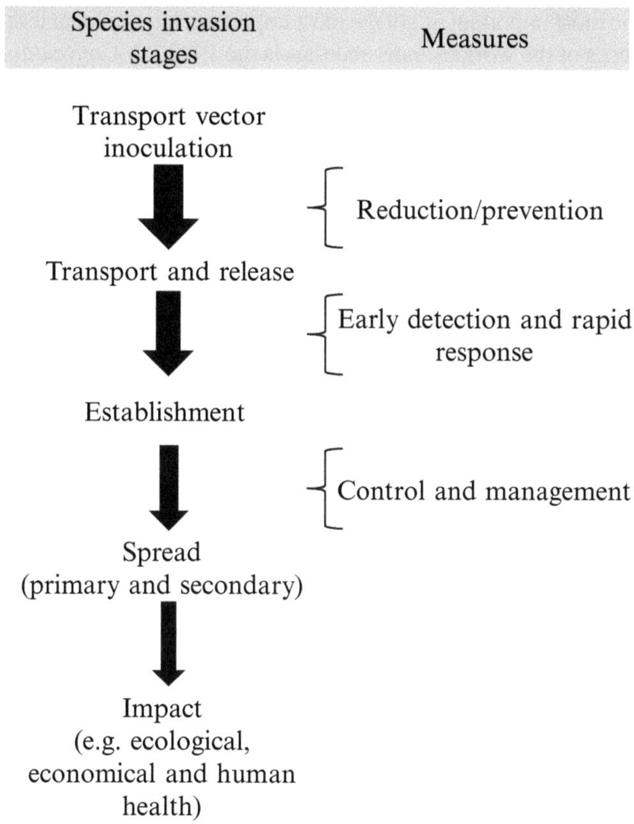

Fig. 1 Stages of non-indigenous species introductions (*left column*), policy and management options (*right column*)

Especially as Waterfowl Habitat (1971).[2] Its Resolution VII.14 requires contracting parties to, wherever possible, address within their jurisdictions the environmental, economic and social impact of invasive species on wetlands. It also suggests to "review existing legal and institutional measures and, where necessary, adopt legislation and programmes to prevent the introduction of new and environmentally dangerous alien species and the movement or trade of such species within their jurisdictions".

The *Bonn Convention for the Conservation of Migratory Species of Wild Animals* (1979) outlines in Article III(4)(c) that contracting parties are "to the extent feasible and appropriate, to prevent, reduce or control factors that are endangering or are likely to further endanger the species, including strictly controlling the introduction of, or controlling or eliminating, already introduced exotic species".

---

[2] http://www.ramsar.org/cda/en/ramsar-tourism-homeindex/main/ramsar/1%5E25816_4000_0__, last accessed 02.10.2012.

One of the most important, if not *the* most important, international legal framework to regulate uses of the world oceans and seas is the 1982 *UN Convention on the Law of the Sea* (UNCLOS) which defines the contamination of the marine environment as the direct or indirect human-mediated introduction of harmful substances or energy into the marine environment, which endangers live creatures, entails a risk to human health and impediments to marine activities, including fishing, and deteriorates the quality of sea water (UNCLOS 1982).[3] UNCLOS defines the obligations of all states concerning the protection and preservation of the marine environment by preventing its contamination, protecting and preserving rare and fragile ecological systems as well as preventing various sources of contamination from destroying plant and animal habitats.[4] Furthermore, it addresses the obligations of states concerning the implementation of all measures necessary for the prevention, reduction and control of environmental pollution from intentional and unintentional introductions of alien and new species to a particular part of the marine environment which may lead to harmful changes. Cooperation is one core mechanism in UNCLOS, especially for the management of enclosed and semi-enclosed seas as well as for research (Suarez de Vivero and Rodriguez Mateos 2002; Hewitt et al. 2009; Pavliha and Martinez Gutierrez 2010).

The 1992 *Convention on Biological Diversity* (CBD) was adopted during the UN Conference on Environment and Development in Rio de Janeiro (CBD 1992). The contracting parties shall preserve indigenous animal and plant species and improve their living conditions.[5] The CBD consists of 27 key principles to provide guidance for the future development of national and international law and decision making and actions to achieve socio-economic development and environmental protection. Among these 27 principles are (McConnell 2002):

- The Prevention Principle;
- The Precautionary Principle; and
- The Polluter-Pays Principle.

In CBD Article 8(h) parties to the Convention are called "as far as possible and as appropriate, (to) prevent the introduction of, control or eradicate those alien species which threaten ecosystems, habitats or species". As adopted in 2002, Decision VI/23 and Guiding Principles describe a "three-stage hierarchical approach" to address invasive alien species (IAS): (1) prevention of IAS introductions as the first line of defence, (2) early detection and rapid response action in cases when prevention fails, (3) eradication as the preferred IAS management option, and containment and long-term control measures as the last option, i.e. should eradication proof to be impossible. The Conference of the Parties (COP)10 of CBD (meeting held in Nagoya, 2010) adopted, for the Strategic Plan 2011–2020, the Aichi target 9: "By 2020, invasive alien species and pathways are identified and prioritized, priority species are controlled or eradicated, and measures are in place

[3] UNCLOS, Official Journal of SFRY – MP, no. 1/86, Article 1, Point 4.

[4] UNCLOS, Articles 192 and 194. Article 1, Point 4.

[5] Official Journal of RS, no. 30/96. *Convention on Biological Diversity.*

to manage pathways to prevent their introduction and establishment." The CBD highlights island biodiversity[6] as vulnerable to IAS and agrees that this aspect should represent a key area for work. A further item of priority should be protected areas and Decision X/31 calls to improve management of IAS in such areas.

Also agreed in 1992, Article 17.30 (a)(vi) of *Agenda 21*[7] provides a special provision directly related to states' international, regional and national commitments to develop governing rules for ballast water discharges to prevent introduction and spread of alien species (McConnell 2002).

In its *Code of Conduct for Responsible Fisheries*, FAO (1995)[8] states that users of living and aquatic resources should conserve aquatic ecosystems and that the right to fish carries with it the obligation to do so in a responsible manner so as to ensure effective conservation and management of the living aquatic resources. Accordingly, the objective of this Code of Practice is to establish best practice principles among nations for responsible management and fishing practices, taking into account all relevant biological, technological, economic, social, cultural and environmental aspects. This EIFAC[9]-agreed voluntary policy document has to fit alongside national legislation and regional best practice guidelines and is designed to prescribe the minimum standards for environmentally friendly, ethically appropriate and socially acceptable recreational fishing.

The *Convention on the Law of Non-Navigational Uses of International Watercourses* (1997)[10] is one of the basic documents of international water related laws. Article 22, Part 4 of the Convention, "Introduction of alien or new species" states: "Watercourse States shall take all measures necessary to prevent the introduction of species, alien or new, into an international watercourse which may have effects detrimental to the ecosystem of the watercourse resulting in significant harm to other watercourse States".

In 2000, the Species Survival Commission (SSC)[11] of the International Union for Conservation of Nature and Natural Resources (IUCN)[12] published guidelines for the prevention of biodiversity loss caused by alien species. These guidelines will support the prevention of biological diversity loss caused by alien invasive species (IUCN 2000). In 2006 IUCN published considerations for responsible use of non-indigenous species in aquaculture (Hewitt et al. 2006) and in 2013 IUCN guidelines for reintroductions and other conservation translocations of species were published (IUCN/SSC 2013).

---

[6] http://www.cbd.int/decision/cop/?id=11013, last accessed 02.04.2013.

[7] http://www.un.org/esa/sustdev/documents/agenda21/english/Agenda21.pdf, last accessed 02.10.2012.

[8] http://www.fao.org/docrep/005/v9878e/v9878e00.HTM, last accessed 02.10.2012.

[9] ftp://ftp.fao.org/fi/DOCUMENT/eifac/eifac23/default.htm, last accessed 02.10.2012.

[10] http://untreaty.un.org/ilc/texts/instruments/english/conventions/8_3_1997.pdf, last accessed 02.10.2012.

[11] http://www.iucn.org/about/work/programmes/species/who_we_are/about_the_species_survival_commission_/, last accessed 02.10.2012.

[12] http://www.iucn.org/, last accessed 02.10.2012.

In addition to the global instruments listed above a high number of regional instruments were developed to regulate non-indigenous species in general or addressing certain intentional and unintentional species introduction vectors, e.g., in aquaculture, stocking and species imports for garden and aquaria. These were not outlined in detail in this chapter because the remainder of this book chapter was written to address ballast water relevant instruments.

Legislation on Ballast Water

The International Maritime Organization (IMO) noted the negative impact of non-indigenous organisms transported in the ballast water of ships as far back as in the early 1970s (IMO 1973a). At the International Conference on Marine Pollution in 1973 the Resolution on the *Research into the Effect of Discharge of Ballast Water Containing Bacteria of Epidemic Diseases* was adopted.[13] After the acknowledgement of the problem in a 1973 resolution, the IMO, through its Marine Environment Protection Committee (MEPC), started to develop an instrument to cope with this problem in the early 1990s (IMO 1993).

The *International Convention for the Prevention of Pollution of Ships.*[14] (MARPOL) adopted by IMO in 1973 contains six regulatory annexes, each relative to a specific source of ship-generated pollution: pollution by oil, pollution by noxious liquid substances, pollution by harmful substances in packaged form, pollution by sewage from ships, pollution by garbage from ships, air pollution from ships (IMO 1973b). Initially the problems concerning ballast water discharge were to be regulated by a new annex for the prevention of uncontrolled ballast water discharge. According to the MARPOL Convention, harmful substances can be defined as any substances dumped into the sea that pose a risk to human health, are noxious to live sea organisms, or disturb any legitimate use of the sea, and should therefore as such be controlled.[15] It was later agreed at IMO that ballast water cannot be viewed as pollutant and could therefore not be covered by MARPOL.

As a first effort, the *International Guidelines for Preventing the Introduction of Unwanted Aquatic Organisms and Pathogens from Ships Ballast Water and Sediment Discharges* were adopted at the 31st Session of MEPC in July 1991. In 1993, the IMO Assembly adopted these Guidelines by Resolution A.774(18) (IMO 1993). It soon became clear thereafter that species' movements in ballast water cannot be completely prevented, and work on this matter continued at IMO. In 1997, the

[13] The Resolution noted that: "ballast water taken in waters which may contain bacteria of epidemic diseases, may, when discharged, cause a danger of spreading of the epidemic diseases to other countries". The Resolution requested IMO and WHO to "initiate studies on that problem on the basis of any evidence and proposals which may be submitted by governments".

[14] MARPOL; International Convention for the Prevention of Pollution of Ships, Official Journal SFRY – MP, no. 2/85.

[15] MARPOL, Article 2.

Guidelines for the Control and Management of Ships' Ballast Water to Minimize the Transfer of Harmful Aquatic Organisms and Pathogens were adopted by Resolution A.868(20) (IMO 1997), which replaced Resolution A.774(18).

The importance of biological invasions was brought into greater focus as several devastating introductions in many countries occurred (e.g., the Atlantic comb jelly, *Mnemiopsis leidyi*, in the Black Sea (Shiganova 1998; Ivanov et al. 2000), the zebra mussel, *Dreissena polymorpha*, in the North American Great Lakes (e.g., Karatayev et al. 2002), the Northern Pacific sea star, *Asterias amurensis*, in Australia and Tasmania (Byrne and Morrice 1997; Rossa et al. 2003), and it was recommended that IMO works towards a stand-alone Convention to address this problem. Consequently the BWM Convention was finalised and adopted at the Diplomatic Conference in London on February 13, 2004 (see chapter "Ballast Water Management Under the Ballast Water Management Convention").

A number of countries have taken the approach to nationally implement BWM requirements of which some have also ratified the BWM Convention. Most of these requirements are based upon the IMO Ballast Water Exchange (BWE) Standard (Regulation D-1), some countries refer to the Ballast Water Performance Standard (D-2 standard) and a minority addresses land-based ballast water reception facilities. Should BWE not be possible due to, e.g., safety reasons, most countries require that the next port of call should be notified that other measures can be taken, which includes a BWE in a designated coastal area or other water treatment (e.g., brine treatment in Canada[16]). Some countries further request ships to have a BWM plan and an up-to-date ballast water record book on board.

The Americas

The most comprehensive BWM requirements are implemented in North America with the USA having the most "diverse" requirements. In Central America ballast water operations in the Panama Canal are restricted and several countries in South America require BWE.

North America

Canada

The Canadian ballast water regulations apply to Canadian vessels everywhere and to vessels that are not Canadian vessels and are in waters under Canadian jurisdiction. Vessels subject to the requirements may use BWE, treatment, land-based

[16] Canada does not have any regulations requiring brine treatment, but inspectors offer it as an option for ships that arrive with tanks having salinity below 30 ppt. It is a good option only if there is a small number of tanks to be treated, and usually for residual ballast (Wang et al. 2012; Bailey personal communication).

reception facilities, no discharge of ballast water in Canadian waters, or treat water to the BWM Convention Regulation D-2 standard. Vessels inbound for the Great Lakes which carry residual amounts of ballast water only and that was not exchanged or treated are required to conduct saltwater flushing (BWCMR 2011).

In addition Canada requires, in close cooperation with the United States, an inspection of those vessels entering the St. Lawrence Seaway arriving from outside Canada's EEZ before they enter the Great Lakes. Researchers concluded that this program seems effective in reducing invasions and recommended it for other freshwater ecosystems world-wide (Bailey et al. 2011).

It is further considered to require a combined BWM measure of BWE and BWT to the D-2 standard for vessels entering the Great Lakes and research works are planned to confirm the assumption that this approach combining two methods represents a better environmental protection than a single measure (Bailey personal communication).

Exceptions from BWM requirements are described in general and for

- similar waters[17];
- areas of exclusive operation[18];
- vessels in transoceanic navigation in Alternate Ballast Water Exchange Areas;
  - Laurentian Channel[19] – east coast; and
  - west coast.[20]

Other exceptions apply in emergency situations in consultation with Transport Canada Marine Safety (BWCMR 2011).

---

[17] Ballast water that is taken on board a vessel in the United States waters of the Great Lakes Basin or in the French waters of the islands of Saint Pierre and Miquelon need not be managed unless it is mixed with other ballast water that was taken on board the vessel in any other area outside waters under Canadian jurisdiction and was not previously subjected to a management process.

[18] Exceptions are also given when vessels are exclusively operated between ports, offshore terminals and anchorage areas on the west coast of North America north of Cape Blanco (Oregon, USA); or between ports, offshore terminals and anchorage areas on the east coast of North America north of Cape Cod (Massachusetts, USA) and ports, offshore terminals and anchorage areas in the Bay of Fundy, on the east coast of Nova Scotia, or on the south or east coast of the island of Newfoundland.

[19] Should the BWE be impossible because of stability or safety reasons, for vessels on a voyage to the Great Lakes Basin, St. Lawrence River or Gulf of St. Lawrence, after notice is provided, an exchange may be conducted between December 1 and May 1, in the Laurentian Channel east of 63° west longitude where the water depth is at least 300 m.

[20] Should on a voyage to a port, offshore terminal or anchorage area on the west coast of Canada BWE as required above be impossible an exchange may be conducted in an area at least 45 NM west of Vancouver Island and the Queen Charlotte Islands and at least 45 NM west of a line extending from Cape Scott to Cape St. James where the water depth is at least 500 m, with the exception of waters within 50 NM of the Bowie Seamount (53°18′north latitude and 135°40′ west longitude).

A general exception applies for

- vessels that operate exclusively in waters under Canadian jurisdiction;
- vessels that operate exclusively in waters under Canadian jurisdiction and in the United States waters of the Great Lakes Basin or the French waters of the islands of Saint Pierre and Miquelon;
- vessels engaged in search and rescue operations that are less than 50 m in overall length and that have a maximum ballast water capacity of 8 m^3;
- pleasure craft that are less than 50 m in overall length and that have a maximum ballast water capacity of 8 m^3;
- vessels that carry permanent ballast water in sealed tanks such that it is not subject to release; or
- vessels that are owned or operated by a state and used only in government non-commercial service.

Vessels entering the Great Lakes not having performed BWE or saltwater flushing have limited alternatives available, which are expensive and time-consuming. Treatment with sodium chloride brine (initial concentration of 230‰) was suggested as an »emergency« BWM option and seems to be effective (Wang et al. 2012). Should ballast water contain lass than 30 psu it is indicated that this was not exchanged at sea and in such cases the addition of brine will be considered, possibly through opened manholes of ballast tanks.

USA

BWM requirements in the U.S. are diverse and are addressed on the federal and state level. Several approaches exist regarding numeric concentrations of organisms in ballast water discharges which include standards adopted at IMO, the U.S. Coast Guard (USCG) and the U.S. Environmental Protection Agency (USEPA), both being federal authorities. Further, selected U.S. coastal states have unilaterally implemented standards.[21] The USEPA and the USCG indicated the likeliness to possibly go beyond the IMO requirements to protect the environment (Albert et al. 2013). In the end of this section we refer to the US "common waters" approach where vessels are exempted from BWM requirements when conducting intra-coastal voyages along parts of the US west coast (Lawrence and Cordell 2010).

United States Federal BWM Regulations

In general, vessels carrying ballast water which enter a U.S. port after operating outside the EEZ have either to conduct BWE prior to entry or use an alternative BWM approach.

[21] The states with ballast water relevant rules include Arizona, California, Connecticut, Hawaii, Illinois, Indiana, Maine, Michigan, Minnesota, New York, Ohio, Rhode Island, Washington and Wisconsin (VGP 2013).

Noting BWE limitations (but see also Costello et al. 2007 who believed it is too early to conclude on its effectiveness) the USCG issued a *Final Rule*, published 2012, establishing a standard for the allowable concentration of living organisms in ships' ballast water discharged in waters of the United States to eventually replace BWE with other BWM measures. This rule requires ballast water discharges to meet the BWM Convention D-2 standard.[22] The standard is subject to a USCG review to evaluate if suitable ballast water management systems (BWMS) are available. The rule also provides details for USCG type-approval of BWMS.

The final rule requires the standard for the following vessels:

- vessels required to conduct BWE which are vessels discharging in U.S. waters with ballast water originating outside the U.S. EEZ; and
- seagoing, coastwise vessels over 1,600 gross register tons discharging ballast water in U.S. waters, but that do not operate beyond the U.S. EEZ.

This final rule also describes requirements how BWMS should be performance tested. BWMS already approved by a foreign administration according to the standards set forth in the International Maritime Organization's BWM Convention, and meeting all applicable requirements, and which are used instead of ballast water exchange, can apply to the USCG for the status of an Alternate Management System (AMS). This status is valid for no longer than 5 years from the date they would otherwise be required to comply with the ballast water discharge standard in accordance with this rule. The presumption is that, during this grace period, the manufacturers of BWMS which have been accepted as AMS will be conducting testing in accordance with USCG Type Approval regulations. Therefore, during the 5 year acceptance period, it is anticipated that the BWMS will receive U.S. Type Approval. Thus, vessels with installed AMS, will then have U.S. type approved systems and be permitted to continue using these systems and be in compliance with both the discharge standard and the requirement for use of a TA system. The implementation schedule of USCG approved BWMS is shown in Table 1.

Table 1 Implementation schedule of USCG approved BWMS

Vessel type	Ballast water capacity	Construction date	Compliance date
New	All	On or after 1st Dec 2013	On delivery
Existing	<1,500 m^3	Before 1st Dec 2013	First scheduled drydocking after 1st Jan 2016
	1,500–5,000 m^3	Before 1st Dec 2013	First scheduled drydocking after 1st Jan 2014
	>5,000 m^3	Before 1st Dec 2013	First scheduled drydocking after 1st Jan 2016

[22] This rule originally consisted of two phases, Phase 1 being the IMO D-2 standard and Phase 2 would have required compliance with a more challenging discharge standard being 1,000 times more stringent than Phase 1.

This applies to all non-recreational vessels, U.S. and foreign, that are equipped with ballast tanks and operate in the waters of the U.S., except

- Department of Defense or Coast Guard vessels,
- Any warship, naval auxiliary, or other vessel owned or operated by a foreign state and used, for the time being, only on government non-commercial service,
- Crude oil tankers[23] engaged in coastwise trade, and
- Vessels that operate exclusively within one Captain of the Port (COTP) Zone.

Some vessels are completely exempt from all BWM requirements, while some are only exempt from meeting the discharge standard. Under certain circumstances, including vessels equipped with USCG approved BWMS, vessels which use only water from a U.S. public water system, vessels with previously cleaned ballast tanks (including sediment) and vessels which discharge to a facility onshore or to another vessel for purposes of treatment, the following vessels are exempted:

- Seagoing vessels that operate in more than one COTP Zone, but do not operate outside of the Exclusive Economic Zone (EEZ), and are less than or equal to 1,600 gross register tons or less than or equal to 3,000 gross tons,
- Non-seagoing vessels,
- Vessels that take on and discharge ballast water exclusively in one COTP Zone, and
- Vessels in innocent passage, i.e. a foreign vessel that is merely traversing the territorial sea of the U.S. (unless bound for, entering or departing a U.S. port or navigating the internal waters of the U.S.).

In summary, the compliance options include USCG approved BWMS, no ballast water discharge, discharge to a reception facility, U.S. Public Water System water for ballast and AMS (as interim solution). The compliance requirements include record keeping, maintain a BWM plan, operate according to BWMPs for uptake and discharge, submit a BW reporting form and continue to meet the discharge standards.

USEPA issued another instrument to manage ballast water which is named *Vessel General Permit* (VGP). Version one was finalised in 2009 and version two was published 2011. In compliance with the provisions of the *Clean Water Act* (CWA), under the VGP, Notices of Intent (NOIs) need to be submitted in advance of discharges from a vessel to USEPA. Although named *Vessel General Permit for Discharges Incidental to the Normal Operation of Vessels* (VGP) ballast water discharges have to be reported, but we consider ballast water discharges as being a regular operations in shipping, i.e. not incidental. However, as per the VGP, it needs to be indicated whether or not a vessel will be using a BWMS and a set of questions needs to be answered including the discharge of residual biocides. This applies to all vessels greater than or equal to 300 gross tonnage or vessels with a ballast water capacity of more than 8 m^3 in ballast tanks. The instrument outlines mandatory BWM practices, including avoidance areas for ballast water uptakes, cleaning of ballast tanks regularly to remove sediments in mid-ocean or under controlled

[23] These vessels are required to meet a different standard.

arrangements in port, or at dry dock, and minimizing the discharge of ballast water essential for vessel operations while in the waters subject to the VGP. A new version of the VGP was issued in March 2013 by the USEPA (VGP 2013). The permit becomes effective 19 December 2013, when the current permit will expire.[24]

The ballast water numeric discharge limitations of VGP are identical with the *Ballast Water Performance Standard* as set out in Regulation D-2 of the BWM Convention (see Table 2). The following vessel types are not required to meet numerical ballast water discharge standards:

- Vessels engaged in short distance voyages, i.e. vessels which operate exclusively in one Coast Guard COTP Zone,
- Vessels which do not travel more than 10 nautical miles (NM) and cross no physical barriers or obstructions (e.g., locks), whether or not they operate within one U.S. Coast Guard COTP zone,
- Unmanned, unpowered barges, such as hopper barges, and
- Existing bulk carriers built before January 1, 2009, confined exclusively to the Great Lakes upstream of the Welland Canal (Lakers).

However, additional requirements exist for Lakers:

- Annual inspections of vessel to assess sediment accumulations. Removal of sediment, if necessary, must be carried out,
- When practical and safe, vessels must minimize the ballast water taken dockside, as to limiting the uptake to the amount of ballast water required to safely depart the dock and then complete ballasting in deeper water, and
- Perform annual inspections of their sea chest screens to assure that they are fully intact and repair or replace deteriorated screens.

Here the same compliance options exist as documented above for the Final Rule, but no AMS provision is given in the current draft. Further, no exemption is given for coastwise tankers as in the Final Rule. The compliance control requirements include Method Detection Limits (biocides and residuals), BWM Plan, BWM practice requirement and to meet the discharge standards.

United States Regulation by Individual States

U.S. states have the authority under the Clean Water Act to impose different ballast water standards and examples are given below. It should be noted that these standards are all subject to change.

Three U.S. states California (see Table 3), Minnesota, and Wisconsin (see Table 4) developed laws, regulations, or permits establishing numeric ballast water discharge standards or issued treatment requirements. Michigan requires the use of a ballast water treatment process approved by the Department of Environmental Quality and sets no numeric standard. Another four states included numeric ballast water discharge limits as part of their Clean Water Act certifications of the VGP,

[24] www.epa.gov/npdes/vessels, last accessed 02.04.2013.

Table 2 Proposed USCG Phase 1 and Phase 2 ballast water discharge standards. The USCG Phase 1 is identical with the requirements of the USEPA proposed VGP. The Phase 2 standard is subject to a USCG review

Standard	Organisms >50 μm	Organisms <50 />10 μm	*Vibrio cholerae* (O1 & O139)	*Escherichia coli*	Intestinal enterococci
USCG Phase 1 = USEPA VGP	<10 ind./m^3	<10 ind./ml	<1 cfu/100 ml	<250 cfu/100 ml	<100 cfu/100 ml
Proposed USCG Phase 2	<1 ind./100 m^3	<1 ind./100 ml	<1 cfu/100 ml	<126 cfu/100 ml	<33 cfu/100 ml

http://www.gpo.gov/fdsys/pkg/FR-2012-03-23/pdf/2012-6579.pdf, last accessed 02.04.2013

Table 3 California's numeric ballast water standards

Organism size class	Performance standards interim	Final standards
Organisms ≥50 μm in minimum dimension	Zero detectable living organisms	Zero detectable living organisms
Organisms ≥10–<50 μm in minimum dimension	<0.01 living organisms per ml	Zero detectable living organisms
Living Organisms <10 μm in minimum dimension:	$<10^4$ viruses/100 ml	Zero detectable living organisms (including viruses)
Escherichia coli	$<10^3$ bacteria/100 ml other than	
Intestinal enterococci	<126 CFU/100 ml	
Toxicogenic *Vibrio cholerae*	<33 CFU/100 ml	
(O1 & O139)	<1 CFU/100 ml	

Table 4 Wisconsin's numeric ballast water standards

Organism size class	Final standards
Organisms ≥50 μm in minimum dimension	<10 viable viable per m^3
Organisms ≥10–<50 μm in minimum dimension	<10 viable viable per ml
Living Organisms <10 μm in minimum dimension:	(viruses not addressed)
Escherichia coli	<126 CFU/100 ml
Intestinal enterococci	<33 CFU/100 ml
Toxicogenic *Vibrio cholerae*	(not addressed)
(O1 & O139)	

Table 5 New York's numeric ballast water standards

Organism size class	Performance standards interim	Final standards
Organisms ≥50 μm in minimum dimension	<1 living organisms per 10 m^3	Zero detectable living organisms
Organisms ≥10–<50 μm in minimum dimension	<1 living organisms per 10 ml	<0.01 living organisms per ml
Living Organisms <10 μm in minimum dimension:	(viruses not addressed)	$<10^4$ viruses/100 ml
Escherichia coli	<126 CFU/100 ml	$<10^3$ bacteria/100 ml other than
Intestinal enterococci	<33 CFU/100 ml	<126 CFU/100 ml
Toxicogenic *Vibrio cholerae*	<1 CFU/100 ml	<33 CFU/100 ml
(O1 & O139)		<1 CFU/100 ml

i.e., Illinois, Indiana, Ohio, and New York (see Table 5). New York has first delayed the standard, then withdrew and exempted all vessels from these requirements stated in the table. However, the table was kept for reasons of comparison. In addition, Pennsylvania (see Table 6), included BWM requirements, however, those were

Table 6 Revoked Pennsylvania's numeric ballast water standards

Organism size class	Performance standards interim	Final standards
Organisms ≥50 μm in minimum dimension	<10 viable organisms per m^3	Zero detectable living organisms
Organisms ≥10–<50 μm in minimum dimension	<10 viable organisms per ml	<0.01 living organisms per ml
Living Organisms <10 μm in minimum dimension:	(viruses not addressed)	$<10^4$ viruses/100 ml
Escherichia coli	<250 CFU/100 ml	$<10^3$ bacteria/100 ml other than
Intestinal enterococci	<100 CFU/100 ml	<126 CFU/100 ml
Toxicogenic *Vibrio cholerae*	<1 CFU/100 ml	<33 CFU/100 ml
(O1 & O139)		<1 CFU/100 ml

subsequently revoked. This table was also kept for comparison reasons. The numeric standards set by the following states are identical with the standards in IMO Regulation D-2: Illinois, Indiana, Michigan, Minnesota and Ohio (Albert et al. 2013). Different numeric ballast water standards of U.S. states are shown in Tables 3, 4, 5, and 6.

We further like to highlight, as an example, the complex ballast water rules of Ohio (VGP 2013). In the chapter Specific Conditions it is stated that "vessels that operate outside the U.S. Exclusive Economic Zone (EEZ) and more than 200 NM from shore, and then enter the Great Lakes via the St. Lawrence Seaway System must conduct salt water flushing of ballast tanks. This condition applies both before and after treatment system deadlines in the VGP; Vessels are prohibited from discharging ballast water sediment" (VGA, part 6.21.4). We understand that this means these vessels need to conduct an exchange of their ballast water with ocean water, which also applies to certain vessels for which ballast water treatment is required so that in this case a double management measure is required (operation of a treatment system plus water exchange). As a consequence of these measures these vessels have marine ballast water on board. In the following part 6.21.5 of this document the following is stated: "It is likely that discharges of ballasted sea water will not meet the toxicity narrative water quality standard if discharged in the relatively shallow water of Ohio's Lake Erie ports, due to the dissolved solids levels in sea water. Discharges in the open waters of the Lake minimize the risk of toxicity, and will allow the standard to be met. In order to prevent toxicity to ambient organisms or rapidly lethal conditions, discharges of ballasted sea water within the breakwalls of Ohio's Lake Erie Ports is prohibited." Does this mean that a ballast water treatment is required for some vessels which is followed by a marine ballast water exchange more than 200 NM from shore, but at the same time these vessels cannot discharge the exchanged oceanic ballast water in Ohio's waters because discharges of marine water are prohibited here? If this is the case what would a vessel operator do? Possibly conduct another ballast water exchange after having entered the Great Lakes, but before arriving in Ohio's waters.

Table 7 Revised implementation schedule of Californian interim BWM standards (*US Senate Bill 814*)

Ballast water capacity of vessel	Standards apply to new vessels in this size class constructed on or after:	Standards apply to all other vessels in this size class beginning on:
<1,500 metric tons	1st January 2016	1st January 2018
1,500–5,000 metric tons	1st January 2016	1st January 2016
>5,000 metric tons	1st January 2016	1st January 2018

California, in addition to its numeric ballast water discharge standards, requires that no unexchanged or untreated ballast water is permitted to be discharged at all, with one exception, i.e. in cases the ballast water originates from within the same port or place where it is intended to be discharged. As a result unmanaged ballast water may, e.g., be transported between different San Francisco Bay ports (Lawrence and Cordell 2010) and some of those are more than 50 NM apart.

In October 2013 the Californian standard implementation schedule was revised due to the lack of available treatment technologies to comply with the California's standards. The *Senate Bill No. 814* delays the implementation of the Californian standards by 2 years (see Table 7).

It should be noted that this delay addresses the interim standard only so that the final discharge standard of zero detectable organisms (see Table 3) is not affected.

The "Same Location" Concept Along the USA West Coast

When conducting intra-coastal voyages along parts of the U.S. west coast vessels are exempted from BWM requirements, but each U.S. Pacific coastal state has unique BWM requirements. In Washington state, ships are excepted from the required BWE when ballast water (and sediments) intended to be discharged originates solely from the "same location". This concept is in-line with the BWM Convention. The "same location" is defined here as the waters of Washington state, the Oregon part of the Columbia River system, and the interior waters of British Columbia south of 50° N, so that the Straits of Georgia and Juan de Fuca are included. Ships arriving from outside of this "same location" are required to complete BWE at least 50 NM offshore (Lawrence and Cordell 2010, Revised *Code of Washington 77.120.030*). In Oregon, the "same location" is larger, and covers the area between latitude 40 and 50° N (Lawrence and Cordell 2010, *Oregon Revised Statutes 783.630*).

In California BWM is required for all vessels. However, BWM exceptions apply for those vessels which intend to discharge ballast water (and sediments) that originates solely from the "same location" or "same port or place", both defined as an area within 1 NM of the berth or within the breakwater of a California port or place at which the ballast water was loaded. The exception also applies to vessels arriving to the "same location" from a port or place outside the Pacific Coast Region provided it can be documented that the ballast water intended to be discharged originates from mid-ocean waters and that it was not mixed with ballast water from another area. For vessels arriving from a port or place within the Pacific Coast Region the exception applies only when ballast water intended to be discharged originates from the "same port or place" and was not mixed with ballast water from another area. Two larger port regions / port complexes have already been identified as a single port or place: (1) San Francisco Bay area east of the Golden Gate Bridge, and (2) Los Angeles, Long Beach and the El Segundo marine terminal (*California's Marine Invasive Species Program 2010*; *California Code of Regulations*, Title 2, Division 3, Chapter 1, Article 4.6). As a result unmanaged ballast water may be transferred between San Francisco Bay ports, which would include ballast water transfer between Oakland and Sacramento, ports which are more than 50 NM apart. In conclusion, the Californian approach follows the "same location" logic of IMO, but the same location is here defined as a wider area. Lawrence and Cordell (2010) recommend to re-examine this exemption approach to evaluate the associated risks. For more information about the BWM Convention's same location concept see Gollasch and David (2012), David et al. (2013) and chapter "Ballast Water Management Under the Ballast Water Management Convention".

Central America

No ballast water operation is permitted in the Panama Canal (Lloyds Register 2011). This requirement is probably more related to avoid blocking of the canal, i.e. not to take a risk of machinery failure which could result in reduced manoeuvrability of vessels or as a worst case scenario in capsizing a vessel. A similar rule exists for the other two major shipping canals in the world – the Kiel and Suez Canals.

South America

BWE is widely required and applies in Argentina, Brazil (with additional requirements for the Amazon and Para River regions), Chile and Peru (Boltovskoy et al. 2011; Lloyds Register 2011). Peru also accepts ballast water treatment as a management measure (Lloyds Register 2011). More recently it became known that Colombia and Ecuador also require BWE.

Europe

The European Maritime Safety Agency (EMSA) is the responsible body for maritime safety and environmental matters also addressing ballast water related issues. In 2009 the EMSA Ballast Water Action Programme was developed and the following objectives were included:

- Development of guidance for sampling for enforcement of the BWM requirements as a basis for global guidance,

- Contributions to the development of the *Mediterranean Ballast Water Action Plan* under the Barcelona Convention/REMPEC/Globallast Partnerships and the Oslo-Paris Commission (OSPAR) Ballast Water programme,
- Information and support to the European Commission and its Member states on issues such as, type approval of BWMS, risk assessment for exemptions from BWM requirements and sampling, to ensure consistency between regional approaches in Europe and help Member states to ratify the BWM Convention, and
- Support IMO working groups and in 2011/2012 chairing the Correspondence Group to finalize the BWM circular on ballast water sampling and analysis.

The key BWM output was the development of guidelines for ballast water sampling for enforcement of BWM Convention standards, which is a key part of the EMSA Ballast Water Action Programme. In addition EMSA developed educational material and held training sessions on Flag state implementation of BWM and practical training sessions on PSC sampling for the BWM Convention were organized.

There is no common EU ballast water policy and no legally binding requirement in place. However, the BWM Convention has been signed by European countries, i.e. Croatia, Denmark, France, Germany, the Netherlands, Spain and Sweden as well as by Norway (as per December 2013). Several EU countries have announced that they are aiming to ratify the BWM Convention soon.

European countries have obligations in relation to alien species and are asked to "strictly control the introduction of non-indigenous species" (*Bern Convention on the Conservation of European Wildlife and Natural Habitats*, see above) and "eradicate those alien species which threaten ecosystems, habitats or species" (*UN Convention on Biological Diversity*, see above). Many European countries unilaterally developed invasive alien species strategies. However, a unified European-wide approach may be beneficial over national measures. The European Commission has recognised the urgent need to address non-indigenous species (*Towards an EU Strategy on Invasive Species* (EU Commission 2008)) and works towards a policy on the issue and further to establish an early warning system of newly found non-indigenous species. In September 2013 the European Commission proposed new legislation with the objective to prevent and manage the rapidly growing problem of invasive species. This proposed regulation addresses prevention, early warning and rapid response as well as management of established invasive alien species of concern. The proposal encourages a shift towards a harmonized and more preventive approach, increasing efficiency and lowering damage costs and the cost of action over time. The proposed regulation is now examined by the European Council and the Parliament. It draws on the *EU Resource Efficiency Roadmap* and the *EU Biodiversity Strategy for 2020* and refers to IMO's biofouling and ballast water instruments: "Action should include voluntary measures, such as the actions proposed by the International Maritime Organization's *Guidelines for the Control and Management of Ships' Biofouling*, and mandatory measures and should build on the experience gained in the Union and in Member states in managing certain pathways, including measures established through the *International Convention for the Control and Management of Ships Ballast Water and Sediments*." The proposed instrument requires Member states to implement action plans on the pathways of invasive alien species. This means to carry out a comprehensive pathway analysis regarding the unintentional introduction and spread of invasive alien species as well as an identification of the pathways which require priority action. This action plan should be designed to include the measures of the BWM Convention (EU Commission 2013).

BWM aspects may also be dealt with under the framework of the new *EU Maritime Policy* and the *EU Marine Strategy Framework Directive*. The *EU Maritime Policy* (EU Commission 2007) has the objective to support maritime transport competitiveness and the protection of the environmental and shipping related matters. Other subjects this instrument addresses include a long-term maritime strategy, the promotion of maritime excellence and innovation, as well as surveillance of maritime transport. Regarding more environmental aspects air pollution and greenhouse gas emissions are in focus (Suarez de Vivero 2007; Suarez de Vivero et al. 2009; Pavliha 2010). Species introduction with ballast water are not explicitly mentioned.

The *EU Marine Strategy Framework Directive* (EU Parliament 2007) requires Member states to "take the necessary measures to achieve or maintain Good Environmental Status in the marine environment" at the latest by the year 2020. The strategies ultimate goal is to reduce pollution. Twelve qualifiers of the Good Environmental Status were developed, one of those are non-indigenous species, but it is not fully clear how non-indigenous species will be addressed when describing the environmental status of the marine environment. However, it becomes clear, although not explicitly stated in the instrument, that ballast water and non-indigenous species are relevant. This Directive further promotes regional cooperation between Member states and also non-EU Member states which share the same marine environment (Suarez de Vivero et al. 2009).

North East Atlantic Ocean, North and Baltic Seas

In 2001, the Baltic area hosted the Baltic Regional Workshop on Ballast Water Management. The workshop was attended by representatives from Estonia, Finland, Germany, Latvia, Lithuania, Poland, Russia, Sweden, the Helsinki Commission, the European Commission, the Global Environment Facility (GEF) and other organizations. The objectives of the workshop were to establish a link between the Baltic and the GloBallast program, present the problem of ballast water, the introductions of alien and harmful marine species, the role of IMO, establish a BWM plan in countries lying on the shores of the Baltic Sea, devise plans for the acquisition of financial support, improve BWM, minimize introductions of harmful substances into Baltic waters, and enhance interregional cooperation. One of the outcomes of the workshop is a project conducted in four states: Estonia, Finland, Lithuania, and Russia – Alien Invasive Species in the North-East Baltic Sea: Monitoring and Assessment of Environmental Impact.[25]

[25] Global Ballast Water Management Programme: Ballast Water News, no. 10, p. 4, and no. 7 p. 6.

The first *Ballast Water Management Strategy for the North Sea* has been developed for the Issue Group on Sustainable Shipping (IGSS) of the Committee of North Sea Senior Officials' (CONSSO) in 2005. The scoping study has been led by the Maritime and Coastguard Agency (MCA) in the United Kingdom, involving also Belgium, Germany, The Netherlands, Norway and Sweden.

This region is geographically fully covered by the Helsinki Commission (HELCOM) and OSPAR. The Contracting Parties of HELCOM and OSPAR developed a voluntary interim application of the Regulation D-1 Ballast Water Exchange Standard of the BWM Convention applicable for shipping in the North-East Atlantic, North and Baltic Seas. General guidance on the voluntary and interim application of the D-1 standard was agreed by HELCOM at the 29th Meeting of the Baltic Marine Environment Protection Commission in March 2008 (HELCOM 2008; IMO 2008, 2009). Since April 1st 2008 HELCOM and OSPAR jointly agreed, also on a voluntary basis and provided safety permits, on requirements for BWM of vessels which enter the following regions of the Atlantic and Arctic Oceans and adjacent seas, including the Baltic Sea, north of 36° north latitude as well as between 42° west longitude and 51° east longitude, and the Atlantic Ocean north of 59° north latitude and between 44° and 42° west longitude. In short, the requirements include:

- Vessels entering the area have to carry a BWM Plan developed in compliance with the relevant IMO Guideline.
- Recording of all ballast water operations on all vessels entering the area.
- Ballast water of all tanks should be exchanged according to the requirements outlined in the D-1 Standard of the BWM Convention, i.e., at least 200 NM from nearest land and in waters of more than 200 m depth. These requirements apply to vessels on trans-Atlantic voyages, and for those entering the region on shipping routes passing the West African coast before entering the North-East Atlantic. In case of non-compliance, vessels are expected to conduct BWE in accordance with the same distance and depth limits within the north-east Atlantic. In those cases where this is also impossible, BWE should be carried out as far as possible from nearest land, but always at least 50 NM away and in depths of at least 200 m.

It should be noted that these guidelines are not meant to replace the requirements as outlined in the BWM Convention, but they represent the first initiative of an interim BWM strategy for this region. All ships already enabled to meet the D-2 standard of the BWM Convention should meet this standard so that these BWE guidelines no longer apply. From entry into force of the BWM Convention these guidelines become mandatory and will consecutively be replaced by the phase-in of the D-2 standard (Gollasch et al. 2007; David and Gollasch 2008). These guidelines are supported by the European Commission and are not relevant for vessels entering the region from the Mediterranean Sea (HELCOM 2008) because further guidance was developed for the Mediterranean Sea (see below).

Two countries in the region have additional BWM regimes. In Norway a ballast water exchange area was designated where ballast water of those vessels should be exchanged which claim that the IMO-required depth and distance limits for BWE could not be met. In the United Kingdom, the Flotta Terminal of the Scapa Flow port on the Orkney Islands provides a shore-based ballast water reception facility (Lloyds Register 2011). However, it is unclear if this facility is to be used for all ballast water discharges or only for ballast water which was transported in cargo holds and which may therefore be contaminated with oil from a previous cargo voyage of this vessel.

Currently HELCOM and OSPAR work towards a jointly agreed BWM related risk assessment method to harmonize the requirements and approaches for exemptions from BWM requirements according to the BWM Convention in the region (see also chapter "Risk Assessment in Ballast Water Management").

At the 11th Trilateral Governmental Wadden Sea Conference (2011) the three countries bordering the Wadden Sea (Denmark, Germany and the Netherlands) decided to develop a common strategy for dealing with alien species in the Wadden Sea. Recommendations how to deal with the species introduction vectors shipping and aquaculture were given. It was also recommended that for all BWM related work, the BWM Convention and it supporting guidelines should be followed as a regional approach (Bouma et al. 2011).

Mediterranean Sea

The Regional Marine Pollution Emergency Response Centre for the Mediterranean Sea (REMPEC) on Malta acts as regional coordinating organization. The Mediterranean strategy on BWM was adopted at the Second GloBallast Regional Task Force Meeting Regional Harmonization Workshop on Management Approaches, Istanbul, Turkey, 1–3 June 2010. Furthermore, harmonized voluntary arrangements for BWM in the Mediterranean region were adopted at the same meeting (REMPEC 2010).

The BWM arrangements are of voluntary interim nature and are applicable from 1 January 2012 until the time that the BWM Convention enters into force. This regime will no longer apply when a ship is in position to meet the ballast water performance standard contained in regulation D-2 of the Convention, or when the Convention comes into force and a ship has to apply the D-2 standard in accordance with the application dates set out in regulation B-3 of the Convention (IMO 2011). Our interpretation of this is that should vessels have BWMS installed, they should use them. However, this is on a voluntary basis.

In short, the requirements include that ships entering the waters of Mediterranean Sea area from the Atlantic Ocean (Straits of Gibraltar), or from the Indian Ocean through the Red Sea (Suez Canal) or leaving the waters of the Mediterranean Sea area to the Atlantic Ocean (Strait of Gibraltar) or to the Indian Ocean through the Red Sea (Suez Canal), should (IMO 2011):

(a) undertake BWE before entering the Mediterranean Sea area, or after leaving the Mediterranean Sea area, as applicable, according to the standard set out in the D-1 standard of the BWM Convention, and at least 200 NM from the nearest land and in waters at least 200 m in depth;
(b) in situations where this is not possible, either due to deviating the ship from its intended voyage or delaying the ship, or for safety reasons, BWE should be undertaken before entering the Mediterranean Sea area, or after leaving the Mediterranean Sea area, as applicable, according to the standard set out in the D-1 standard of the BWM Convention, as far from the nearest land as possible, and in all cases in waters at least 50 NM from the nearest land and in waters of at least 200 m depth.

From the 1st of October 2012, vessels leaving the Mediterranean Sea and sailing to destinations in North-East Atlantic or the Baltic Sea, and the vessels sailing from these areas to the Mediterranean Sea, should conduct BWE in the North-East Atlantic area and at least 200 NM from the nearest land and in waters at least 200 m in depth. If this is not possible for operational reasons, then BWE should be conducted as far as possible from the nearest land, and in all cases in waters at least 50 NM from the nearest land and in waters of at least 200 m depth (IMO 2012).

The Exceptions from BWM requirements under the Regulations A-3 may apply and the Exemptions from BWM requirements under the regulation A-4 may be granted. BWE should not in any way jeopardise vessels safety. Vessels should have onboard a Ballast Water Management Plan and keep record of all ballast operations (IMO 2011, 2012).

In the Adriatic Sea sub-region, the initiative on establishing common BWM measures is being carried through the Commission for the Protection of the Adriatic (members Croatia, Italy, Montenegro and Slovenia, and observers Bosnia and Hercegovina and Albania). In 2003, the Ballast Water Management Task Force was established to cope with the ballast water issue at the Adriatic level, which in 2004 developed into a more formal Ballast Water Management Sub-Commission (BWMSC). BWMSC is the formal body for BWM issues in the Adriatic, wherein experts and government representatives work on plans and proposal developments. There are also other regional frameworks for ballast water policies (e.g., Adriatic-Ionian Initiative, Adriatic Partnership) (David and Gollasch 2008).

In November 2013 the strategic project Ballast Water Management System for Adriatic Sea Protection (BALMAS) started. The overall BALMAS objective is to establish a common cross-border system linking all Adriatic research, experts and national responsible authorities to avoid the unwanted risks to the environment and humans from the transfer of HAOP, through the control and management of ships' ballast waters and sediments. Further, developments will be encouraged in related knowledge and technology at cross-border level for a long-term effective BWM in the Adriatic according to the BWM Convention, Europe wide developments and local specifics. BALMAS partners are research institutions and government bodies/ministries from all Adriatic countries (Italy, Slovenia, Croatia, Bosnia and Hercegovina, Montenegro and Albania). The BALMAS project will end in March 2016 (Matej David personal communication).

Black Sea

As a result of activities during the first GloBallast Programme[26] and as recommended by the Black Sea conferences on ballast water control and management a Regional Ballast Water Task Force was established to minimize the transfer of harmful aquatic organisms and pathogens in ships' ballast water.

In 2008, at an OSCE International Expert Conference on the Safety of Navigation and Environmental Security in a Transboundary Context in the Black Sea Basin, which was held in Odessa, one session dealt with BWM. Although best efforts were made to agree on a regional approach, it became clear that the Black Sea countries have divided positions regarding BWM requirements. Further, already existing national BWM requirements vary substantially within the Black Sea countries, i.e. a harmonized and agreed upon uniform approach is lacking (Kideys 2008).

The Black Sea countries BWM requirements include:

- Ballast water reporting, follow the IMO Assembly Resolution 868(20) which contains a Ballast Water Reporting Form. This is required by Bulgaria, Georgia, Turkey and Ukraine;
- Ballast water reception facilities are available in the Georgian Ports of Batumi and Poti, but it remains unclear if these are only in use for ballast water carried in empty cargo holds of oil tankers, i.e. oil-contaminated ballast water;
- A BWM scenario and requirements for BWE of ballast water originating outside the Black Sea are implemented in Bulgaria, Georgia, Ukraine and the Russian Federation (Bashtannyy et al. 2001; Georgian Ballast Water Decree 2002, Velikova personal communication);
- Georgia accepts also releases of treated ballast water (Lloyds Register 2011);
- Ballast water is monitored for chemical contamination in the Russian port Novorossiysk (BSC 2007) and non-compliance with the BWE requirements in place for this port (Lloyds Register 2011) may cause delay and/or penalties; and
- Ukraine prohibits vessels to enter their territorial waters with unexchanged ballast water (Aquatic Code of the Ukraine 1995; Berdnikov 2008). Further, Ukrainian authorities sample ballast water to assess possible chemical contamination and ballast water discharge fees apply when sampling results proof chemical contamination above certain limits (Beken et al. 2007, Velikova personal communication, Cabinet Ministers of Ukraine 2002). Ballast water needs to be exchanged and treated prior discharge (Lloyds Register 2011).

A BWM strategy is under development by Turkish authorities and Bulgaria works towards regional cooperation concerning the designation of a BWE area (Kideys 2008). All Black Sea countries are planning to harmonize procedures on a regional level, but funding is critical (Velikova personal communication, Bashtannyy et al. 2001; BSC 2007; Kideys 2008; Matheickal 2008). BWM was also incorporated in the revised *Strategic Action Plan of the Black Sea* which was adopted at a ministerial meeting in 2008 (Velikova personal communication).

---

[26] http://globallast.imo.org/index.asp?page=reports/report1.htm, last accessed 02.04.2013.

Caspian Sea

In 2005 IMO and the Caspian Environmental Programme Coordination Unit jointly organized a workshop on ballast water sampling where BWM matters were also discussed (Gollasch 2005). One consideration at the workshop was that vessels originating from outside the Caspian Sea may treat or conduct BWE prior of having left the Black/Azov and Baltic Seas. With this measure ballast water would be managed before inbound vessels enter the water-ways connecting to the Caspian Sea. The workshop participants concluded that land-based ballast water reception facilities at major Caspian Sea ports may also be a realistic BWM option.

A regional road map for the development of a regional BWM scenario in the Caspian Sea was prepared through a series of workshops facilitated by IMO and the Caspian Sea Environmental Programme (Matheickal personal communication).

Persian Gulf Region

For all ports in the Persian Gulf region BWE or treatment with a certified BWMS is required. BWE should be conducted as outlined in the BWM Convention (Lloyds Register 2011).

Malaysia

The Maritime Department Malaysia has issued their *Shipping Notice* (*MSN 04/2012*) to inform the maritime community that BWM requirements were implemented for ships constructed in or after 1st June 2012, which have a ballast water capacity of 5,000 cubic meters or more. These vessels shall conduct BWM to, at least, meet the D-2 standard. Although the BWM Convention is not implemented, the above requirement still applies. The Maritime Department further states that this notice applies to ships calling at Malaysian ports inside of its Exclusive Economic Zone after operation on the waters beyond its Exclusive Economic Zone during any part of its voyage (Malaysia Shipping Notice 2012).[27]

China

Should any ballast water discharges be planned in Chinese waters an application shall be made to the Harbour Superintendency Administration for approval (Ministry of Commerce, People's Republic of China 1979[28]). In current practice vessel

[27] http://www.ombros-consulting.com/?p=1164, last accessed 02.10.2012.

[28] http://english.mofcom.gov.cn/aarticle/lawsdata/chineselaw/200211/20021100050602.html, assessed 13.06.2013.

captains avoid any ballast water discharges in Chinese waters. In situations where ballast water operations are unavoidable, a tank to tank transfer is used (Kumar personal communication).

Australia

The Department of Agriculture, Fisheries and Forestry (DAFF) is the responsible agency within the Australian Government for the regulation of BWM on internationally arriving vessels.

DAFF is currently redrafting the legislation that governs BWM, as part of a broader reform of the biosecurity activities undertaken by the Department.

Biosecurity is protecting the economy, environment and people's health from pests and diseases. It includes trying to prevent new pests and diseases from arriving in Australia, and helping to control outbreaks when they do arrive. This terminology and the reform agenda represents a move away from the concept of "quarantine" (actions at the border) to a broader "biosecurity" framework of actions taken pre-border, at the border and post-border.

The new legislation will include requirements consistent with the BWM Convention. Under the new legislation it will be an offence for a vessel to discharge ballast water in Australian seas unless the discharge is covered by an exemption or the ballast water has been appropriately managed by conduct of an acceptable BWE or by using an approved method of BWM. An approved method of BWM includes the use of type approved BWMSs and approved prototype treatment technologies.

Until the new legislation comes into force, internationally arriving vessels will need to meet the current regulations, which prohibit the discharge of high risk ballast water inside Australia's territorial seas. High risk ballast water is defined as "all salt water from ports and coastal waters outside Australia's territorial sea". Vessels are required to manage high risk ballast water by BWE, retaining high risk ballast water on board or using fresh potable water that has been municipally sourced. Written approval must be obtained from DAFF prior to any discharge.[29]

Exceptions may apply under certain circumstances, e.g., when BWE was impossible due to safety reasons (DAFF 2011; Lloyds Register 2011).

The current regulations do not include any numerical ballast water discharge standards, however the new legislation will include reference to the D-1 and D-2 standards.

Victoria, one of seven Australian states/territories, has additional requirements for the management of domestic ballast water. Vessels visiting Victorian ports must undertake a ballast water risk assessment on a voyage by voyage basis, to assess whether their domestic ballast water is either high or low risk. Ballast water identified as high risk must be managed by conduct of a BWE outside of Australia's

[29] www.daff.gov.au/aqis/avm/vessels/quarantine_concerns/ballast/australian-ballast-water-management-requirements, last accessed 07.02.2013.

territorial sea. Vessels have to submit a ballast water report form and ballast water log to the Victorian Authority and must not discharge ballast water until written permission has been granted to do so (EPA VIC 2012).[30]

New Zealand

Vessels which uptake ballast water outside New Zealand waters can only discharge in New Zealand waters with the approval of an inspector. New Zealand accepts discharges of ballast water which was either

- exchanged at sea in areas free from coastal influences, preferably 200 NM from the nearest land and in water over 200 m in depth;
- is fresh water (not more than 2.5 ppt sodium chloride);
- treated with a shipboard treatment system approved by the relevant New Zealand authorities; or
- discharged in an onshore treatment facility approved by the relevant New Zealand authorities.

Should BWE be impossible to conduct, e.g., due to safety reasons, exceptions apply. However, vessels having uptaken ballast water in areas assessed as high risk are prohibited to discharge ballast water. It should be noted that presently no treatment facilities are approved by the New Zealand authorities nor land-based ballast water reception facilities. However, systems that have been type approved in accordance with the requirements of IMO are taken as being approved by the New Zealand authorities (MAF 2010).

As in Australia, no numerical ballast water discharge standard applies in New Zealand. However, as above, if a ship entering New Zealand waters has a type approved BWMS on board the D-2 standard would apply to the discharge. The New Zealand government has agreed to accede to the BWM Convention, and has passed the necessary primary legislation to enable this. New regulations need to be put in place under the amended primary legislation, following which New Zealand will be in a position to accede to the Convention.

Acknowledgements We express our grateful thanks to Brian Elliott, Arnaud Leroy and Mirja Ikonen (EMSA) for their updated information on BWM developments in EU as well as to Violeta Velikova, and Ahmet Kideys (Black Sea Commission, Istanbul, Turkey), for providing insights of the BWM situation in the Black Sea countries. We also thank Jose Matheickal (IMO) for his comments on the Caspian Sea chapter. We further like to thank two anonymous reviewers of the U.S. chapter and Maurya Falkner (California State Lands Commission) for her review of the U.S. same location concept. Part of this publication has been produced with the financial assistance of the IPA Adriatic Cross-Border Cooperation Programme – strategic project Ballast Water Management System for Adriatic Sea Protection (BALMAS). The contents of this publication are the sole responsibility of authors and can under no circumstances be regarded as reflecting the position of the IPA Adriatic Cross-Border Cooperation Programme Authorities.

[30] www.epa.vic.gov.au/en/your-environment/water/ballast-water, last accessed 07.02.2013.

References

Albert R, Lishman J, Saxena J (2013) Ballast water regulations and the move towards concentration-based numeric discharge limits. Ecol Appl 23:289–300

Aquatic Code of the Ukraine (1995) Article 67. In: Proceedings of the Supreme Rada (PSR), 1995, # 24. (Put in force by the SR resolution # 21495-BP of 06.06.95, BBP 1995, # 24, p 190), p 189

Bailey SA, Deneau MG, Jean L, Wiley CJ, Leung B, MacIsaac HJ (2011) Evaluating efficacy of an environmental policy to prevent biological invasions. Environ Sci Technol 45(7):2554–2561

Bashtannyy R, Webster L, Raaymakers S (2001) In: Proceedings of the first Black Sea conference on Ballast water control and management. Globallast monograph series no. 3

Beken SCP, Tolun LG, Murat S (2007) Regional gap analysis of the Black Sea strategic action plan. Final report. Institute of Environment and Chemistry, Marmara Research Center, The Scientific and Technological Research Council of Turkey, Ankara

Berdnikov V (2008) Ballast water management in the Black Sea. In: Proceedings of the international OSCE expert conference on "the safety of navigation and environmental security in a transboundary context in the Black Sea Basin", Odessa

Boltovskoy D, Almada P, Correa N (2011) Biological invasions: assessment of threat from ballast-water discharge in Patagonian (Argentina) ports. Environ Sci Policy 14(5):578–583

Bonn Convention (1979) Convention on the conservation of migratory species of wild animals. Available at: www.wcmc.org.uk/cms/cms_conv.htm. Accessed 12 Aug 2013

Bouma S, Gollasch S, Lengkeek W (2011) Neobiota in the Wadden Sea including recommendations for a trilateral strategy. Report nr 11-097. Prepared for Programma Rijke Waddenzee and the Common Wadden Sea Secretariat Wilhelmshaven

BSC (2007) Black Sea Commission (Commission on the Protection of the Black Sea Against Pollution). Final Report on Policy Measures Development and Data Analysis for 2006/2007. Istanbul

BWCMR (2011) Ballast water control and management regulations, SOR/2011-237, Canada. Available at: http://laws-lois.justice.gc.ca/eng/regulations/SOR-2011-237/index.html. Last accessed 12 July 2012

Byrne M, Morrice MG (1997) Introduction of the northern Pacific asteroid *Asterias amurensis* to Tasmania: reproduction and current distribution. Mar Biol 127(4):673–685

Cabinet Ministers of Ukraine (2002) Decree of cabinet ministers of Ukraine no. 269 dated 19.02.1996 with amendments no. 431 dated 29.03.2002

Campbell ML, Grage A, Mabin CJT, Hewitt CL (2009) Conflict between international treaties: failing to mitigate the effects of introduced marine species. Academy of the Social Sciences in Australia. Canberra. Dialogue 28(1):46–56

CBD (1992) Convention on Biological Diversity (CBD). http://www.cbd.int. Accessed 20 Oct 2013

Convention on the Law of Non-Navigational Uses of International Watercourses (1997) Adopted by the general assembly of the United Nations on 21 May 1997. United Nations Treaty Series. Accessed 20 Oct 2013

Costello C, Drake JM, Lodge DM (2007) Evaluating an invasive species policy: ballast water exchange in the Great Lakes. Ecol Appl 17(3):655–662

DAFF (2011) Australian ballast water management requirements: version 5. Department of Agriculture, Fisheries, and Forestry. http://www.daff.gov.au/__data/assets/pdf_file/0004/713884/Ballast-Water-Mgmt-Requirements.pdf. Accessed 12 July 2012

David M, Gollasch S (2008) EU shipping in the dawn of managing the ballast water issue. Mar Pollut Bull 56(12):1966–1972

David M, Gollasch S, Pavliha M (2013) Global ballast water management and the "same location" concept: a clear term or a clear issue? Ecol Appl 23(2):331–338

EPA VIC (2012) Environment protection authority, Victoria, Australia. Ballast water regulations. www.epa.vic.gov.au/en/your-environment/water/ballast-water. Accessed 7 Feb 2013

EU Commission (2007) An integrated maritime policy for the European Union. Commission of the European Communities, Communication from the Commission to the European Parliament. European Economic and Social Committee and the Committee of the Regions, Brussels, October 10, 2007. COM (2007) 575 final

EU Commission (2008) European Parliament, the European Economic and Social Committee and the Committee of the Regions. Towards an EU Strategy on Invasive Species, Brussels, December 3, 2008. COM (2008) 789 final

EU Commission (2013) Proposal for a regulation of the European parliament and of the council on the prevention and management of the introduction and spread of invasive alien species, Brussels, September 9, 2013. COM(2013) 620 final

EU Parliament (2007) Protection and conservation of the marine environment: marine strategy directive. EU Parliament, Strasbourg, 11 Dec, 2007. Provisional edition, P6_TA-PROV(2007)0595

Georgian Ballast Water Decree (2002) The order on approval of the regulations on ballast waters management procedures in Georgia. Based on the decree of the president of Georgia No. 227 "On Ship Ballast Waters Management Control" of June, 2001. Ministry of the Environment and Natural Recourses of Georgia and Ministry of Justice of Georgia State Register of Normative Acts

Gollasch S (2005) Workshop report of the IMO-UNOPS project on ballast water management in the Caspian region, Baku, November 10–11, 2005

Gollasch S, David M (2012) A unique aspect of ballast water management requirements—the same location concept. Mar Pollut Bull 64:1774–1775

Gollasch S, David M, Voigt M, Dragsund E, Hewitt CL, Fukuyo Y (2007) Critical review of the IMO international convention on the management of ships' ballast water and sediments. Harmful Algae 6:585–600

HELCOM (2008) General guidance on the voluntary interim application of the D-1 ballast water exchange standard in the north east Atlantic and the Baltic Sea. Agenda Item 2 HELCOM Baltic Sea Action Plan. Document code 2/8, 26.2.2008. Submitted by the Executive Secretary. In: Proceedings of the HELSINKI COMMISSION, Baltic Marine Environment Protection Commission, Twenty-ninth Meeting, Helsinki

Hewitt CL, Campbell ML, Gollasch S (2006) Alien species in aquaculture. Considerations for responsible use. IUCN, Gland

Hewitt CL, Everett RA, Parker N (2009) Examples of current international, regional and national regulatory frameworks for preventing and managing marine bioinvasions. In: Rilov G, Crooks J (eds) Biological invasions in marine ecosystems ecological, management and geographic perspectives. Ecol Studies 204. Springer, Berlin, pp 335–352

IMO (1973a) Conference to adopt the international convention for the prevention of pollution from ships. Resolution 18

IMO (1973b) International convention for the prevention of pollution of ships (MARPOL). www.imo.org. International Maritime Organization, London. Accessed 20 Oct 2013

IMO (1993) Guidelines for preventing the introduction of unwanted aquatic organisms and pathogens from ships' ballast water and sediment discharges. Resolution A.774(18). International Maritime Organization, London

IMO (1997) Guidelines for the control and management of ship's ballast water to minimize the transfer of harmful aquatic organisms and pathogens. Resolution A.868(20). International Maritime Organization, London

IMO (2004) International convention for the control and management of ships' ballast water and sediments, London 2004. International Maritime Organization, London

IMO (2008) International convention for the control and management of ships' ballast water and sediments, 2004, Communication received from the Administration of the United Kingdom. General guidance on the voluntary interim application of the D-1 Ballast Water Exchange Standard in the North-East Atlantic and the Baltic Sea. BWM.2/Circ.14, International Maritime Organization, London

IMO (2009) International convention for the control and management of ships' ballast water and sediments, 2004, Communication received from the Administration of Sweden. General guidance on the voluntary interim application of the D-1 Ballast Water Exchange Standard by vessels leaving the Baltic Sea and transiting the North-East Atlantic to other destinations. BWM.2/Circ.22, International Maritime Organization, London

IMO (2011) International convention for the control and management of ships' ballast water and sediments, 2004, communication received from the Regional Marine Pollution Emergency Response Centre for the Mediterranean Sea (REMPEC). Harmonized voluntary arrangements for ballast water management in the Mediterranean region. BWM.2/Circ.35, International Maritime Organization, London

IMO (2012) International convention for the control and management of ships' ballast water and sediments, 2004, Communication received from the Administration of Croatia. General guidance on the voluntary interim application of the D-1 Ballast Water Exchange Standard by vessels operating between the Mediterranean Sea and the North-East Atlantic and/or the Baltic Sea. BWM.2/Circ.39, International Maritime Organization, London

IUCN (2000) Guidelines for the prevention of biodiversity loss caused by alien invasive species. International Union for Conservation of Nature and Natural Resources, Gland

IUCN/SSC (2013) Guidelines for reintroductions and other conservation translocations. Version 1.0. International Union for Conservation of Nature and Natural Resources, Species Survival Commission, Gland

Ivanov VP, Kamakin AM, Ushivtzev VB, Shiganiva T, Zhukova O, Aladin N, Wilson SI, Harbison RG, Dumont H (2000) Invasion of the Caspian Sea by the comb jellyfish *Mnemiopsis leidyi* (Ctenophora). Biol Invasion 2:255–258

Karatayev AY, Burlakova LE, Padilla DK (2002) Impacts of zebra mussels on aquatic communities and their role as ecosystem engineers. In: Leppäkoski E, Gollasch S, Olenin S (eds) Invasive aquatic species of Europe: distribution, impacts and management. Kluwer Academic Publishers, Dordrecht, pp 433–446

Kideys A (2008) Towards ballast water management in the Black Sea. In: Proceedings of the international OSCE expert conference on "The Safety of Navigation and Environmental Security in a Transboundary Context in the Black Sea Basin". Odessa

Lawrence DJ, Cordell JR (2010) Relative contributions of domestic and foreign sourced ballast water to propagule pressure in Puget Sound, Washington, USA. Biol Conserv 143:700–709

Lloyds Register (2011) National ballast water management requirements. Lloyds Register, London

MAF (2010) Guide to New Zealand ballast water controls. Ministry of Agriculture and Forestry New Zealand. Available at: http://www.biosecurity.govt.nz/enter/ships/ballast and http://www.biosecurity.govt.nz/files/ihs/ballastwater.pdf. Accessed 12 July 2012

Malaysia Shipping Notice (2012) Notice to shipowners, ship agents, Masters, seafarers, port operators and the maritime industry. Installation of ballast water treatment system. Notics, MSN 04/2012

Matheickal J (2008) Progressing the IMO BWM convention implementation in the Black Sea Region. In: Proceedings of the international OSCE expert conference on "The Safety of Navigation and Environmental Security in a Transboundary Context in the Black Sea Basin". Odessa

McConnell M (2002) GloBallast; legislative review – final report. GloBallast monograph series no. 1. International Maritime Organization, London

Ministry of Commerce, People's Republic of China (1979) Regulations governing supervision and control of foreign vessels by the People's Republic of China. Chapter VII, Prevention of pollution, Article 36. Available at: http://english.mofcom.gov.cn/aarticle/lawsdata/chineselaw/200211/20021100050602.html. Assessed 13 June 2013

Pavliha M (2010) New European maritime policy for cleaner oceans and seas. In: Martinez Gutierrez NA (ed) Serving the rule of international maritime law: essays in honour of professor David Joseph Attard (Imli Studies in International Maritime Law). Routledge (Francis & Tailor Group), Milton Park, pp 22–31

Pavliha M, Martinez Gutierrez NA (2010) Marine scientific research and the 1982 United Nations convention on the Law of the sea. Ocean Coast Law J 16(1):115–133

RAMSAR Convention or convention on wetlands of international importance especially as waterfowl habitat (1971) Available at: www.ramsar.org. Accessed 11 Aug 2013

REMPEC (2010) Second GloBallast regional task force meeting report. REMPEC/WG.31/8. Istanbul

Rossa DJ, Johnsona CR, Hewitt CL (2003) Variability in the impact of an introduced predator (*Asterias amurensis*: Asteroidea) on soft-sediment assemblages. J Exp Mar Biol Ecol 288:257–278

Shiganova TA (1998) Invasions of the Black Sea by the ctenophore *Mnemiopsis leidyi* and recent changes in pelagic community structure. Fish Oceanogr 7:305–310

Suarez de Vivero JL (2007) The european vision for oceans and seas – social and political dimensions of the Green Paper on maritime policy for the EU. Mar Policy 31:409–414

Suarez de Vivero JL, Rodrıguez Mateos JC (2002) The Mediterranean and Black Sea: regional integration and maritime nationalism. Mar Policy 26:383–401

Suarez de Vivero JL, Rodrıguez Mateos JC, Florido del Corral D (2009) Geopolitical factors of maritime policies and marine spatial planning: State, regions, and geographical planning scope. Mar Policy 33:624–634

UNCLOS (1982) UN convention on the law of the sea (UNCLOS). United Nations Treaty Series. Accessed 20 Oct 2013

VGP (2013) Vessel general permit for discharges incidental to the normal operation of vessels (VGP). Final version

Wang TN, Bailey SA, Reid DF, Johengen TH, Jenkins PT, Wiley CJ, MacIsaac HJ (2012) Efficacy of NaCl brine for treatment of ballast water against freshwater invasions. J Great Lakes Res 38:72–77

Ballast Water Management Under the Ballast Water Management Convention

Matej David, Stephan Gollasch, Brian Elliott, and Chris Wiley

Abstract The importance of ballast water as a vector for moving non-indigenous species was initially addressed in a 1973 International Maritime Organization (IMO) resolution. Subsequently IMO worked towards the finalization of the *International Convention for the Control and Management of Ships' Ballast Water and Sediments* (BWM Convention) which was adopted in February 2004 at a diplomatic conference in London. The BWM Convention's main aim is to prevent, minimize and ultimately eliminate the risks to the environment, human health, property and resources which arise from the transfer of harmful aquatic organisms and pathogens via ships' ballast waters and related sediments. It should be noted that harmful aquatic organisms in this context are not limited to non-indigenous species, but covers all aquatic species irrespective of their origin. As defined at IMO "Ballast Water Management means mechanical, physical, chemical, and biological processes, either singularly or in combination, to remove, render harmless, or avoid the uptake or discharge of Harmful Aquatic Organisms and Pathogens within Ballast Water and Sediments." The BWM Convention and its supporting guidelines are described in this chapter, outlining the ballast water exchange and performance standards, warnings concerning ballast water uptake in certain areas, ballast water reception facilities, sediment management as well as exemptions and exceptions from ballast water management requirements. This chapter ends with the description of implementation options of the BWM Convention.

M. David (✉)
Dr. Matej David Consult, Korte, Izola, Slovenia
e-mail: matej.david@siol.net

S. Gollasch
Gollasch Consulting (GoConsult), Hamburg, Germany
e-mail: sgollasch@aol.com

B. Elliott
European Maritime Safety Agency, Lisbon, Portugal
e-mail: brian.elliott@emsa.europa.eu

C. Wiley
Transport Canada Marine Safety, Sarnia, Canada
e-mail: chris.wiley@dfo-mpo.gc.ca

M. David, S. Gollasch (eds.), *Global Maritime Transport and Ballast Water Management*, Invading Nature - Springer Series in Invasion Ecology 8,
DOI 10.1007/978-94-017-9367-4_5

Keywords Ballast water management • Ballast water management convention • Ballast water guidelines • Ballast water exchange standard • Ballast water performance standard • Reception facilities • Sediment management • Exemptions • Exceptions • Ballast water management convention implementation

The Ballast Water Management Convention

The importance of ballast water as a vector for moving non-indigenous species was initially addressed in a 1973 International Maritime Organization (IMO) resolution (IMO 1973). Subsequently IMO worked towards the finalization of the *International Convention for the Control and Management of Ships' Ballast Water and Sediments* (BWM Convention) which was adopted in February 2004 at a diplomatic conference in London (IMO 2004). This Convention's aim is to prevent, minimize and ultimately eliminate the risks to the environment, human health, property and resources which arise from the transfer of harmful aquatic organisms and pathogens (HAOP) via ships' ballast waters and related sediments. It should be noted that harmful aquatic organisms in this context are not limited to non-indigenous species, but covers all species irrespective of their origin.

The BWM Convention consists of 22 Articles followed by five sections with Regulations. In addition, two Appendices provide standard formats and requirements regarding the form of International Ballast Water Management Certificates as well as recording operations for reporting and verification in a Ballast Water Record Book.

The Regulations for the control and management of ships' ballast water and sediments are presented in five sections:

- Section A: General provisions: Definitions, General applicability, Exceptions, Exemptions, Equivalent Compliance;
- Section B: Management and control Requirements for Ships: Ballast Water Management;
- Section C: Special Requirements in Certain Areas;
- Section D: Standards for Ballast Water Management; and
- Section E: Survey and Certification requirements for Ballast Water Management.

Certain obligations are to be met by all stakeholders including the ship, the Administrations, i.e., both in their capacity as Flag state, Port State, and as the representative of a Party, and IMO.

The BWM Convention enters into force 12 months after the date on which more than 30 states, with combined merchant fleets not less than 35 % of the gross tonnage of the world's merchant shipping, have signed this Convention. As of December 2013, 38 states ratified the BWM Convention, representing 30.38 % of the world merchant shipping gross tonnage (for an update visit Status of Conventions at www.imo.org).

Table 1 Guidelines to the BWM Convention and their development status. MEPC = IMO Marine Environment Protection Committee

Title	Work progress
Guidelines for Sediment Reception Facilities (G1)	Adopted at MEPC 55, Oct. 2006 (IMO 2006a)
Guidelines for Ballast Water Sampling (G2)	Adopted at MEPC 58, Oct. 2008 (IMO 2008a)
Guidelines for Ballast Water Management Equivalent Compliance (G3)	Adopted at MEPC 53, Jul. 2005 (IMO 2005a)
Guidelines for Ballast Water Management and Development of Ballast Water Management Plans (G4)	Adopted at MEPC 53, Jul. 2005 (IMO 2005b)
Guidelines for Ballast Water Reception Facilities (G5)	Adopted at MEPC 55, Oct. 2006 (IMO 2006b)
Guidelines for Ballast Water Exchange (G6)	Adopted at MEPC 53, Jul. 2005 (IMO 2005c)
Guidelines on Risk Assessments under Regulation A-4 (G7)	Adopted at MEPC 56, Jul. 2007 (IMO 2007a)
Guidelines for the Approval of Ballast Water Management Systems (G8)	Adopted at MEPC 53, Jul. 2005, amended at MEPC 58, Oct. 2008 (IMO 2008b)
Procedure for Approval of Ballast Water Management Systems that make use of Active Substances (G9)	Adopted at MEPC 53, Jul. 2005, amended at MEPC 57, Apr. 2008 (IMO 2008c)
Guidelines for Approval and Oversight of Prototype Ballast Water Treatment Technology Programmes (G10)	Adopted at MEPC 54, Mar. 2006 (IMO 2006c)
Guidelines for Ballast Water Exchange Design and Construction Standard (G11)	Adopted at MEPC 55, Oct. 2006 (IMO 2006d)
Guidelines for Sediment Control on Ships (G12)	Adopted at MEPC 55, Oct. 2006 (IMO 2006e)
Guidelines for Additional Measures Including Emergency Situations (G13)	Adopted at MEPC 56, Jul. 2007 (IMO 2007b)
Guidelines on Designation of Areas for Ballast Water Exchange (G14)	Adopted at MEPC 55, Oct. 2006 (IMO 2006f)
Guidelines for Port State Control	In preparation

In total 15 guidelines support the uniform implementation of the BWM Convention (see Table 1) by providing technical guidance to implement the BWM Convention principles. The majority of these guidelines (G1–G14) have already been adopted, however the *Guidelines for Port State Control* that have the purpose of harmonizing port State control activities and to define criteria for a detailed inspection of the ship (Article 9 of the BWM Convention) are still not yet finalised.

Guidelines at IMO are intended to be at high level, providing an overall structure for the implementation of the BWM Convention. However, because of the highly complex nature of the subject matter and the sophistication of the technology, many Guidelines have become quite specific and detailed.

Agreements reached on a global level usually represent a compromise, and the BWM Convention is not an exception. During the BWM Convention and over the BWM Convention's Guidelines negotiations many issues were controversial and in certain cases it proved extremely hard and difficult to reach agreements. In order to explain all the concepts, controversial views and agreements reached in its entirety a separate book of its own would be needed. Therefore, the focus of this chapter will remain with the requirements of the BWM Convention, as well as the availability and feasibility of ballast water management (BWM) options. Compliance control measures are also found to be closely related to the BWM requirements and options, hence these are presented in chapters "Ballast Water Sampling and Sample Analysis for Compliance Control" and "Ballast Water Management Decision Support System".

What Is Ballast Water Management?

As defined at IMO: "Ballast Water Management means mechanical, physical, chemical, and biological processes, either singularly or in combination, to remove, render harmless, or avoid the uptake or discharge of Harmful Aquatic Organisms and Pathogens within Ballast Water and Sediments."

BWM in its core sense means the prevention, minimization and ultimate elimination of the transfer of HAOP via vessels' ballast waters and sediments. In light of this, BWM cannot only be understood as mechanical, physical, chemical, and biological processes preventing the transfer of HAOP, because the process includes also different precautionary measures to minimize the uptake of HAOP and sediments. Those include the avoidance of ballast water uptake, where practicable,

- in areas identified by the port State in connection with advice provided by ports;
- in darkness when the organism concentration in upper water layers increases;
- in areas with outbreaks, infestations or known populations of HAOPs;
- in very shallow water because it is more likely to pump in bottom living organisms;
- where propellers may stir up sediment;
- where dredging is or recently has been carried out; and
- nearby sewage outfalls.

Furthermore, no mixing of ballast water should occur and additional management practices may apply, e.g., risk assessment (RA) (see chapter "Risk Assessment in Ballast Water Management"), decision support system (see chapter "Ballast Water Management Decision Support System"). Hence BWM should be understood as a complex, multi-facetted process of all precautionary measures, preventive and treatment procedures, as well as additional measures taken to prevent, minimize and ultimately eliminate the transfer of HAOP via ballast water and sediments.

Vessels should also, whenever possible, implement precautionary practices, i.e., avoid the unnecessary discharge of ballast water. Should it be necessary to take on and discharge ballast water in the same port to facilitate safe cargo operations, unnecessary discharge of ballast water that has been taken up in another port should be avoided. Managed ballast water which is mixed with unmanaged ballast water is no longer in compliance with Regulations D-1 and D-2.

Ballast Water Management Requirements

By the basic principle, vessels (not ports) are required to conduct BWM according to the requirements of the BWM Convention. However, port reception facilities are also considered by the BWM Convention as a BWM option, i.e., Regulation B-3.6 and Guidelines for ballast water reception facilities (G5) (G5 Guidelines) (IMO 2006b). During the BWM Convention negotiations ballast water reception facilities were considered as the primary BWM measure. However, as ships may need to conduct ballast water operations also outside ports (see chapter "Vessels and Ballast Water"), such reception facilities would not cover all ballast water discharges. Therefore, treatment on board ship before ballast water discharge is required.

Standards for BWM are dealt with by the BWM Convention in Regulations D-1 and D-2. The BWM Convention introduces these two different protective regimes as a sequential implementation regime:

- *Ballast Water Exchange Standard* (Regulation D-1, so called D-1 standard) requiring ships to exchange a minimum of 95 % ballast water volume;
- *Ballast Water Performance Standard* (Regulation D-2, so called D-2 standard) requires that the discharge of ballast water have the number of viable organisms below the specified limits.

The D-2 standard is based on a limited number of organisms that can be discharged with ballast water. The phase-in of the D-2 standard was originally planned gradually, based on the vessels total ballast tanks capacity and if these vessels are existing or are new builds (see Fig. 1). When the phase-in dates were set, the expectation was that technology and manufacturing capacity would be first available for vessels with lower ballast water capacities and flow rates. As such dates were set to allow a gradual maturity of the technology with the expectation that the very high flow rates would come later due to the technical challenges. These include that on smaller vessels due to engine room limited space it might be difficult to install ballast water management systems (BWMS) at that time. Higher flow rates were considered difficult as the first generation of BWMS was not able to meet these flow requirements.

However, the BWM Convention has not come into force and certain phase-in dates have already passed. This resulted in a debate at IMO and Marine Environment Protection Committee (MEPC) at its 65th session (May 2013) approved a draft IMO Assembly resolution on the application of Regulation B-3 of the BWM Convention, which addresses the fixed dates, to ease and facilitate the smooth implementation of the BWM Convention. This was approved at the 28th session of the IMO Assembly (25 November to 4 December 2013). This resolution recommends that ships

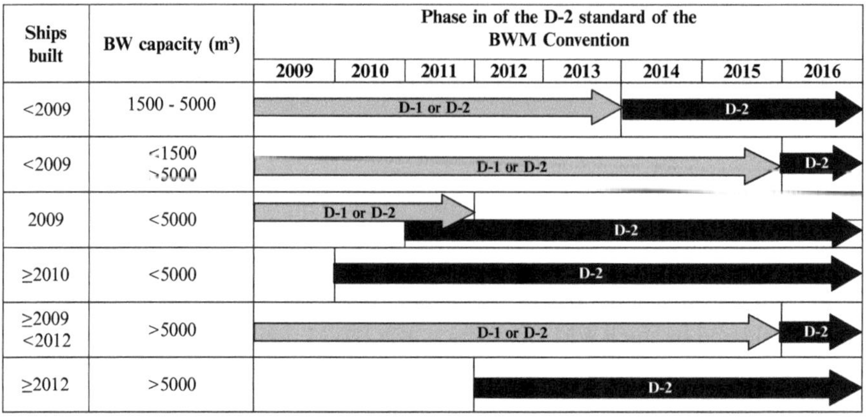

Fig. 1 The original phase-in plan of the ballast water performance standard (Regulation D-2) in relation to the ballast water exchange standard (Regulation D-1) (David and Gollasch 2008) (Reprinted from David and Gollasch 2008, copyright 2008, with permission from Elsevier)

constructed before the entry into force of the BWM Convention will not be required to comply with Regulation D-2 until their first renewal survey following the date of entry into force of the BWM Convention. The aim of the resolution is to clarify that although the BWM Convention itself cannot be changed prior to entry into force, Regulation B-3 may be enforced on a realistic timeline upon entry into force of the BWM Convention. This needs consensus amongst all IMO Member states. One issue that was not anticipated was that the term "renewal survey" is not specifically tied to any statutory requirement. That was solved by using the requirements for the date of the issuance of the International Oil Pollution Prevention (IOPP) certificate as the trigger for the renewal survey.

Several Delegations at MEPC65 expressed their concerns regarding this approach because, due to the reduced urgency to implement BWM methods on board, it may result in a relaxation of efforts to ratify the BWM Convention. It was further assumed that this new approach would negatively impact the developers of BWMS as sales of their units may be delayed.

Ballast Water Exchange Standard: D-1 Standard

Approximately 10 years ago when the D-2 standard was negotiated at IMO no BWMS was readily available. In the absence of full scale BWMS to be installed on vessels, it was suggested by MEPC that ballast water exchange (BWE) at sea may reduce the risk of species introductions. Most vessels are enabled to conduct a BWE without needing extra installations.

The reasoning behind BWE is that coastal organisms pumped on board during ballast water uptake, when discharged at sea are unlikely to survive due to, e.g., salinity issues and the lack of a hard substrate to complete their life cycle. In addition, high sea organisms when pumped on board during the BWE will unlikely survive when released in coastal waters also due to possible salinity changes and the lack of suitable habitats. Further, it is well-known that organism concentrations are much lower in high seas compared to coastal waters which reduces the risk of species introductions. However, sampling studies on board of commercial vessels have shown that in certain instances after BWE a higher concentration of organisms was found in the ballast water (e.g., Macdonald and Davidson 1998; McCollin et al. 2001). This specifically occurred when the BWE was undertaken in shallower seas or during high organism concentrations, such as algal blooms, which are also known to occur in the high seas.

Other BWE efficiency limitations include that, due to ballast tank design, a certain amount of unpumpable ballast water and sediments always remains inside the tank on almost all ships (see chapter "Vessels and Ballast Water"). As a result a one time BWE will not be sufficient to reduce the organism load. IMO noted this and therefore Regulation D-1 of the BWM Convention requires at least a 95 % water exchange. This may be met by emptying and refilling the tank or by pumping through three times the tank volume (Rigby and Hallegraeff 1994). However, when Gollasch and David conducted shipboard tests of different BWM methods it was noticed that on vessels which were trimmed ahead, about 15 % and more of unpumpable water remained in the tanks during the empty-refill (sequential) BWE. Furthermore, a 95 % volumetric BWE is unlikely equivalent with a 95 % organism removal because the organisms are not homogeneously distributed in a tank (e.g., Murphy et al. 2002). In contrast, under certain circumstances, the 95 % volumetric exchange may result in an even higher than 95 % organism removal. In conclusion, pumping through less than three times the volume may also be acceptable provided the ship can demonstrate that at least 95 % volumetric exchange limit is met.

When conducting BWE Guidelines for Ballast Water Exchange (G6) are to be considered. Three methods are accepted to conduct BWE and can be described as follows (IMO 2005c):

Sequential method – is a process by which a ballast tank is first emptied and then refilled with replacement ballast water to achieve at least a 95 % volumetric exchange.

Flow-through method – is a process by which replacement ballast water is pumped into a ballast tank, allowing water to flow through an overflow on deck or other arrangements.

Dilution method – is a process by which replacement ballast water is filled through the top of a ballast tank with simultaneous discharge from the bottom at the same flow rate so that a constant water level is maintained in the tank throughout the BWE.

In addition to the requirements to be met in relation to the BWE methods used, a ship should also consider requirements regarding where BWE shall, whenever possible, be conducted. In the first place, this is at least 200 nautical miles from nearest land and in water depths of at least 200 m. If this is impossible, then the

BWE should be conducted as far from nearest land as possible, and in all cases at least 50 nautical miles from nearest land and in waters of at least 200 m depth (IMO 2004).

Ballast Water Exchange Areas

In sea areas where these BWE depth and distance requirements cannot be met, the port State may designate a ballast water exchange area (BWEA). This should be done in consultation with adjacent or other states, as applicable. Any such designation should follow the principles of Guidelines on Designation of Areas for Ballast Water Exchange (G14).

However, a ship shall not be required to deviate from its intended voyage, or delay the voyage to conduct BWE. In contrast, a port State may require a ship to deviate from its intended route or delay its voyage in case a designated BWEA has been established. The BWE activity for each tank should not start if the process cannot be fully completed.

In general, ships should follow the G6 Guidelines and shall only be required to comply with any BWE requirements if those would not threaten the safety or stability of the ship, its crew, or its passengers because of, e.g., adverse weather, ship design or stress, equipment failure, or any other extraordinary condition.

Vessels operating in coastal areas are unlikely to meet the distance (200 nm or 50 nm distance from nearest land) and water depth (200 m depth) requirements of the BWM Convention. Further, routes may be too short to conduct a complete BWE of all ballast tanks intended to be discharged in the port of call. Management options for those vessels may therefore be based on a selective approach, i.e., use a designated BWEA or by granting exemptions based on RA (see chapter "Risk Assessment in Ballast Water Management").

The rationale for the BWEA designation is that it provides an area where ships can safely exchange ballast water as a risk reducing measure while at the same time minimising harmful environmental effects. However, next to shipping and nautical aspects, the challenge is to identify such areas from a biological perspective. It is understood that coastal BWEA pose a higher risk of species introductions compared to mid-ocean exchange, but at the same time it may be preferred to use specially designated BWEA rather than to discharge unmanaged ballast water in a port or across the entire coastal area.

Strong concerns have already been voiced that the designation of near-shore BWEA may expose certain regions to additional ballast water discharges, which may pose a risk to those ballast water receiving environments. This is why BWEA need to be selected very carefully using RA to prove it is environmentally safe. Ideal would be a BWEA with off-shore directed water currents, it should be as far from nearest land and as deep as possible, free of pollution or HAOP. When these requirements are met the BWEA may be considered environmentally safe and effective. When considering shipping aspects, the BWEA needs to be designed as large as possible and as close as possible to shipping routes (David 2007).

In practice this implies difficulties especially for the designation of BWEA in shallow seas (e.g., North Sea, Baltic Sea) or semi-enclosed seas (e.g., Adriatic). Considerations should be given to the trade-offs between (a) additional ballast water discharges in such areas, (b) the dimension of the BWEA to allow complete BWE and (c) to its location to avoid major deviations from the vessels' intended routes. To meet the requirements vessels with bigger ballast water capacities may slow down when sailing through BWEA to gather extra time to complete the BWE operation or to exchange just the "critical" (i.e., assessed as highest risk ballast) ballast water. A decision on the minimum management measure required should be taken according to the level of RA (see chapter "Risk Assessment in Ballast Water Management").

BWEA should be biologically monitored frequently to document the presence/absence of introduced species or other HAOP. A worst case scenario may be that HAOP become introduced and established in such an area and are rapidly spread unnoticed due to the ongoing BWE activities in this area.

A unique situation occurs in e.g. Europe and USA as some of the busiest ports are located in estuaries with brackish or even freshwater conditions (e.g., Antwerp, Hamburg and parts of Rotterdam, inner parts of Chesapeake Bay and San Francisco Bay). A high risk for a species introduction occurs when freshwater organisms (e.g., the zebra mussel) are transported in ballast tanks between two freshwater ports, but these two ports are separated by marine water conditions, which poses a natural migration barrier so that the freshwater organisms cannot spread by their natural means between these freshwater ports. In those instances BWE in higher saline waters, also in coastal waters (i.e., <50 NM from the nearest land and <200 m depth), may be a risk reducing measure. However, some organisms show a very wide salinity tolerance, i.e., BWE alone will not completely eliminate the risk of species introductions.

We therefore recommend that freshwater ballast should be exchanged in marine waters even if this is in coastal waters provided that the voyage is sufficiently long to complete BWE en-route in marine waters for the ballast water intended for discharge.

Undue Delay and Deviation from Planned Route

As per the BWM Convention vessels should not be forced to deviate or be unduly delayed by BWM requirements. The BWM Convention gives the vessel a right for compensation when it has been unduly delayed. However, the term "undue delay" has never clearly been defined by IMO in relation to the BWM Convention or other IMO applications.

The designation of BWEA should not require major vessel deviations. However, a cost/benefit analysis considering the costs caused by negative impacts of introduced species vs. re-routing costs for shipping may reveal that a slight re-routing of vessels may be considered. Similarly, if a RA identifies that a vessel carries ballast water with an unacceptable risk, then the reasoning for a deviation may apply and it is therefore not "undue". It may therefore be considered that vessels use specific routes even if this results in a delay of a few hours.

Ballast Water Performance Standard: D-2 Standard

The *Ballast Water Performance Standard* as outlined in Regulation D-2 stipulates that ships meeting the requirements of the BWM Convention must discharge:

- less than 10 viable organisms per cubic meter greater than or equal to 50 µm in minimum dimension, and
- less than 10 viable organisms per millilitre less than 50 µm in minimum dimension and greater than or equal to 10 µm in minimum dimension, and
- less than the following concentrations of indicator microbes, as a human health standard:
  - Toxigenic *Vibrio cholerae* (serotypes O1 and O139) with less than 1 colony forming unit (cfu) per 100 ml or less than 1 cfu per 1 g (wet weight) of zooplankton samples,
  - *Escherichia coli* less than 250 cfu per 100 ml, and
  - Intestinal *Enterococci* less than 100 cfu per 100 ml.

This standard formed the basis for significant discussions and continuing controversy at IMO. The acceptable organism numbers and the method to determine their size classes were debated intensively. This compromise was reached through negotiations by various countries which ranged from an acceptable number of organisms above 50 µm in minimum dimension between 100 and 0.01 per cubic meter. The current version of the D-2 standard is seen as a considerable reduction compared to the amount of organisms discharged in unmanaged ballast water or even that obtained by BWE.

The D-2 standard for both organism groups greater than or equal to 10 µm in minimum dimension refers to all organisms, not per species, and not only for non-indigenous or harmful organisms. As a result the individual taxonomic species identification is not required for purposes of compliance testing.

Also of note is the inclusion of a discharge limit for "indicator microbes" with a human health impact in the D-2 standard. A number of delegations insisted on incorporating these bacteria as they had specific issues, hoping this would result in a strong signal to R&D interests. Existing and developing ballast water treatment technologies are able to meet these standards using a combination of treatment methods (see chapter "Ballast Water Management Systems for Vessels").

Although the D-2 standard results in a considerable reduction in organisms being released we note that vessels carry up to 100,000 tonnes of ballast water or more so that still a high number of organisms may be discharged with ballast water being in compliance with this Convention. Assuming that 10,000 tonnes of ballast water are discharged, the acceptable D-2 standard organism concentration for individuals greater than or equal to 50 µm in minimum dimension is less than 100,000, which theoretically means 99,999. The number of organisms to establish a founder population in new environments is largely unknown, but we suspect that an inoculation of approximately 100,000 individuals (although of different species) may not *eliminate* the risk of species introductions in all cases. Another weak point regarding the D-2 standard is that it does not address organisms below 10 μm (in minimum dimension), but a considerable number of species, including bloom forming harmful algae, are smaller than 10 μm (e.g., *Phaeocystis* spp., *Pfiesteria* spp. and *Chrysochromulina* spp.).

How to Achieve Compliance with the D-2 Standard?

The D-2 standard is based on a limited number of organisms that can be discharged with ballast water, and is not considering only non-indigenous or harmful organisms, but all viable organisms in relevant size classes, or limited number of cfu per indicator microbes. Indicator microbes are in general present only in coastal environments, into which these may be discharged with untreated river run-offs contaminated with human influence or due to improper sewage treatment plants. Therefore BWE may still be efficient to manage ballast water according to the D-2 standard in terms of indicator microbes as in open ocean these organisms are absent. However, the open ocean concentration of viable organisms greater than or equal to 10 μm in minimum dimension, and especially those greater than or equal to 50 μm in minimum dimension, may be higher in BWEA than the D-2 standard (Gollasch and David, own observations). Consequently BWE is not an option to manage ballast water to comply with the D-2 standard. With this the on board installation of ballast water treatment systems, so called BWMS, became a viable option and requirement. It is interesting to note that a recent summary of existing and developing BWMS revealed more than 100 such systems. However, some of these are not considered realistic, but if only half of those make it to the market, a large variety of BWMS becomes available so that all vessel types with their specific BWM requirements can be equipped with BWMS. As of the December 2013, 33 BWMS have been type approved. Details about BWMS are given in chapter "Ballast Water Management Systems for Vessels".

Issues which further may need to be considered are the possible regrowth of organisms in ballast tanks after treatment and also that organisms may remain in the tank from previous ballast water operations and may become re-suspended during ballast water operations (Murphy et al. 2008). Consequently, upon discharge, treated water may contain unacceptably high organism numbers although the treatment systems proved that the D-2 standard was met during water uptake. To ensure that ballast water discharges always meet the D-2 standard it is recommended to treat the water during uptake and discharge and also to develop BWMS which by far exceed the standards set forth in the BWM Convention.

In the case of fresh water ecosystems, some countries such as Canada are examining the possibility of continuing the use of BWE to take advantage of the salinity shock imposed on fresh water organisms when vessels travel between freshwater donor and freshwater recipient ports, i.e., in cases when vessels ballast in freshwater, a marine water BWE would provide a salinity shock to the originally pumped in freshwater organisms. At the same time, marine organisms pumped on board during

BWE would be exposed to a salinity shock when released in a recipient freshwater port. Land-based trials have indicated an up to tenfold reduction of risk compared to the use of BWMS alone (Briski et al. 2013).

Warnings Concerning Ballast Water Uptake in Certain Areas

The BWM Convention encourages Administrations to conduct monitoring programmes in their coastal waters, i.e. typical ballast water uptake zones, and further to notify mariners if ballast water uptake restrictions are necessary. Such notifications may include suggestions for alternative ballast water uptake areas. Ballast water uptake warnings are useful e.g., in cases of outbreaks of toxic algal blooms (e.g., Hallegraeff 1998), in the presence of human pathogens, or other (potentially) harmful organisms. Ballast water uptake should also be avoided near sewage outfalls and when tidal flushing is poor. Relevant notifications should be communicated to IMO and potentially affected states.

These monitoring activities may be conducted within the framework of a regional cooperation. One key problem is that in most countries existing monitoring programmes were created for other purposes and lack sampling sites in ports or port regions, i.e. in ballast water uptake areas.

Ballast Water Reception Facilities

BWM requirements in the BWM Convention do not apply to ships which intend to discharge ballast water to a reception facility. If available, such facilities should be designed according to the G5 Guidelines. A ballast water reception facility may be a good solution for a vessel that didn't manage ballast water properly and would need to discharge it. This would be especially important when the ballast water is posing a high risk to the recipient environment (see chapter "Risk Assessment in Ballast Water Management").

Reception facilities may be land based or floating, e.g., barges, tankers (IMO 2013). Reception facilities may have a capacity to receive ballast water and treat it later before the discharge into the environment, or the treatment process is applied directly during the discharge to the environment. Where ballast water is discharged into the aquatic environment it should at least meet the D-2 standard of the BWM Convention (IMO 2006b).

A reception facility should provide adequate pipelines, manifolds, reducers, equipment and other resources to enable, ships wishing to discharge ballast water in a port to use the facility (IMO 2006b). However, today ships are lacking a (standardised) pipework connection, which would enable the discharge of ballast water to reception facilities. Tankers have standardized piping and manifolds for cargo transfers and the concept of standard fittings is embedded in ship design and construction. Therefore, for these vessels ballast water transfer to a reception facility could easily be achieved provided the cargo transfer pipes may be used for ballast water discharge. Hence, ships planning to use this option need to have adequate equipment installed.

It should be noted that prior to the introduction of double hulls and segregated ballast tanks, designed to minimize the threat of oil pollution to the environment, tankers pumped their ballast ashore. Refineries worldwide have ballast water reception facilities. Major crude oil exporting ports, such as Valdez (Alaska, USA) and Scapa Flow (Orkney Islands, United Kingdom) still use these shore-based facilities for the reception and treatment of oily ballast from crude oil tankers. This proves that the engineering, pumping, storage etc. of massive quantities of ballast is technically possible and economically feasible within the operating cost structures of modern shipping and ports. Adapting this approach to include biological treatment to remove or render harmless the ballast water organisms is unlikely to be any more challenging or less feasible than the original development of these facilities – especially as technology has advanced.

Land-based ballast water reception facilities may also be used to provide biologically clean ballast water at the source ports, which prevents the problem already at ballast water uptake.

Sediment Management

Regulation B-5 of the BWM Convention requires that all ships shall remove and dispose ballast water related sediments in accordance with the vessels' ballast water management plan.

All possible practical steps should be taken during ballast uptake to avoid sediment accumulation, but it is known that it cannot be avoided to take sediment on board and this will settle on tank surfaces and bottoms. The sediment amount in a ballast tank should be monitored on a regular basis. When sediment has accumulated, tank bottoms and other surfaces should be flushed when in suitable areas, i.e. areas complying with the minimum depth and distance requirements as described for BWE.

The frequency and timing of sediment removal depends on several factors, including dimension of sediment build up, ship's trading pattern, availability of reception facilities, work load of the ship's personnel and safety issues.

The removal of sediment should preferably be undertaken under controlled conditions in a port, at a repair facility or in a dry dock. The removed sediment should be disposed of in a sediment reception facility in line with the waste disposal requirements of the coastal state. Regulation B-5 further requires that ships constructed in or after 2009 should, without compromising safety or operational efficiency, be designed and constructed to minimize the sediment uptake and entrapment, facilitate removal of sediments, and to provide safe access for sediment removal and sampling, taking into account the Guidelines for sediments control on ships (G12). This also applies to ships constructed prior to 2009, to the extent practicable.

Exemptions from BWM and Additional Measures

Some ships may be exempted from BWM requirements provided that the risk level of such a discharge is acceptable based on Guidelines on Risk Assessments under Regulation A-4 (G7). In other cases, when the risk is identified as (very) high, such ships may be required to take additional measures based on Guidelines for Additional Measures Including Emergency Situations (G13). The level of risk is a result of RA (see chapter "Risk Assessment in Ballast Water Management").

The BWM Convention addresses the selective BWM approach in Article 4.2. This article requests a party to develop BWM policies, strategies or programs regarding to its particular conditions and capabilities. It was understood that no "one size fits all" approach is available because different states may have different geographical, environmental, socio-economic, organizational, political and other conditions as well as different shipping patterns. In light of RA based exemptions from BWM requirements, these can be given on the basis of Regulation A-4, while additional measures may be introduced based on Regulation C-1 (see Fig. 2).

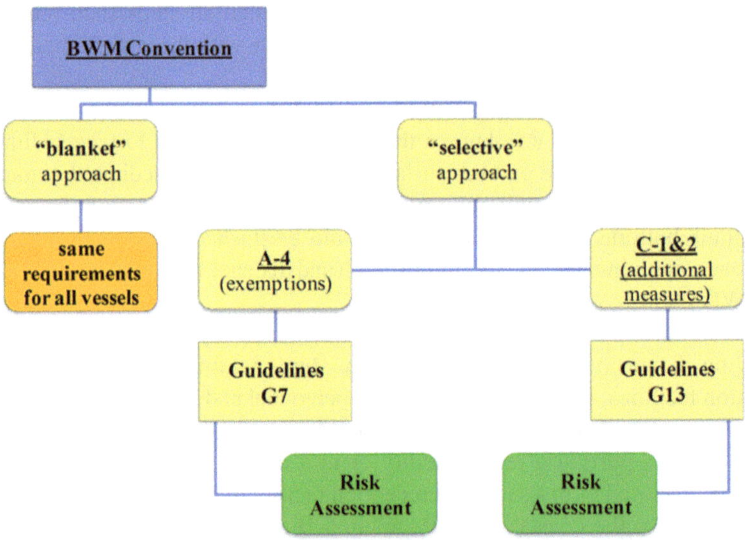

Fig. 2 Risk assessment procedures according to the BWM Convention (Enhanced after Gollasch et al. 2007) (Reprinted from Gollasch et al. 2007, copyright 2007, with permission from Elsevier)

Exceptions from BWM

Further to the above mentioned exemptions, the BWM Convention also includes provisions for cases where vessels do not need to manage their ballast water at all. This refers to vessels being in line with the Regulation A-3 *Exceptions*. Exceptions are identified for specific cases including (IMO 2004):

1. ballast water uptake or discharge is needed for ensuring the safety of a ship in emergency situations;
2. accidental discharge results from damage to a ship or its equipment;
3. uptake or discharge of ballast water is used to avoid or minimize pollution incidents;
4. uptake and discharge of the same ballast water is conducted on the high seas; or
5. uptake and discharge occurs at the same location, provided no mixing occurs with other locations.

The "high seas" and "same location" exceptions may apply permanently if this is a regular vessel operation. Granting an exemption or a permanent exception means that a vessel is not required to install a ballast water treatment system with the clear benefit of avoiding capital and operational costs as well as burdens associated with the certification and inspections. However, the BWM Convention is not specific in defining the term "same location" (IMO 2004; Gollasch and David 2012; David et al. 2013). Therefore the concept is subject to different interpretations which depend on the interpreters' approach and this may be based on one or a combination of the following: environmental parameters, hydrological regimes, biological meaningful parameters, or political aspects. The shipping industry would benefit from a larger "same location", as it avoids ballast water management requirements on voyages inside each such location. In contrast maximizing environmental protection requires that a "same locations" should be as small as possible. As a result, the "same location" may be of different dimensions, including a mooring, port basin, port, anchorage, part of a sea, or even an entire sea with numerous ports. These different interpretations introduce difficulties in the uniform implementation of the BWM Convention, including an opportunity for the secondary transfer of organisms between ports within a large "same location" (Gollasch and David 2012; David et al. 2013).

In light of the above the identification of a "same location" for ballast water management is not an easy task. This should be port specific and each port has its unique peculiar situation regarding the number of port basins, it may extend over waters of different salinity regimes, and ports likely have different cargo patterns resulting in different ballast water operation profiles. The issue becomes more complex when the same location needs to be explained in biologically meaningful terms addressing aquatic species invasions. To biologically identify a "same location" the species diversity and their abundance may be considered. This assessment should include indicator microbes and human pathogens as listed in the D-2 standard. Should all species, including indicator microbes and human pathogens, be identical and their abundance is very similar, this area could be considered as the same location (Gollasch and David 2012; David et al. 2013).

National authorities responsible for the BWM Convention implementation may receive applications from shipping companies for permanent exceptions based on the "same location" concept. Consequently the authorities will need to decide, on a case-by-case basis, how the term should be applied. We recommend that "same location" means the smallest practicable unit, i.e., the same harbour, mooring or anchorage, as stated in IMO Guidelines G3. When considering the diversity of ships ballast operations and ports, as well as possible differences in environmental conditions and species compositions among port terminals or basins, we recommend that an entire smaller port, possibly also including the anchorage, should be considered as "same location". For larger ports with a gradient of environmental conditions, the "same location" should mean a terminal or a port basin. We further suggest that IMO considers the preparation of a guidance document to include concepts, criteria and processes how to identify a "same location", which limits should be clearly identified. Large areas encompassing more ports should not be identified as a "same location" as this would seriously undermine purpose of the BWM Convention, as unmanaged ballast water would be transferred in this area (Gollasch and David 2012; David et al. 2013). (see also the U.S. same port or place concept in chapter "Policy and Legal Framework and the Current Status of Ballast Water Management Requirements").

Compliance Monitoring

In accordance with Article 9.1, ships to which the BWM Convention applies may be subject to inspections for the purpose of revealing violations of the provisions of the BWM Convention. These inspections shall:

- Verify that the ship is carrying a valid Ballast Water Management Certificate;
- Verify that a Ballast Water Management Plan specific to the ship and approved by the Flag state is onboard;
- Undertake an inspection of the Ballast Water Record Book.

As a part of the Port State Control and to demonstrate compliance with the D-2 standard, port authorities may consider sampling ballast water for subsequent analyses. IMO provided guidance on sampling ballast water in Guidelines for Ballast Water Sampling (G2). We have summarised the state of knowledge regarding ballast water sampling in chapter "Ballast Water Sampling and Sample Analysis for Compliance Control".

Implementation of the Ballast Water Management Convention

A Blanket or a Selective Approach?

The BWM Convention incorporates two different basic BWM regimes; i.e., the "blanket" and the "selective" approach. A blanket approach results in a situation where all ships intending to discharge ballast water in a port are required by the port

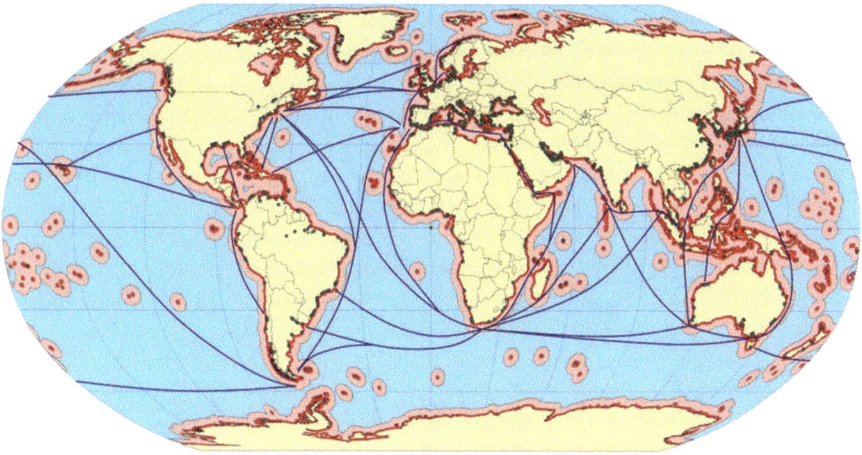

Fig. 3 World map indicating the main intercontinental shipping routes (*blue lines*) and BWE areas according to the BWM Convention (*red shading* = 50 NM and *pink shading* = 200 NM limit to nearest land and >200 m water depth) (After David et al. 2005)

State to conduct BWM. The selective approach means that the appropriate BWM measures to take vary depending on the different levels of risk posed by the intended ballast water discharge, which also depends on the BWM feasibility in certain circumstances.

Ballast Water Management Feasibility

Whenever possible and until the D-2 standard is required, BWE should be undertaken as a risk reducing measure. Provided safety permits, it is assumed that most vessels operating on oceanic voyages are enabled to undertake BWE that meets the IMO water depth and distance to nearest land limits (see Fig. 3).

However, there are limitations in BWE applications, which are primarily due to shipping patterns of a port (e.g., shipping routes, length of voyages) and local specifics regarding the required/available conditions according to the BWM Convention (i.e., distance from nearest land, water depth, BWEA). BWE has also substantial limitations in its biological effectiveness especially in semi-enclosed or enclosed areas. Ships in these areas usually sail within 50 nautical miles distance from the nearest land, and therefore, according to the BWM Convention, cannot meet the requirements to conduct BWE. Because of geographical specifics, not only ships in Short-Sea-Shipping fall into this category (see Fig. 4).

Hence, from the most effective BWM perspective worldwide, the use of BWMS would be essential.

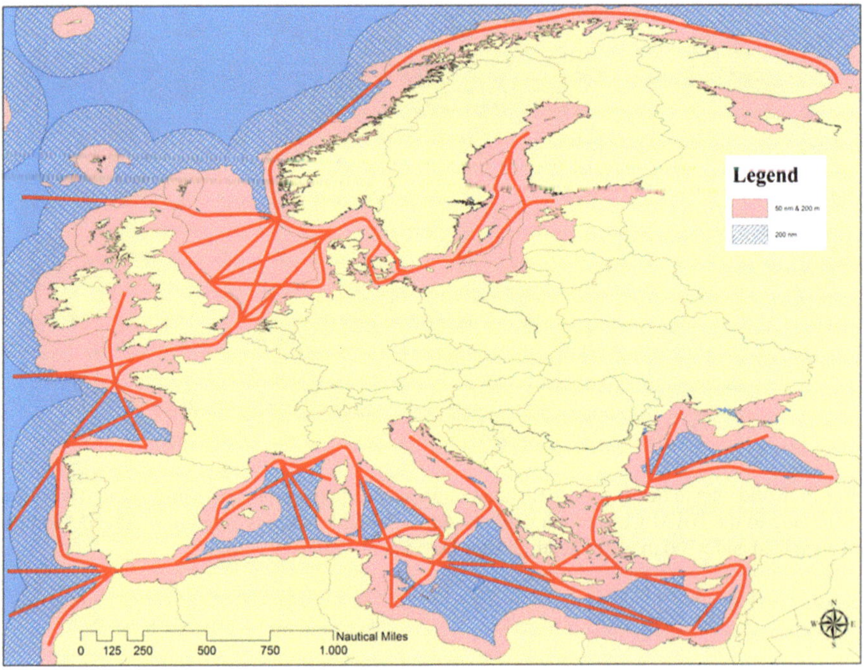

Fig. 4 The seas surrounding Europe with the 50 nautical miles and 200 m depth limit shown in *pink*, and *pink shaded* the 200 nautical miles limit. The *red lines* show the main shipping routes (After David and Gollasch 2008) (Reprinted from David and Gollasch 2008, copyright 2008, with permission from Elsevier)

Acknowledgements The research leading to part of these results has received funding from the European Community's Seventh Framework Programme (FP7/2007–2013) under Grant Agreement No. [266445] for the project Vectors of Change in Oceans and Seas Marine Life, Impact on Economic Sectors (VECTORS). Part of this publication has been produced with the financial assistance of the IPA Adriatic Cross-Border Cooperation Programme – strategic project Ballast Water Management System for Adriatic Sea Protection (BALMAS). The contents of this publication are the sole responsibility of authors and can under no circumstances be regarded as reflecting the position of the IPA Adriatic Cross-Border Cooperation Programme Authorities.

Disclaimer The content of this chapter does not necessarily reflect the official opinion of the European Maritime Safety Agency or the European Commission. Responsibility for the information and views expressed in this chapter lies entirely with the authors.

References

Briski E, Allinger LE, Balcer M, Cangelosi A, Fanberg L, Markee TP, Mays N, Polkinghorne CN, Prihoda KR, Reavie ED, Regan DH, Reid DM, Saillard HJ, Schwerdt T, Schaefer H, TenEyck M, Wiley CJ, Bailey SA (2013) Multidimensional approach to invasive species prevention. Environ Sci Technol 47(3):1216–1221

David M (2007) A decision support system model for ballast water management of vessels: doctoral dissertation. University of Ljubljana, Portorož

David M, Gollasch S (2008) EU shipping in the dawn of managing the ballast water issue. Mar Pollut Bull 56(12):1966–1972

David M, Gollasch S, Hewitt C, Jakomin L (2005) Ballast water management – blanket vs. flexible approach. Poster presented at the Fourth International Conference on Marine Bioinvasions, Wellington, 24–26 August 2005

David M, Gollasch S, Pavliha M (2013) Global ballast water management and the same location concept – a clear term or a clear issue? Ecol Appl 23(2):331–338

Gollasch S, David M (2012) A unique aspect of ballast water management requirements – the same location concept. Mar Pollut Bull 64:1774–1775

Gollasch S, David M, Voigt M, Dragsund E, Hewitt C, Fukuyo Y (2007) Critical review of the IMO international convention on the management of ships' ballast water and sediments. Harmful Algae 6:585–600

Hallegraeff GM (1998) Transport of toxic dinoflagellates via ships' ballast water: bioeconomic risk assessment and efficacy of possible ballast water management strategies. Mar Ecol Prog Ser 168:297–309

IMO (1973) Conference to adopt the international convention for the prevention of pollution from ships. Resolution 18

IMO (2004) International convention for the control and management of ships' ballast water and sediments. International Maritime Organization, London

IMO (2005a) Guidelines for Ballast Water Management Equivalent Compliance (G3). Marine Environment Protection Committee, Resolution MEPC.123(53), 22 July 2005. International Maritime Organization, London

IMO (2005b) Guidelines for ballast water management and development of ballast water management plans (G4). Marine Environment Protection Committee, Resolution MEPC.127(53), 22 Jul 2005. International Maritime Organization, London

IMO (2005c) Guidelines for ballast water exchange (G6). Marine Environment Protection Committee, Resolution MEPC.124(53), 22 Jul 2005. International Maritime Organization, London

IMO (2006a) Guidelines for Sediment Reception Facilities (G1). Marine Environment Protection Committee, Resolution MEPC.152(55), 13 October 2006. International Maritime Organization, London

IMO (2006b) Guidelines for ballast water reception facilities (G5). Marine Environment Protection Committee, Resolution MEPC.153(55), 13 October 2006. International Maritime Organization, London

IMO (2006c) Guidelines for approval and oversight of prototype ballast water treatment technology programmes (G10). Marine Environment Protection Committee, Resolution MEPC.140(54), 24 March 2006. International Maritime Organization, London

IMO (2006d) Guidelines for ballast water exchange design and construction standards (G11). Marine Environment Protection Committee, Resolution MEPC.149(55), 13 October 2006. International Maritime Organization, London

IMO (2006e) Guidelines on design and construction to facilitate sediment control on ships (G12). Marine Environment Protection Committee, Resolution MEPC.150(55), 13 October 2006. International Maritime Organization, London

IMO (2006f) Guidelines on designation of areas for ballast water exchange (G14). Marine Environment Protection Committee, Resolution MEPC.151(55), 13 October 2006. International Maritime Organization, London

IMO (2007a) Guidelines for Risk Assessment under Regulation A-4 of the BWM Convention (G7). Marine Environment Protection Committee, Resolution MEPC.162(56), 13 July 2007. International Maritime Organization, London

IMO (2007b) Guidelines for additional measures regarding ballast water management including emergency situations (G13). Marine Environment Protection Committee, Resolution MEPC.161(56), 13 July 2007. International Maritime Organization, London

IMO (2008a) Guidelines for ballast water sampling (G2). Marine Environment Protection Committee, Resolution MEPC.173(58), 10 October 2008. International Maritime Organization, London

IMO (2008b) Guidelines for approval of ballast water management systems (G8). Marine Environment Protection Committee, Resolution MEPC.174(58), 10 October 2008. International Maritime Organization, London

IMO (2008c) Procedure for approval of ballast water management systems that make use of active substances (G9). Marine Environment Protection Committee, Resolution MEPC.169(57), 4 April 2008. International Maritime Organization, London

IMO (2013) Port-based mobile ballast water treatment facility (BWTBOAT) as another method of ballast water management for regional and coastal trading ships. Submitted by India. Marine Environment Protection Committee, 26 December 2013. MEPC 66/2/8. International Maritime Organization, London

Macdonald EM, Davidson RD (1998) The occurrence of harmful algae in ballast water discharges in Scottish ports and the effects of mid-water exchange in regional seas. In: Reguera B, Blanco J, Fernandez ML, Wyatt T (eds) Harmful Algae. Xunta de Galicia and Intergovernmental Oceanographic Commission of UNESCO, pp 220–223

McCollin T, Macdonald EM, Dunn J, Hall C, Ware S (2001) Investigations into ballast water exchange in European regional seas. In: Proceedings of international conference on marine bioinvasions, New Orleans April 9–11 2001, pp 94–95

Murphy KR, Ritz D, Hewitt CL (2002) Heterogeneous zooplankton distribution in a ship's ballast tanks. J Plankton Res 24(7):729–734

Murphy KR, Field MP, Waite TD, Ruiz GM (2008) Trace elements in ships' ballast water as tracers of mid-ocean exchange. Sci Total Environ 393:11–26

Rigby GR, Hallegraeff GM (1994) The transfer and control of harmful marine organisms in shipping ballast water: behaviour of marine plankton and ballast water exchange on the MV "Iron Whyalla". J Mar Environ Eng 1:91–110

Ballast Water Management Systems for Vessels

Matej David and Stephan Gollasch

Abstract In this chapter we focus on ballast water management systems (BWMS) which are currently in use as well as treatment approaches manufacturers have chosen for the development of future BWMS. The main purpose of this review is to identify the current availability of BWMS technologies worldwide. Until January 2014 we brought together information of 104 different BWMS. To achieve the ballast water discharge standards, different water treatment technologies are used, mostly in combination, and applied in different stages of the ballasting process. In general, the treatment processes can be split in three stages, i.e., pre-treatment, treatment and residual control (neutralisation). Among the 104 BWMS identified, 100 apply their treatment at the uptake, some of those BWMS require also a treatment during ballast water discharge (in-line treatment) and three BWMS apply treatment only during the voyage (in-tank treatment). The majority of BWMS use filtration or a combination of hydrocyclone and filtration as pre-treatment separation step. The dominating treatment processes are to use an active substance, mostly generated on board by electrolysis/electrochlorination. The second most frequent treatment process is UV. BWMS to be installed for operation on vessels need to be type approved by a state. By the writing of this chapter more than 30 BWMS were type approved. It should be noted that the development of BWMS is a very dynamic market with newly proposed BWMS appearing almost on a monthly basis. The chapter also outlines how BWMS are applied on vessels, their capacities and installation requirements, which BWMS were type approved, and what projected global market for BWMS may exist. A recent calculation on the estimated value of the global market for purchasing and installing BWMSs resulted in an estimated turnover of possibly $50–74 billion. The chapter ends with a list of manufacturers, commercial names of their BWMS, applied treatment technologies used and links to BWMS web pages where available.

M. David (✉)
Dr. Matej David Consult, Korte, Izola, Slovenia
e-mail: matej.david@siol.net

S. Gollasch
Gollasch Consulting (GoConsult), Hamburg, Germany
e-mail: sgollasch@aol.com

M. David, S. Gollasch (eds.), *Global Maritime Transport and Ballast Water Management*, Invading Nature - Springer Series in Invasion Ecology 8,
DOI 10.1007/978-94-017-9367-4_6

Keywords Ballast water management systems • Type approval process • Treatment technologies • Global market • On board installation

Introduction

As the entry into force of the *International Convention for the Control and Management of Ships' Ballast Water and Sediments* (BWM Convention; IMO 2004) is approaching rapidly the industry is more and more aware and considers ballast water management a good business. This becomes obvious when noting the high number of vessels which need to be equipped with treatment systems.

This comprehensive review of ballast water management systems (BWMS) was conducted until January 2014. In this chapter we focus on BWMS which are currently in use as well as treatment approaches the manufacturers have chosen for future BWMS.

The main purpose of this review is to identify the current availability of BWMS technologies worldwide, to briefly introduce these and their use on vessels, identify their timely availability in relation to the BWM Convention requirements, and to identify the prospects of the global BWMS market. At the beginning of this chapter the requirements that BWMS need to comply with are presented, followed by an introduction of BWMS identified, which technologies different BWMS use, how are BWMS applied on vessels, what are BWMS capacities and their installation requirements, what is the situation with BWMS testing and approvals and what is the foreseen global market for BWMS. At the end, names of manufacturers, commercial names of their BWMS, treatment technologies used and links to BWMS web pages are provided, where available (see Table 2).

IMO Requirements

With the *Guidelines for Approval of Ballast Water Management Systems* (*G8*) (G8 Guidelines), IMO has in 2008 adopted requirements for a comprehensive test programme to evaluate the performance and suitability of BWMS. This includes performance tests in larger scale on land under controlled conditions as well as shipboard tests to show the efficiency and seaworthiness of BWMS. Noting some shortcomings in these test requirements some countries have developed their own requirements and test protocols, which set more stringent standards than the G8 Guidelines. One example is the USA with its Environmental Technology Verification (ETV) Program developed for the U.S. Environmental Protection Agency and the U.S. Coast Guard Shipboard Technology Evaluation Program (STEP) (NSF International 2010; STEP 2010).

At present there are many different treatment technologies available, and most of those were previously developed for municipal and other industrial applications. However, when applying those without modifications and improvements to the ballast water treatment purpose, none of these technologies have shown the capability to treat the ballast water to the level required by the BWM Convention D-2 *Ballast Water Performance Standard* (see chapter "Ballast Water Management Under the Ballast Water Management Convention").

The setting of these proposed regulations is an important driving force for ballast water treatment technology developments worldwide. As a result, it was expected that the development and implementation of these systems will proceed at a greatly accelerated rate. However, the ambitious phase-in of the D-2 standard as shown in chapter "Ballast Water Management Under the Ballast Water Management Convention" was modified at MEPC65 (in May 2013) and approved by the IMO Assembly in December 2013. The required starting times were now set as in maximum 5 years after the entry into force of the BWM Convention because the time limits as agreed earlier (see chapter "Ballast Water Management Under the Ballast Water Management Convention") are valid for so many vessels that timely retrofitting may become very difficult or impossible because of BWMS manufacturing and dockyard limitations (IMO 2010g, h; IMO 2011a, z).

Ballast Water Management Systems

World-wide available information about 104 different BWMS was collected and is presented in this chapter (IMO 2005a, b, 2006a, b, c, d, e, f, g, h, i, j, 2007a, b, c, d, e, f, g, h, i, j, 2008a, b, c, d, e, f, g, h, i, j, k, l, m, n, o, p, q, r, s, t, u, v, w, 2009a, b, c, d, e, f, g, h, i, j, k, l, m, n, o, p, q, r, 2010a, b, c, d, e, f, g, h, i, j, k; 2011a, b, c, d, e, f, g, h, i, j, k, l, m, n, o, p, q, r, s, t, u, v, w, x, y, z, za; 2012a, b, c, d, e, f, g, h, i, j, k, l, m, n, o, p, q, r, s, t, u, v, w, x, y, z, za, 2013a, b, c, d, e, f, g, h, i, j, k; Mesbahi 2004; Köster 2010; Shiferaw 2012, Stephan Gollasch pers. comm.). Some of these systems are in the (early) development stage, hence information about these is limited or not available due to confidentiality reasons or patents pending. It is further assumed that not all systems listed will reach full commercially ready development and some manufacturers have stopped the further development of the systems under consideration or withdrawn the system from the market. It should be noted that the development of BWMS is a very dynamic market with newly proposed BWMS appearing almost on a monthly basis.

Ballast Water Management Systems Treatment Technologies

To be able to achieve the requirements of the ballast water discharge standards, different water treatment technologies are used, mostly in combination, and applied in different stages of the ballasting process. In general, the treatment processes can be split

Table 1 Generic treatment process and some main BWMS technologies

	Treatment			
Pre-treatment	Chemical	Physical	Biological	Residual control
Filtration	Chlorination	UV radiation	Bioaugmentation with microorganisms	Chemical reduction (Neutralisation)
Hydrocyclone	Electrochlorination	Deoxygenation		
Coagulation	Ozonation	Inert gas or Nitrogen injection		
Flocculation	Chlorine dioxide	Ultrasonic treatment		
	Peracetic acid	Cavitation		
	Other active substances	Fine filtration		
		Heat		

in three stages, i.e., pre-treatment, treatment and residual control (neutralisation). In the pre-treatment stage the main focus is to exclude as much as possible solid material and bigger organisms, and with this helping the other treatment process(es) to be more effective, e.g., UV efficacy is limited if there are many solid particles in suspension because organisms survive when being in shadows of such particles, and the more solids and organisms are in the water, the more active substances are needed to achieve the same lethal effect. The residual control stage (neutralisation) is needed if there are any substances left in the ballast water after the treatment process is completed that could cause harm when being discharged from a vessel, e.g., residual toxicity from the use of active substances and their by-products (see Table 1).

In the following paragraphs we describe some of the main working principles of BWMS components.

Filtration

Filtration of ballast water seems to be the most environmentally sound method, but the amounts of ballast water that have to be treated are immense. Different filter technologies are in use, e.g., disk filters, mesh and wedge-wire filters. Ultra-filtration methods have not yet been tested or proven to work with large volumes of ballast water and high loads of sediments. The efficacy of removing particles larger than the mesh-size of these filter units is with 95–98 % very effective. In addition some percentage of the smaller particles may also be removed. Some systems use a combination of two filters where the first removes very large particles, which enhances the efficiency of the second finer filter. However, although the organism removal rate is high the D-2 standard is unlikely to be met with filtration as a stand-alone technology. Most filters used include an automatic backwash mechanisms for self-cleaning to ensure continuous operation. Overboard disposal of the collected residues as filter backwash would occur at the location of ballasting rather than at the destination port, thereby avoiding the transfer of non-native species with the filter backwash.

Hydrocyclone

Cyclonic separation has been proposed as a relatively simple and inexpensive way of removing larger particles and organisms from ballast water. Water and particles enter the hydrocyclone tangentially, thus setting up a circular flow. They are then drawn through tangential slots and are accelerated into the separation chamber. Centrifugal action tosses particles heavier than the water to the perimeter of the separation chamber. The solids gently drop along the perimeter and end up in the calm collection chamber of the separator. Solids may be periodically purged or continuously extracted from the separator. However, cyclonic separation of organisms with a specific gravity similar to that of water is limited which is valid for many plankton taxa. Therefore, some BWMS use the hydrocyclone as a pre-separator which is followed by a filtration unit thereby enhancing the performance of the filtration unit.

Ultraviolet Radiation

Ultraviolet (UV) radiation is commonly used for sterilising potable or waste water and for the purification in aquaculture and fisheries. UV radiation operates by causing photochemical reactions of biological components such as nucleic acids (DNA and RNA) and proteins. The lower UV wavelengths are generally more effective. However, radiation at these wavelengths shows a lower transmission in water. It's performance may further be affected by organic material, particles or bubbles. The effectiveness of UV treatment depends also largely upon the pigmentation, size, morphology of organisms (surface/volume ratio). Viruses require similar dosages to bacteria. Algae require larger dosages than bacteria due to their size and their pigmentation. Disadvantaging is the effect that some smaller organisms could pass the UV unit in the shadow of larger organisms/particles with reduced treatment and the reduced transmission of UV-radiation in turbid waters. It was observed in tests that some organisms have a self-repair mechanism so that re-growth of organisms after UV treatment occurred. This is (partly) overcome by applying the UV exposure also during ballast water discharge. Another and unsolved problem is that the UV effect on organisms is not immediately observed (Liebich et al. 2012; Martínez et al. 2012) so that compliance with the D-2 standard is difficult to show when the water is treated during discharge.

Electrochlorination

The use of electrochlorination as a means of preventing marine growth is well known. Electrochlorination is used on board so that the active substances are generated from the ballast water taken on board (no storage tank of chemicals) and this may either be done in a side stream or in the full ballast water stream of a vessel. Electrolyzers usually consist of a number of reactor cells arranged in series.

A minimum salinity is needed for its efficient use, in freshwater and lower brackish ballast water uptake zones marine water may be pumped into the line from a previously filled ballast tank to reach the required minimum salinity.

Chemical Dosing

A large number of chemical disinfectants are commercially available. These have been used successfully for many years in land-based potable and wastewater treatment applications. For the purpose of ballast water treatment several substances and formulations were considered, e.g., Chlorine dioxide, PeracleanOcean and SeaKleen. These systems have in common that an on board storage is needed and it would be beneficial that a supply of additional such substances is available in all ports the vessel is calling which may be logistically challenging. Further, ozone, generated on board from ambient air, is used in several BWMS. Most chemicals are usually applied during ballast water uptake with a mixing device to allow efficient treatment.

Neutralisation

The vast majority of ballast water treatment systems which make use of active substances add a neutralization substance. Such a neutralization step may not always be needed as e.g. on longer voyages the active substance may be (bio-)degraded before the ballast water discharge occurs. It seems most useful to apply the neutralising substance during the ballast water discharge. Proper mixing should occur so that the neutralizer is well circulated in the ballast water and that its neutralizing power is applied before the ballast water has left the vessel. Our review has shown that Sodium Thiosulphate is the most frequently used neutralizer today (see Table 2).

Application of Ballast Water Management Systems Technologies on Vessels

Different vendors developed different BWMS combining different technologies. Different systems (or parts of these) have their application in different stages of the ballasting process, i.e., at the uptake of ballast water, during holding the ballast water in tanks during navigation, and/or at discharge.

Among the 104 BWMS identified (see Table 2), 100 apply some treatment at the uptake, of these four apply treatment at the uptake and during the voyage (Table 2, nr. 17, 61, 74 and 95), and three are known to apply the treatment only during the voyage (nr. 12, 38 and 58). 29 BWMS treat the ballast water at uptake and discharge.

Some pre-treatment technology is used by 80 BWMS, of these 70 use filtration, three use a combination of hydrocyclone and filtration (nr. 16, 28 and 32), one uses a combination of flocculation and filtration (nr. 39), four use a hydrocyclone (nr. 35, 90, 95, and 97), and the remaining two use different other methods (nr. 5 and 17). It is interesting to note that 24 systems do not have a pre-treatment separation step.

Most of BWMS identified are regarded as BWMS that make use of an active substance (58). The most frequently used technique seems to be electrolysis/electrochlorination (35), and is applied as stand-alone treatment method by 28 BWMS, and by seven in combination with other techniques. The remaining 24 BWMS use dosing of different active substances, e.g., chlorine, PeraClean, SeaKleen and Akrolein.

In the second place is UV treatment with 34 BWMS, 24 of these use UV as stand-alone treatment method, while ten systems use UV in combination with one or more other techniques, i.e., TiO_2, ultrasound, ozonation, electrolysis, plasma.

In total 26 BWMS use two or more treatment techniques in combination as the main treatment method, while 75 rely on one treatment technique, no information was available for three BWMS (see Table 2).

One BWMS (Table 2, nr. 74) is the only system which makes use of vacuum deoxygenation and bioaugmentation. Bioaugmentation is a mechanism to, e.g., start activated sludge bioreactors in municipal wastewater treatment plants. In this BWMS microorganisms will be used to treat living organisms.

The application of BWMS that make use of active substances may result in residual active substances above the maximum allowable level (TRO[1] 0.2 mg L^{-1}) when this is to be discharged into the surrounding waters, hence they need to neutralise these before the discharge. The BWMS without neutralization will depend on a longer holding time of ballast water in the tanks during which the chlorine will breakdown to uncritical substances. Chlorine dioxide has a half-life of approximately 6–12 h (according to the suppliers and Olivieri et al. 1986), but at the concentrations at which it is employed it can be safely discharged after a maximum of 24 h. However, this relates also to water salinity and temperature and both should be taken into account when evaluating the minimum retention time before discharge.

Thirty-four BWMS that make use of active substances have included also an obligatory neutralisation process at discharge, and further three have this as an option. The most frequently used neutralisation is by Sodium Thiosulphate (24 BWMS), Sodium Sulphite use five BWMS, three use Sodium Biosulphite, one uses Activated Carbon, one uses Thiosulphate, and for three BWMS the substance is unknown (see Table 2). Most chlorination systems are applying a dose which results in approx. 10 mg L^{-1} chlorine during treatment, which has proven to be effective to kill organisms, but less than 0.2 mg L^{-1} residual chlorine in the ballast water discharges has proven to be environmentally acceptable to the recipient waters (see various references of Final Approvals of BWMS and GESAMP BWWG reports (IMO 2005–2012)). Most ozonation suppliers are using an ozone dose of 1–2 mg L^{-1} which has proven to be effective (Lloyds Register 2011a).

[1] TRO = total residual oxidants

According to the Lloyd's Register review of BWMS (Lloyds Register 2011a, b), technical features of the products are not necessarily common to all of them and are specific to generic types of process technologies. Deoxygenation is effective because the deoxygenated water is stored in sealed ballast tanks. However the process takes between 1 and 4 days to take effect, and thus represents the only type of technology where longer voyage length is a factor in process efficacy. This type of technology is also the only one where, technically, a decrease in corrosion propensity would be expected (and, according to one supplier, has been recorded as being suppressed by 50–85 %), since oxygen is a key component in the corrosion process. The water is re-aerated on discharge to avoid any unwanted effects to the recipient environment. However, the efficiency of deoxygenation is of concern as some organism can change their metabolism to another source than oxygen and other organisms are not dependent on oxygen at all.

Essentially most UV systems operate using the same type of medium pressure amalgam lamps. A critical aspect of UV effectiveness is the applied UV dose/power of the lamp. This information has not been given by all suppliers. Another aspect of UV effectiveness is the clarity of the water. In waters with a high turbidity or colloidal content, UV would not expected to be as effective as in very clear waters, but it was shown that UV systems also under these conditions meet the D-2 standard. Most of the busy ports in Europe (e.g., Rotterdam, Antwerp, Felixstowe and Hamburg) are located in estuaries with high sediment content.

Ballast Water Management Systems Capacities and Installation Requirements

Different BWMS have different capacities and technical profiles, which are mainly related to the aspects of appropriate capacity of the ballast water system of a vessel, as well as to the system space requirement and power consumption. For many BWMS the information available was very limited, and for some BWMS no information became known at all.

BWMS capacities range from 50 $m^3 h^{-1}$ to more than 10,000 h^{-1}, while five manufacturers informed that their systems are (will be) able to treat 20,000 and more h^{-1}. In terms of footprint space requirements the systems with the capacity 200 h^{-1} could occupy from even less than 1 m^2 and up to 30 m^2, while the systems with the capacity 2,000 $m^3 h^{-1}$ would occupy from 1 m^2 and up to 145 m^3. Systems operate also with no electricity requirement, and others may consume up to 200 kW per 1,000 $m^3 h^{-1}$ water to be treated.

Chemical dosing systems such as PeracleanOcean, SeaKleen and chlorine dioxide have low capital costs because only a dosing/mixing pump is required but these systems require chemical storage facilities and availability of chemicals in all ports visited. Should the active substance be transported in higher concentrations, as during shipment to the vessel, some special regulations regarding the transport of dangerous goods may apply in certain ports due to safety concerns.

The biggest operating cost for most systems is power and for large power consumers (electrolytic, advanced oxidation processes and UV) the availability of shipboard power will be a factor which may limit its installation and operation. For chemical dosing systems, power consumption is very low and chemical costs are the major factor. For these reasons chemical addition may be better suited to treat small ballast capacities.

Although the BWMS operate at generally low pressure and thus do not require additional ballast water pumping pressure, those employing Venturi devices (for exerting shear forces and proper mixing of chemicals) incur pressure losses of up to 2 bar.

For most systems it is recommended that the installation takes place in the engine/machine room near the existing ballast water pumps, although installation on deck may also be possible if appropriate precautions are taken. If the location is in an explosion zone, then the installation will need explosion proofing. Some of the technologies can be provided as explosion-proof products, but there is a cost factor for this. The generation of hydrogen by the electrolytic technologies is not considered an issue, provided the gas is vented and diluted with air to safe levels.

Whilst disinfection by-products are an issue, and central to the approval of ballast water management systems that make use of active substances, suppliers are confident that the levels of active substances and by-products generated are unlikely to be problematic. There is a large amount of scientific and technical information on disinfection by-products formation that is likely to support this. However, all systems using active substances will be reviewed by an independent expert group of GESAMP to assess the environmental acceptability of the treated water at discharge.

Ballast Water Management Systems Testing and Approvals

All systems need to be type approved by a Flag state before being sold to a client. Systems that use Active Substances by the definition in the BWM Convention have to undergo a more thorough certification process and obtain Basic and Final Approvals by IMO MEPC. This process was initiated to proof the environmental acceptability of treated ballast water when discharged from a vessel.

All systems are tested in a land-based setting with challenging water conditions (different water parameters and high organism numbers) to show that the D-2 standard is met. Ten test cycles need to be carried out in minimum. In addition, at least three test cycles need to be undertaken over a period of at least 6 months on board of commercial vessels to document that they meet the D-2 standard and are seaworthy. These tests are addressed in the IMO Guidelines G8. Currently a harmonization of sampling methods and sample analysis options is ongoing with all test facilities and shipboard sampling teams being involved (GloBalTestNet) and in October 2013 a Memorandum of Understanding of these was signed to achieve these goals. Test facilities are located in China, Denmark, Germany (Stephan Gollasch for shipboard

BWMS	**BWMS approval according to G9 (GESAMP and MEPC)**	**BWMS approval according to G8 (Flag state)**		**BWMS approval according to G9 (GESAMP and MEPC)**	**BWMS Type Approval Certificate (Flag state)**
using active substance(s) →	Basic Approval →	Land based tests →	Shipboard tests →	Final Approval →	Type Approval
without using active substance(s) →	→	Land based tests →	Shipboard tests →	→	Type Approval

Fig. 1 The approval process of BWMS according to the IMO requirements

tests), Japan, Korea, The Netherlands, Norway, Singapore, Slovenia (Matej David for shipboard tests), UK and the USA, and others are further planned in India and South Africa (Gollasch 2010 and pers. comm.).

After all these tests the system gets eventually type approved by a Flag state. This comprehensive approval process usually takes 1.5 years or longer. The duration of a type approval depends on many factors, including the test requirements, the availability of land and shipboard test facilities, the success of BWMS performance test runs and whether or not a system makes use of active substances. When active substances are used comprehensive basic and final approval dossiers need to be prepared, which requires additional tests. These dossiers are evaluated by the Ballast Water Working Group of GESAMP (see Fig. 1).

At present BWMS are in different stages of development, testing and approval processes, while 33 were already type approved by different administrations (IMO 2013k, see grey shading in Table 2). The authors have further obtained information that German authorities have issued an additional certificate for the Aquaworx BWMS (nr. 6) (Clason, personal communication) which brings the total number of type approved BWMS to 34. We expect this number to rise soon as several other BWMS are in the final phase of the approval process.

The Global Market for Ballast Water Management Systems

Japanese experts calculated the number of vessels to which Regulation D-2 would have applied if it would have been implemented as originally planned from 2009 to 2020. The number of vessels would have totalled to more than 75,000 vessels, with the highest annual number in 2017, i.e., more than 16,000 vessels. Divided by 365 this results in an installation demand of ca. 45 BWMS per day. The number of vessels required to install BWMS was expected to rapidly increase in 2015 and sharply drop in 2020, because the vessels constructed before 2009 should have installed BWMS between 2015 and 2019. The number of existing vessels that would need to retrofit would be in total approximately 34,000 vessels and the number of vessels, which are required to retrofit BWMS is estimated at 2,500 vessels in 2015 and 2016, 11,000 vessels in 2017, and 9,000 vessels in 2018 and 2019. The phase-in of the vessels to meet the D-2 standard was recently time-wise relaxed (see chapter "Ballast Water Management Under the Ballast Water Management Convention"), which will likely result in a longer high demand of BWMS to be installed on board vessels (IMO 2010g, h).

A recent calculation on the estimated value of the global market for purchasing and installing BWMSs was conducted by IMarEST (IMO 2011a, z) and the estimations resulted a turn-over between 2011 and 2016 of possibly $50–74 billion.

As per the original IMO requirements more than 21,000 vessels were subject to the first round of BWMS retrofits. This would have included vessels with a ballast water capacity of 1,500–5,000 m^3. With 16,000 out of these 21,000 vessels, the majority of those vessels would have been general cargo ships. IMarEST analysed the "delivered" vessels by type and it was estimated that more than 68,000 vessels would need to install on board BWMS before 2020 (IMO 2011a, z).

Fishing vessels are a special case and only those of >300 gross tons were included in the analysis of IMarEST (IMO 2011a, z). Considering the tight profit range of especially smaller fishing vessels, it is unlikely that they will include the installation of BWMS in their business plans. Other limitations for those vessels may be the lack of space to install BWMS so that those vessels may have to find another way to comply with ballast water management regulations.

According to IMarEST estimates the cost range of BWMS across system types and categories of ship was estimated to be between $640,000 and $947,000 per vessel, however the authors in direct contact with BWMS vendors received information that the system prices would start from approximately 250,000 Euro. It should also be noted that installation costs will vary to a great extend which is related to the BWMS and ship characteristics and the footprint and other requirements. In some cases, depending on the number of ballast pumps aboard, more than one BWMS may have to be installed.

BWMS manufacturers and shipowners assume that minimal or even no lost profit may occur due to the retrofitting of BWMS provided the installation time does not extend the normal shipyard time. Alternatively the BWMS may be installed during navigation, but cabin and lifeboat limitations may occur when planning to accommodate the installation crew (IMO 2011a, z).

Ballast Water Management Systems Information

Table 2 BWMS manufacturers (in alphabetical order), commercial names of their BWMS, technologies used and available web pages (last accessed January 2014). Type approved BWMS are shown with grey shading

Nr.	Manufacturer	System name	Pre-treatment	Treatment	Residual control	Web site
1	21st Century Shipbuilding Co., Ltd	ARA Ballast (Blue Ocean Guardian BOG)	Filtration	Plasma+UV	–	www.21csb.com/eng /sub04_02.html
2	Akballast	Akballast	Filtration	UV	–	–
3	Alfa Laval Tumba AB	PureBallast (3.0)	Filtration	UV + TiO_2	–	www.alfalaval.com /campaigns/pureballast3/Documents/index.htm
4	Alfa Laval Tumba AB	PureBallast (3.0 Ex)	Filtration	UV + TiO_2	–	www.alfalaval.com /campaigns/pureballast3/Documents/index.htm
5	Aquaeng Co. Ltd.	AquaStar BWMS	Smart pipe unit	Electrolysis/electrochlorination	Sodium thiosulphate	www.aquaeng.kr
6	Aquaworx ATC GmbH as original developer, now with GEA Westfalia	AquaTriComb, new name: BallastMaster ultraV	Filtration	UV + ultrasound	–	www.westfalia -separator.com/applications/marine/ballast-water-treatment/gea-westfalia -separator-ballastmaster-ultrav.html
7	atgUVTechnology (ATG Willand)	–	Filtration	UV	–	www.atguv.com /marine-shipping
8	ATLAS-DANMARK	ATLAS-DANMARK ABTS	Filtration	Electrochemical (Anolyte)	–	www.atlas -denmark.com
9	Auramarine	CrystalBallast	Filtration	UV	–	www.auramarine.com/news/auramarine-_new_challenger_in_the_market_for_ballast_water_treatment_systems
10	Azienda Chimica Genovese	ECOLCELL BTs	Filtration	Electrolysis/electrochlorination	Sodium thiosulphate	www.acgmarine.com/en/products/ecolcell -bts
11	BaWaPla (stopped development)	–	Filtration	Electrolysis/electrochlorination	–	www.bawapla.com
12	Bawat	–	–	Inert gas+heat	–	www.bawat.dk
13	Bio-UV	Bio-SEA	Filtration	UV	–	www.ballast -water-treatment.com
14	Brillyant Marine	–	–	Elelctric pulse	–	www.brillyantinc.com
15	Cavipure (old name: Jetsam)	–	Filtration	UV+ultrasound	–	–
16	China Ocean Shipping Company (COSCO)	Blue Ocean Shield	Hydrocyclone + Filtration	UV	–	www.cosco.com/en
17	Coldharbour Marine	Coldharbour BWT	Cavitation	Deoxigenation	–	www.coldharbourmarine.com
18	Dalian Maritime University	DMU OH BWMS	Filtration	hydroxyl radicals, ozone and hydrogen peroxide	Sodium thiosulphate	www.dlmu.edu.cn
19	DESMI OceanGuard AS	DESMI OceanGuard BWMS	Filtration	Ozonation+UV	–	www.desmioceanguard.com
20	Dow Chemical Pacific (Singapore) Pte Ltd.	Dow-Pinnacle BWMS	Filtration	Ozonation	Sodium thiosulphate (optional)	–
21	Ecochlor Inc	Ecochlor	Filtration	Chlorination (ClO_2)	–	www.ecochlor.com
22	Ecologiq	BallaClean	Filtration	Electrolysis/electrochlorination	–	www.ecologiq.us
23	Electrichlor Inc	Electrichlor	Filtration	Electrolysis/electrochlorination	–	www.electrichlor.com
24	EltronWaterSystems	PeroxEgen	–		–	www.eltronwater.com
25	Environmental Technologies Inc	ETI	Filtration	Ozonation+ultrasound	–	www.tlmcos.com
26	Envirotech and Consultancy PTE ltd.	BlueSeas BWMS	Filtration (microsized strainer)	Electrochemical disinfection	Sodium thiosulphate	www.blueseas.com.sg (under construction)
27	Envirotech and Consultancy PTE ltd.	BlueWorld BWMS	Filtration (microsized strainer)	Electrochemical disinfection	Sodium thiosulphate	www.blueseas.com.sg (under construction)
28	Erma First SA	Erma First BWMS	Hydrocyclone	Electrolysis/electrochlorination	Sodium bisulphite	www.ermafirst.com /ballast-water
29	Ferrate Treatment Technologies (stopped development)	Ferrator BW	–	Fe6+	–	www.ferratetreatment.com /ballast water.htm
30	Gauss (stopped developement)	–	Filtration	UV	–	–

(continued)

Table 2 (continued)

Nr.	Manufacturer	System name	Pre-treatment	Treatment	Residual control	Web site
31	GEA Westfalia	BallastMaster ecoP	Filtration	Electrolysis/elec trochlorination	Sodium thiosulphate	www.westfalia - separator.com/applications/marine /ballast-water-treatment/gea - westfalia -separator-ballastmaster - ecop.html
32	Hamann AG (Degussa) (withdrawn from market)	Sedna using Peraclean Ocean	Hydrocyclone + Filtration	Paracetic acid	–	www.haman nag.com
33	Hamworthy	Aquarius UV	Filtration	UV	–	www.wartsila.com /en/ballast - water-management - system/hamworthy/aquarius -uv
34	Hamworthy	Aquarius EC	Filtration	Electrolysis/elec trochlorination	Sodium bisulphite	www.wartsila.com /en/ballast - water-management- system/hamworthy/aquarius -ec
35	Hamworthy Greenship (stopped development)	Greenship Sedinox	Hydrocyclone	Electrolysis/elec trochlorination	–	–
36	Hanla IMS Co., Ltd.	EcoGuardian	Filtration	Electrolysis/elec trochlorinationc	Sodium thiosulphate	www.hanlaims.com/
37	Headway Technology Co., Ltd.	OceanGuard	Filtration	Electrolysis/elec trochlorination +ultrasonic treatment (EUT)	Sodium thiosulphate (optional)	www.headwaytech.com/en
38	Hi Tech Marine Pty Ltd	Ballast water disinfection	–	Heating	–	www.htmarine.com.au
39	Hitachi	ClearBallast	Flocculation (magnetic particles) + Filtration		–	www.hitachi - pt.com/products/es/ballast
40	HWASEUNG R&A Co., Ltd.	HS-Ballast	–	Electrolysis/elec trochlorination	Sodium thiosulphate	www.hsrna.com/eng/main
41	HyCa Technologies Pvt. Ltd.	HyCator	Filtration	Electrolysis/elec trochlorination	Sodium thiosulphate	www.hycator.com/index.php/ourp roduct/hycator-bwt
42	Hyde Marine Inc	Hyde Guardian Gold	Filtration	UV	–	www.hydemarine.com /ballast_wa ter
43	Hyde Marine Inc	Seakleen ™(Vitamar)	–	SeaKleen	–	www.hydemarine.com /ballast_wa ter
44	Hyundai Heavy Industries	EcoBallast	Filtration	UV	–	http://english.hhi.co.kr
45	Hyundai Heavy Industries	HiBallast	Filtration	Electrolysis/elec trochlorination	Sodium thiosulphate	http://english.hhi.co.kr
46	JFE Engineering Corporation	JFE BallastAce BWMS (uses TG Ballastcleaner)	Filtration	Chlorination+re sidual Clorine+cavitati on (TG BallastCleaner)	Sodium sulphite (TG Environmental guard)	www.jfe - eng.co.jp/en /products/machine/ma rine/mar01.html
47	JFE Engineering Corporation	JFE BallastAce BWMS (uses NEO-CHLOR MARINE)	Filtration	Chemical injection (Neo-Chlor Marine)	Sodium sulphite	www.jfe - eng.co.jp/en /products/machine/ma rine/mar01.html
48	Jiujiang Precision Measuring Technology Research Institute	OceanDoctor	Filtration	UV+photocataly tic reaction	–	–
49	Kashiwa Kuraray Co.Ltd. (ref doc 61/2/6)	Microfade	Filtration	Chlorination (Cl_2)	Sodium sulphite	www.kuraray.co.jp
50	Katayama Chemical inc.	Sky-System using PeracleanOcean	–	PeracleanOcean	Sodium sulphite	www.nipponyuka.jp
51	Knutsen Ballast Vann AS	KBAL	–	vacuum+UV	–	www.knutsenoas.com/knutsen-technology/knutsen-ballast-water-treatment-technology-kbal%C2%AE/
52	Korea Top Marine (KT Marine) Co., Ltd.	KTM-BWMS (Plankill pipe™)	–	Electrolysis/elec trochlorination	Sodium thiosulphate	–
53	Kwang San Co., Ltd.	En-Ballast	Filtration	Electrolysis/elec trochlorination	Sodium thiosulphate	www.kwangsan.com
54	Mahle NFV GmbH	Ocean Protection System OPS	Filtration	UV	–	www.mahle-industrialfiltration.com
55	Marenco Technology Group Inc	Marenco BWTS	Filtration	UV	–	–
56	Maritime Solutions Inc.	–	Filtration	UV	–	www.maritimesolutionsinc.com
57	Mexel Industries	–	–	Chemical treatment	–	www.mexelusa.com
58	MH Systems Inc	MH Systems BWTS	–	Deoxigenation	–	www.mhsystemscorp.com
59	Mitsui Engineering & Shipbuilding	FineBallast® OZ (Special Pipe SP-Hybrid BWMS with ozone)	–	Ozonation+cavi tation	Activated carbon	www.mes.co.jp/english /business/ship/ship_13.html

(continued)

Table 2 (continued)

Nr.	Manufacturer	System name	Pre-treatment	Treatment	Residual control	Web site
60	MMC Green Technology AS	MMC	Filtration	UV	–	www.mmcgt.no/
61	NEI Treatment Systems LLC (two independent type approvals, i.e., Marshall Islands and Malta)	Venturi Oxygen Stripping	–	Cavitation+deoxigenation	–	www.nei-marine.com/en/about-us
62	NK Company	NK-03 BlueBallast	–	Ozonation	Sodium thiosulphate	http://nk-eng.nkcf.com
63	Nutech 03	Mark III	–	Ozonation	–	www.nutech-o3.com
64	Oceansaver AS (MetaFil AS)	OceanSaver	Filtration	Cavitation+electrolysis/electrochlorination+deoxigenation	Sodium thiosulphate	www.oceansaver.com
65	Oceansaver AS (MetaFil AS)	OceanSaver with optional N2 supersaturation	Filtration	Electrolysis/electrochlorination	Sodium thiosulphate	www.oceansaver.com
66	Optimarin AS	OptiMarin Ballast System OBS	Filtration	UV	–	www.optimarin.com
67	Panasia	GloEn-Patrol	Filtration	UV	–	www.pan -asia.co.kr
68	Panasia	GloEn-Saver	Filtration	Electrolysis/electrochlorination	Sodium thiosulphate	www.pan -asia.co.kr
69	Qwater	–	Filtration	Ultrasound	–	www.qwatercorp.com
70	REDOX Maritime Technologies (RMT) AS	REDOX AS	Filtration	ozone+UV	Sodium thiosulphate	www.redoxmaritime.no/uk/index.html
71	RWO GmbH Marine Water Technology	CleanBallast	Filtration	Electrochlorination+OH	Substance unknown	www.rwo.de /en/technologies_products_and_Solutions/Ballast_Water_Treatment /
72	Samsung Heavy Indistries	PuriMar	Filtration	Electrolysis/electrochlorination	Sodium thiosulphate	www.shi.samsung.co.kr/Eng/default.aspx
73	Samsung Heavy Indistries	NEO-PuriMar	Filtration	Electrolysis/electrochlorination	Sodium thiosulphate	www.shi.samsung.co.kr/Eng/default.aspx
74	Sea Knight Corporation	Insitu™	–	Chemical tr.+vacuum deoxigenation + bioaugmentation	–	www.seaknight.net
75	Sea Reliance Marine Services	–	Filtration	UV	–	–
76	Seair	–	Filtration	Ozonation	–	www.seair.ca
77	Sembawang	Semb -Eco	Filtration	UV incl. LED-UV	–	www.sembship.com
78	Severn Trent De Nora	BalPure	optional	Electrolysis/electrochlorination +residual Clorine	Sodium bisulphite or Sodium sulphite or Sodium thiosulphate	http://www.severntrentdenora.com/Products-and-Services/Ballast-Water-Treatment-Systems/
79	Shanghai Cyeco Environmental Technology Co. Ltd.	Cyeco	Filtration	UV	–	www.cyecomarine.com/product2.html
80	Siemens (now as Evoqua Water Technologies)	SiCURE	Filtration	Electrolysis/electrochlorination	Sodium sulphite (optional)	www.water.siemens.com
81	Sincerus	Sincerus maritime	Filtration	Electrolysis/electrochlorination	–	–
82	SPO System	Special Pipe Hybrid BWMS with PeracleanOcean	–	Cavitation+peracleanOcean	–	–
83	STX Metal Co. Ltd.	Smart Ballast	–	Electrolysis/electrochlorination	Sodium thiosulphate	www.stxmetal.co.kr
84	Sumetomo Electric Industries Ltd.	Ecomarine	Filtration	UV	–	http://global-sei.com/news/press/11/11_35.html
85	SUNBO Industries Co., Ltd.	Blue Zone	–	ozone	Thiosulphate	http://sunboind.en.ec21.com
86	Sunrui Corrosion and Fouling Control Company (Sunrui CFCC)	BalClor BWMS (Sunrui BWMS)	Filtration	Electrolysis/electrochlorination	Sodium thiosulphate	www.sunrui.net /Products/BalClorTMBallastWaterManagementSystem/
87	Techcross	Electro-Cleen System ECS	–	Electrolysis/electrochlorination	Sodium thiosulphate	www.techcross. com /new/main/main.asp

(continued)

Table 2 (continued)

Nr.	Manufacturer	System name	Pre-treatment	Treatment	Residual control	Web site
88	The Ship Stability Research Centre (SSRC), University of Strathclyde	ClearBal	–	–		www.sumobrain.com/patents/wipo/Ballast-water-treatment-system/WO2010086604.html
89	Trojan	Trojan Marinex	Filtration	UV	–	www.trojanmarinex.com
90	University of Dubrovnik, Croatia	–	Hydrocyclone	High pressure	–	–
91	Van Oord B.V.	VO-BWMS	–	Drinking water+chlorine	Sodium bisulphite	www.vanoord.com/
92	Wilhelmsen Technical Solutions/Resource (withdrawn from market, relaunch planned)	Unitor Resource BWTS	Filtration	Cavitation+electrolysis/electrochlorination+ozonation	–	www.resource-technology.com
93	Wuxi Brightsky Electronic Co., Ltd.	BSKY	Filtration	UV	–	www.bsky.cn
94	Confidential	Thermal Aqua Filtration (TAF)	Filtration	Heat	–	–
95	Confidential	–	Hydrocyclone	–	–	–
96	Confidential	–	Filtration	Cavitation+Electrochemical treatment	–	–
97	Confidential	–	Hydrocyclone	Chlorination	–	–
98	Confidential	–	Filtration	Akrolein	Not decided	–
99	Confidential	–	Filtration?	Electrolysis/electrochlorination	–	–
100	Confidential	–	Filtration	UV	–	–
101	Confidential	–	Filtration	Electrolysis/electrochlorination	Substance unknown	–
102	Confidential	–	Filtration	UV+microwaves+ozone	–	–
103	Confidential	–	Filtration	UV	–	–
104	Confidential	–	Filtration	UV	–	–

Acknowledgements The research leading to part of these results has received funding from the European Community's Seventh Framework Programme (FP7/2007–2013) under Grant Agreement No. [266445] for the project Vectors of Change in Oceans and Seas Marine Life, Impact on Economic Sectors (VECTORS). Part of this publication has been produced with the financial assistance of the IPA Adriatic Cross-Border Cooperation Programme – strategic project Ballast Water Management System for Adriatic Sea Protection (BALMAS). The contents of this publication are the sole responsibility of authors and can under no circumstances be regarded as reflecting the position of the IPA Adriatic Cross-Border Cooperation Programme Authorities.

References

Gollasch S (2010) Global expert workshop on harmonization of methodologies for test facilities of ballast water management systems. Global Ballast Water Management Programme (GloBallast). World Maritime University, Malmö, 24–25 January 2010

IMO (2004) International convention for the control and management of ships' ballast water and sediments. International Maritime Organization. Adopted on 13 February 2004, IMO, London

IMO (2005a) Harmful aquatic organisms in ballast water: information to be considered by the Review Group. Submitted by Sweden. International Maritime Organization, Marine Environment Protection Committee, MEPC 53/2/6. 15 April 2005. IMO, London

IMO (2005b) Application for basic approval of active substances used by Electro-Clean (electrolytic disinfection) ballast water management system. Submitted by Republic of Korea. International Maritime Organization, Marine Environment Protection Committee, MEPC 54/2/3. 16 December 2005. IMO, London

IMO (2006a) Report of the first meeting of the GESAMP-BWWG. Note by the Secretariat. International Maritime Organization, Marine Environment Protection Committee, MEPC 54/2/12. 28 February 2006. IMO, London
IMO (2006b) Report of the Marine Environment Protection Committee on its fifty-fourth session. International Maritime Organization, Marine Environment Protection Committee, MEPC 54/21. 27 March 2006. IMO, London
IMO (2006c) Basic approval of active substances used by Special Pipe Ballast Water Management System (combined with Ozone treatment). Submitted by Japan. International Maritime Organization, Marine Environment Protection Committee, MEPC 55/2. 12 April 2006. IMO, London
IMO (2006d) Basic approval of active substances used by NK Ballast Water Treatment System. Submitted by the Republic of Korea. International Maritime Organization, Marine Environment Protection Committee, MEPC 55/2/3, 20 April 2006. IMO, London
IMO (2006e) Basic approval of active substances used by EctoSys electrochemical system. Submitted by Sweden. International Maritime Organization, Marine Environment Protection Committee, MEPC 55/2/4, 21 April 2006. IMO, London
IMO (2006f) Basic approval of active substances used by Pure Ballast management system. Submitted by Sweden. International Maritime Organization, Marine Environment Protection Committee, MEPC 55/2/5, 21 April 2006. IMO, London
IMO (2006g) Review documentation of ballast water treatment systems in Germany, Information (Update of MEPC 53/2/11 Annex 1) provided by Elga Berkefeld GMBH, Lückenweg, 5, 29227 Celle, Germany and its subsidiary RWO Marine Water Technology, Leerkämpe 3, 29259, Bremen, Germany. International Maritime Organization, Marine Environment Protection Committee, MEPC 55/2/17, Annex 1. 7 July 2006. IMO, London
IMO (2006h) Report of the second meeting of the GESAMP-Ballast Water Working Group (GESAMP-BWWG). Note by the Secretariat. International Maritime Organization, Marine Environment Protection Committee, MEPC 55/2/16. 7 July 2006. IMO, London
IMO (2006i) Report of the Marine Environment Protection Committee on its fifty-fifth session. International Maritime Organization, Marine Environment Protection Committee, MEPC 55/23. 16 October 2006. IMO, London
IMO (2006j) Application for final approval of ballast water management system using Active Substances. Submitted by Norway. International Maritime Organization, Marine Environment Protection Committee, MEPC 56/2/1. 15 December 2006. IMO, London
IMO (2007a) Basic approval of active substances used by Resource Ballast Technologies Systems (Cavitation combined with Ozone and Sodium Hypochlorite treatment). Submitted by South Africa. International Maritime Organization, Marine Environment Protection Committee, MEPC 56/2/3. 6 April 2007. IMO, London
IMO (2007b) Report of the third meeting of the GESAMP-Ballast Water Working Group (GESAMP-BWWG). Note by the Secretariat. International Maritime Organization, Marine Environment Protection Committee, MEPC 56/2/2. 13 April 2007. IMO, London
IMO (2007c) Review documentation of ballast water treatment technologies in Germany. Submitted by Germany. International Maritime Organization, Marine Environment Protection Committee, MEPC 56/2/10. 4 May 2007. IMO, London
IMO (2007d) Report of the Marine Environment Protection Committee on its fifty-sixth session. International Maritime Organization, Marine Environment Protection Committee, MEPC 56/23. 30 July 2007. IMO, London
IMO (2007e) Application for basic approval of active substances used by Hitachi Ballast Water Purification System (ClearBallast). Submitted by Japan. International Maritime Organization, Marine Environment Protection Committee, MEPC 57/2/2. 7 September 2007. IMO, London
IMO (2007f) Application for final approval of a ballast water management system using active substances. Submitted by Germany. International Maritime Organization, Marine Environment Protection Committee, MEPC 57/2/3. 7 September 2007. IMO, London
IMO (2007g) Application for final approval of a ballast water management system using active substances. Submitted by Germany. International Maritime Organization, Marine Environment Protection Committee, MEPC 57/2/5. 7 September 2007. IMO, London

IMO (2007h) Basic approval of active substance used by GloEn-Patrol™ Ballast Water Management System. Submitted by the Republic of Korea. International Maritime Organization, Marine Environment Protection Committee, MEPC 57/2/4. 7 September 2007. IMO, London

IMO (2007i) Report of the fourth meeting of the GESAMP-Ballast Water Working Group (GESAMP-BWWG). Note by the Secretariat. International Maritime Organization, Marine Environment Protection Committee, MEPC 57/2. 19 December 2007. IMO, London

IMO (2007j) Application for basic approval of a combined ballast water management system consisting of sediment removal and an electrolytic process using seawater to produce active substances (Greenship Ltd). Submitted by the Netherlands. International Maritime Organization, Marine Environment Protection Committee, MEPC 57/2/7. 20 December 2007.

IMO (2008a) Report of the fifth meeting of the GESAMP-Ballast Water Working Group (GESMP-BWWG). Note by the Secretariat. International Maritime Organization, Marine Environment Protection Committee, MEPC 57/2/10. 25 January 2008. IMO, London

IMO (2008b) Report of the Marine Environment Protection Committee on its fifty-seventh session. International Maritime Organization, Marine Environment Protection Committee, MEPC 57/21. 7 April 2008. IMO, London

IMO (2008c) Application for final approval of the OceanSaver® Ballast Water Management System (OS BWMS). Submitted by Norway. International Maritime Organization, Marine Environment Protection Committee, MEPC 58/2/1. 19 March 2008. IMO, London

IMO (2008d) Application for final approval of the Electro-Clean System (ECS). Submitted by the Republic of Korea. International Maritime Organization, Marine Environment Protection Committee, MEPC 58/2. 20 March 2008. IMO, London

IMO (2008e) Application for basic approval of the Ecochlor® Ballast Water Treatment System. Submitted by Germany. International Maritime Organization, Marine Environment Protection Committee, MEPC 58/2/2. 20 March 2008. IMO, London

IMO (2008f) Application for final approval of the NK-O3 Blue Ballast System (Ozone). Submitted by the Republic of Korea. International Maritime Organization, Marine Environment Protection Committee, MEPC 58/2/3. 21 March 2008. IMO, London

IMO (2008g) Procedure for approval of ballast water management systems that make use of active substances (G9). MEPC 57/21. Annex 1. Resolution MEPC.169(57). Adopted on 4 April 2008. IMO, London

IMO (2008h) Report of the sixth meeting of the GESAMP-Ballast Water Working Group. Note by the Secretariat. International Maritime Organization, Marine Environment Protection Committee, MEPC 58/2/7. 14 July 2008. IMO, London

IMO (2008i) Report of the seventh meeting of the GESAMP-Ballast Water Working Group. Note by the Secretariat. International Maritime Organization, Marine Environment Protection Committee, MEPC 58/2/8. 28 July 2008. IMO, London

IMO (2008j) Information on the Type Approval Certificate of the SEDNA® Ballast Water Management System using Peraclean® Ocean (SEDNA® 250). Submitted by Germany. International Maritime Organization, Marine Environment Protection Committee, MEPC 58/INF.17. 31 July 2008. IMO, London

IMO (2008k) Guidelines for approval of ballast water management systems (G8). International Maritime Organization, Marine Environment Protection Committee, MEPC 58/23. Annex 4. Resolution MEPC.174(58). Adopted on 10 October 2008. IMO, London

IMO (2008l) Report of the Marine Environment Protection Committee on its fifty-eighth session. International Maritime Organization, Marine Environment Protection Committee, MEPC 58/23. 16 October 2008. IMO, London

IMO (2008m) Application for final approval of the RWO Ballast Water Management System (CleanBallast) Submitted by Germany. International Maritime Organization, Marine Environment Protection Committee, MEPC 59/2. 28 November 2008. IMO, London

IMO (2008n) Application for final approval of the Special Pipe Hybrid Ballast Water Management System (combined with Ozone treatment). Submitted by Japan. International Maritime Organization, Marine Environment Protection Committee, MEPC 59/2/1. 4 December 2008. IMO, London

IMO (2008o) Application for basic approval of the Blue Ocean Shield Ballast Water Management System. Submitted by China. International Maritime Organization, Marine Environment Protection Committee, MEPC 59/2/2. 5 December 2008. IMO, London
IMO (2008p) Application for final approval of the NK-O3 Blue Ballast System (Ozone). Submitted by the Republic of Korea. International Maritime Organization, Marine Environment Protection Committee, MEPC 59/2/3. 8 December 2008. IMO, London
IMO (2008q) Application for basic approval of the HHI Ballast Water Management System (EcoBallast). Submitted by the Republic of Korea. International Maritime Organization, Marine Environment Protection Committee, MEPC 59/2/4. 9 December 2008. IMO, London
IMO (2008r) Application for final approval of the Hitachi Ballast Water Purification System (ClearBallast). Submitted by Japan. International Maritime Organization, Marine Environment Protection Committee, MEPC 59/2/5. 11 December 2008. IMO, London
IMO (2008s) Application for final approval of the Greenship Sedinox Ballast Water Management System. Submitted by the Netherlands. International Maritime Organization, Marine Environment Protection Committee, MEPC 59/2/6. 12 December 2008. IMO, London
IMO (2008t) Application for final approval of the GloEn-Patrol™ Ballast Water Treatment System. Submitted by the Republic of Korea. International Maritime Organization, Marine Environment Protection Committee, MEPC 59/2/7. 16 December 2008. IMO, London
IMO (2008u) Application for basic approval of the AquaTriComb™ Ballast Water Treatment System. Submitted by Germany. International Maritime Organization, Marine Environment Protection Committee, MEPC 59/2/8. 16 December 2008. IMO, London
IMO (2008v) Application for final approval of the Resource Ballast Technologies System (Cavitation combined with Ozone and Sodium Hypochlorite treatment). Submitted by South Africa. International Maritime Organization, Marine Environment Protection Committee, MEPC 59/2/10. 19 December 2008. IMO, London
IMO (2008w) Application for basic approval of the Siemens SiCURE™ Ballast Water Management System. Submitted by Germany. International Maritime Organization, Marine Environment Protection Committee, MEPC 59/2/11. 19 December 2008. IMO, London
IMO (2009a) Information on the type approval certificate of the Electro-Cleen™ System (ECS). Submitted by the Republic of Korea. International Maritime Organization, Marine Environment Protection Committee, MEPC 59/INF.6. 20 January 2009. IMO, London
IMO (2009b) Report of the eighth meeting of the GESAMP-Ballast Water Working Group. Note by the Secretariat. International Maritime Organization, Marine Environment Protection Committee, MEPC 59/2/16, Annex 4. 8 April 2009. IMO, London
IMO (2009c) Information on the type approval certificate of the OceanSaver® Ballast Water Management System. Submitted by Norway. International Maritime Organization, Marine Environment Protection Committee, MEPC 59/INF.17. 6 May 2009. IMO, London
IMO (2009d) Type approval of the Hyde GUARDIAN™ Ballast Water Management System. Submitted by the United Kingdom. International Maritime Organization, Marine Environment Protection Committee, MEPC 59/INF.20. 7 May 2009. IMO, London
IMO (2009e) Report of the Marine Environment Protection Committee on its fifty-ninth session. International Maritime Organization, Marine Environment Protection Committee, MEPC 59/24. 27 July 2009. IMO, London
IMO (2009f) Application for basic approval of the ATLAS-DANMARK Ballast Water Treatment System (with onboard generated ANOLYTE oxidant). Submitted by Denmark. International Maritime Organization, Marine Environment Protection Committee, MEPC 60/2. 10 August 2009. IMO, London
IMO (2009g) Application for basic approval of the DESMI Ocean Guard Ballast Water Management System. Submitted by Denmark. International Maritime Organization, Marine Environment Protection Committee, MEPC 60/2/4. 19 August 2009. IMO, London
IMO (2009h) Application for final approval of HHI Ballast Water Management System EcoBallast. Submitted by the Republic of Korea International Maritime Organization, Marine Environment Protection Committee, MEPC 60/2/1. 20 August 2009. IMO, London

IMO (2009i) Application for final approval of the JFE Ballast Water Management System (JFE-BWMS) that makes use of "TG Ballastcleaner® and TG Environmentalguard®." Submitted by Japan. International Maritime Organization, Marine Environment Protection Committee, MEPC 60/2/2. 20 August 2009. IMO, London
IMO (2009j) Application for basic approval of the Sunrui ballast water management system. Submitted by China. International Maritime Organization, Marine Environment Protection Committee, MEPC 60/2/3. 24 August 2009. IMO, London
IMO (2009k) Application for basic approval of Blue Ocean Guardian (BOG) Ballast Water Management System. Submitted by the Republic of Korea. International Maritime Organization, Marine Environment Protection Committee, MEPC 60/2/5. 24 August 2009. IMO, London
IMO (2009l) Application for basic approval of the Hyundai Heavy Industries Co., Ltd. (HHI) Ballast Water Management System (HiBallast). Submitted by the Republic of Korea. International Maritime Organization, Marine Environment Protection Committee, MEPC 60/2/6. 24 August 2009. IMO, London
IMO (2009m) Application for basic approval of Kwang San Co., Ltd. (KS) Ballast Water Management System "En-Ballast." Submitted by Korea. International Maritime Organization, Marine Environment Protection Committee, MEPC 60/2/7. 25 August 2009
IMO (2009n) Application for basic approval of the OceanGuard™ Ballast Water Management System. Submitted by Norway. International Maritime Organization, Marine Environment Protection Committee, MEPC 60/2/8. 26 August 2009. IMO, London
IMO (2009o) Application for basic approval of the Severn Trent DeNora BalPure® Ballast Water Management System. Submitted by Germany. International Maritime Organization, Marine Environment Protection Committee, MEPC 60/2/9. 28 August 2009. IMO, London
IMO (2009p) Draft MEPC resolution on the installation of ballast water management systems on new ships in accordance with the application dates contained in the BWM Convention. Note by the Secretariat. International Maritime Organization, Marine Environment Protection Committee, MEPC 60/2/10. 23 September 2009. IMO, London
IMO (2009q) Report of the tenth meeting of the GESMP-Ballast Water Working Group. Note by the Secretariat. International Maritime Organization, Marine Environment Protection Committee, MEPC 60/2/11. 30 October 2009. IMO, London
IMO (2009r) Report of the eleventh meeting of the GESMP-Ballast Water Working Group. Note by the Secretariat. International Maritime Organization, Marine Environment Protection Committee, MEPC 60/2/12, Annex 4. 1 December 2009. IMO, London
IMO (2010a) Information on the type approval certificate of the NK-O3 BlueBallast System (Ozone). Submitted by the Republic of Korea. International Maritime Organization, Marine Environment Protection Committee, MEPC 60/INF.14. 15 January 2010. IMO, London
IMO (2010b) Report of the twelfth meeting of the GESAMP – Ballast Water Working Group. Note by the Secretariat. International Maritime Organization, Marine Environment Protection Committee, MEPC 60/2/16. 8 February 2010. IMO, London
IMO (2010c) Report of the marine environment protection committee on its sixtieth session. International Maritime Organization, Marine Environment Protection Committee, MEPC 60/22. 14 April 2010. IMO, London
IMO (2010d) Information on the type approval of the OptiMarin Ballast System. Submitted by Norway. International Maritime Organization, Marine Environment Protection Committee, MEPC 61/INF.4. 21 June 2010. IMO, London
IMO (2010e) Information on the type approval certificate of the Hitachi Ballast Water Management System (ClearBallast). Submitted by Japan. International Maritime Organization, Marine Environment Protection Committee, MEPC 61/INF.21. 23 July 2010. IMO, London
IMO (2010f) Report of the thirtieth meeting of the GESAMP – Ballast Water Working Group. Note by the Secretariat. International Maritime Organization, Marine Environment Protection Committee, MEPC 61/2/15. 26 June 2010. IMO, London
IMO (2010g) Information for examination and analysis of the applicable requirements for vessels described in regulation B-3.1. Submitted by Japan. International Maritime Organization, Marine Environment Protection Committee, MEPC 61/2/17. 23 July 2010. IMO, London

IMO (2010h) Information for examination and analysis of the applicable requirements for vessels described in regulation B-3.1. Submitted by Japan. Corrigendum. International Maritime Organization, Marine Environment Protection Committee, MEPC 61/2/17/Corr.1. 11 August 2010. IMO, London

IMO (2010i) Information of type approval certificate of the GloEn-Patrol™ Ballast Water Management System. Submitted by the Republic of Korea. International Maritime Organization, Marine Environment Protection Committee, MEPC 61/2/19. 6 August 2010. IMO, London

IMO (2010j) Report of the Marine Environment Protection Committee on its sixty-first session. International Maritime Organization, Marine Environment Protection Committee, MEPC 61/24. 6 October 2010. IMO, London

IMO (2010k) Report of the fourteenth meeting of the GESAMP – Ballast Water Working Group. Note by the Secretariat. International Maritime Organization, Marine Environment Protection Committee, MEPC 62/2/21. 23 August 2010. IMO, London

IMO (2011a) Report of the fifteenth meeting of the GESAMP – Ballast Water Working Group. Note by the Secretariat. International Maritime Organization, Marine Environment Protection Committee, MEPC 62/2/11. 24 February 2011. IMO, London

IMO (2011b) Information on the type approval of the PureBallast 2.0 and PureBallast 2.0 Ex Ballast Water Management System. Submitted by Norway. International Maritime Organization, Marine Environment Protection Committee, MEPC 62/INF.14. 8 April 2011. IMO, London

IMO (2011c) Information on the type approval of the OceanSaver® Ballast Water Management System. Submitted by Norway. International Maritime Organization, Marine Environment Protection Committee, MEPC 62/INF.15. 8 April 2011. IMO, London

IMO (2011d) Report of the sixteenth meeting of the GESAMP – Ballast Water Working Group. Note by the Secretariat. International Maritime Organization, Marine Environment Protection Committee, MEPC 62/2/12. 28 April 2011. IMO, London

IMO (2011e) Information on the type approval of the Resource Ballast Technologies System (Cavitation combined with ozone and sodium hypochlorite treatment). Submitted by South Africa. International Maritime Organization, Marine Environment Protection Committee, MEPC 62/INF.18. 6 May 2011. IMO, London

IMO (2011f) Information on the type approval of the JFE Ballast Water Management System (JFE BallastAce) Submitted by Japan. International Maritime Organization, Marine Environment Protection Committee, MEPC 62/INF.25. 6 May 2011. IMO, London

IMO (2011g) Information on the type approval of the Blue Ocean Shield Ballast Water Management System. Submitted by China. International Maritime Organization, Marine Environment Protection Committee, MEPC 62/INF.28. 6 May 2011. IMO, London

IMO (2011h) Information on the type approval of the BalClor™ Ballast Water Management System. Submitted by China. International Maritime Organization, Marine Environment Protection Committee, MEPC 62/INF.29. 6 May 2011. IMO, London

IMO (2011i) Information on the type approval of the BSKYTM Ballast Water Management System. Submitted by China. International Maritime Organization, Marine Environment Protection Committee, MEPC 62/INF.30. 6 May 2011. IMO, London

IMO (2011j) Report of the seventeenth meeting of the GESAMP – Ballast Water Working Group. Note by the Secretariat. International Maritime Organization, Marine Environment Protection Committee, MEPC 62/2/18. 13 June 2011. IMO, London

IMO (2011k) Application for basic approval of DMU·OH Ballast Water Management System. Submitted by China. International Maritime Organization, Marine Environment Protection Committee, MEPC 63/2. 21 July 2011. IMO, London

IMO (2011l) Application for final approval of ERMA FIRST Ballast Water Management System. Submitted by Greece. International Maritime Organization, Marine Environment Protection Committee, MEPC 63/2/1. 25 July 2011. IMO, London

IMO (2011m) Report of the Marine Environment Protection Committee on its sixty-second session. International Maritime Organization, Marine Environment Protection Committee, MEPC 62/24. 26 July 2011. IMO, London

IMO (2011n) Application for final approval of the AquaStar™ Ballast Water Management System. Submitted by the Republic of KoreaInternational Maritime Organization, Marine Environment Protection Committee, MEPC 63/2/3. 2 August 2011. IMO, London
IMO (2011o) Application for final approval of MICROFADE™ Ballast Water Management System. Submitted by Japan. International Maritime Organization, Marine Environment Protection Committee, MEPC 63/2/2. 5 August 2011. IMO, London
IMO (2011p) Application for basic approval of the EcoGuardian™ Ballast Water Management System. Submitted by the Republic of Korea. International Maritime Organization, Marine Environment Protection Committee, MEPC 63/2/4. 8 August 2011. IMO, London
IMO (2011q) Application for basic approval of the HS-BALLAST Ballast Water Management System. Submitted by the Republic of Korea. International Maritime Organization, Marine Environment Protection Committee, MEPC 63/2/5. 8 August 2011. IMO, London
IMO (2011r) Application for final approval of Samsung Heavy Industries Co., Ltd. (SHI) Ballast Water Management System (Neo-Purimar™). Submitted by the Republic of Korea. International Maritime Organization, Marine Environment Protection Committee, MEPC 63/2/6. 19 August 2011. IMO, London
IMO (2011s) Application for basic approval of the KTM-BWMS Ballast Water Management System. Submitted by the Republic of Korea. International Maritime Organization, Marine Environment Protection Committee, MEPC 63/2/8. 31 August 2011. IMO, London
IMO (2011t) Application for basic approval of the Hamworthy AQUARIUS™-EC Ballast Water Management System (Hamworthy AQUARIUS™-EC BWMS). Submitted by the Netherlands. International Maritime Organization, Marine Environment Protection Committee, MEPC 63/2/9. 2 September 2011. IMO, London
IMO (2011u) Report of the eighteenth meeting of the GESAMP-Ballast Water Working Group. International Maritime Organization, Marine Environment Protection Committee, MEPC 63/2/10. 11 November 2011. IMO, London
IMO (2011v) Information on the type approval of the HiBallast™ Ballast Water Management System. Submitted by the Republic of Korea. International Maritime Organization, Marine Environment Protection Committee, MEPC 63/INF.4. 25 November 2011. IMO, London
IMO (2011w) Information on the type approval of the EcoBallast™ Ballast Water Management System. Submitted by the Republic of Korea. International Maritime Organization, Marine Environment Protection Committee, MEPC 63/INF.5. 25 November 2011. IMO, London
IMO (2011x) Information on the type approval of the Samsung Heavy Industries Co., Ltd. (SHI) Ballast Water Management System (Purimar™). Submitted by the Republic of Korea. International Maritime Organization, Marine Environment Protection Committee, MEPC 63/INF.6. 25 November 2011. IMO, London
IMO (2011y) Report of the nineteenth meeting of the GESAMP – Ballast Water Working Group. Note by the Secretariat. International Maritime Organization, Marine Environment Protection Committee, MEPC 63/2/11. 14 December 2011. IMO, London
IMO (2011z) Information on the type approval of FineBallast® OZ (the Special Pipe Hybrid Ballast Water Management System combined with Ozone treatment version). Submitted by Japan. International Maritime Organization, Marine Environment Protection Committee, MEPC 63/INF.12. 23 December 2011. IMO, London
IMO (2012a) Report of the twentieth meeting of the GESAMP – Ballast Water Working Group. Note by the Secretariat. International Maritime Organization, Marine Environment Protection Committee, MEPC 63/2/21. 24 January 2012. IMO, London
IMO (2012b) Application for basic approval of the OceanDoctor Ballast Water Management System. Submitted by China. International Maritime Organization, Marine Environment Protection Committee, MEPC 64/2. 13 March 2012. IMO, London
IMO (2012c) Report of the Marine Environment Protection Committee on its sixty-third session. International Maritime Organization, Marine Environment Protection Committee, MEPC 63/23. 14 March 2012. IMO, London
IMO (2012d) Application for final approval of JFE BallastAce that makes use of NEO-CHLOR MARINE™. Submitted by Japan. International Maritime Organization, Marine Environment Protection Committee, MEPC 64/2/1. 14 March 2012. IMO, London

IMO (2012e) Application for final approval of the ballast water management system (Smart Ballast BWMS). Submitted by the Republic of Korea. International Maritime Organization, Marine Environment Protection Committee, MEPC 64/2/2. 15 March 2012. IMO, London
IMO (2012f) Application for basic approval of the HS-BALLAST ballast water management system. Submitted by the Republic of Korea. International Maritime Organization, Marine Environment Protection Committee, MEPC 64/2/3. 15 March 2012. IMO, London
IMO (2012g) Application for basic approval of the GloEn-Saver™ Ballast Water Management System. Submitted by the Republic of Korea. International Maritime Organization, Marine Environment Protection Committee, MEPC 64/2/4. 15 March 2012. IMO, London
IMO (2012h) Application for basic approval of Dow-Pinnacle Ballast Water Management System. Submitted by Singapore. International Maritime Organization, Marine Environment Protection Committee, MEPC 64/2/5. 16 March 2012. IMO, London
IMO (2012i) Report of the twenty-first meeting of the GESAMP-Ballast Water Working Group. Note by the Secretariat. International Maritime Organization, Marine Environment Protection Committee, MEPC 64/2/6. 11 June 2012. IMO, London
IMO (2012j) Report of the twenty-second meeting of the GESAMP-Ballast Water Working Group. Note by the Secretariat. International Maritime Organization, Marine Environment Protection Committee, MEPC 64/2/7. 29 June 2012. IMO, London
IMO (2012k) Information on the type approval of the OceanSaver® Ballast Water Management System. Submitted by Norway. International Maritime Organization, Marine Environment Protection Committee, MEPC 64/INF.4. 29 June 2012. IMO, London
IMO (2012l) Report of the twenty-third meeting of the GESAMP-Ballast Water Working Group. Note by the Secretariat. International Maritime Organization, Marine Environment Protection Committee, MEPC 64/2/19. 27 July 2012. IMO, London
IMO (2012m) Information on the type approval of the Cyeco™ Ballast Water Management System. Submitted by China. International Maritime Organization, Marine Environment Protection Committee, MEPC 64/INF.12. 27 July 2012. IMO, London
IMO (2012n) Information on the type approval of MICROFADE Ballast Water Management System. Submitted by Japan. International Maritime Organization, Marine Environment Protection Committee, MEPC 64/INF.17. 27 July 2012. IMO, London
IMO (2012o) Information on the type approval of AquaStar™ Ballast Water Management System. Submitted by the Republic of Korea. International Maritime Organization, Marine Environment Protection Committee, MEPC 64/INF.18. 27 July 2012. IMO, London
IMO (2012p) Information on the type approval certificate for the BalPure® BP-500 Ballast Water Management System. Submitted by Germany. International Maritime Organization, Marine Environment Protection Committee, MEPC 64/INF.20. 27 July 2012. IMO, London
IMO (2012q) Information on the type approval of the ERMA FIRST BWTS Ballast Water Management System. Submitted by Greece. International Maritime Organization, Marine Environment Protection Committee, MEPC 64/INF.26. 27 July 2012. IMO, London
IMO (2012r) Information on the type approval of ARA PLASMA BWTS Ballast water management system. Submitted by the Republic of Korea. International Maritime Organization, Marine Environment Protection Committee, MEPC 64/INF.33. 27 July 2012. IMO, London
IMO (2012s) Report of the Marine Environment Protection Committee on its Sixty-Fourth Session. International Maritime Organization, Marine Environment Protection Committee, MEPC 64/23. 11 October 2012. IMO, London
IMO (2012t) Application for final approval of the AQUARIUS® EC Ballast Water Management System. Submitted by the Netherlands. International Maritime Organization, Marine Environment Protection Committee, MEPC 65/2/1. 12 October 2012. IMO, London
IMO (2012u) Application for basic approval of the REDOX AS Ballast Water Management System. Submitted by Norway. International Maritime Organization, Marine Environment Protection Committee, MEPC 65/2/3. 12 October 2012. IMO, London
IMO (2012v) Application for basic approval of the Van Oord Ballast Water Management System. Submitted by the Netherlands. International Maritime Organization, Marine Environment Protection Committee, MEPC 65/2/2. 19 October 2012. IMO, London

IMO (2012w) Application for final approval of the EcoGuardian™ Ballast Water Management System. Submitted by the Republic of Korea. International Maritime Organization, Marine Environment Protection Committee, MEPC 65/2/4. 26 October 2012. IMO, London

IMO (2012x) Application for basic approval of the Blue Zone™ Ballast Water Management System. Submitted by the Republic of Korea. International Maritime Organization, Marine Environment Protection Committee, MEPC 65/2/5. 26 October 2012. IMO, London

IMO (2012y) Application for final approval of the OceanDoctor Ballast Water Management System. Submitted by China. International Maritime Organization, Marine Environment Protection Committee, MEPC 65/2/6. 26 October 2012. IMO, London

IMO (2012z) Application for basic approval of the HyCator®: BWT Reactor System. Submitted by India. International Maritime Organization, Marine Environment Protection Committee, MEPC 65/2/7. 26 October 2012. IMO, London

IMO (2013a) Information on type approval of the DESMI Ocean Guard OxyClean Ballast Water Management System. Submitted by Denmark. International Maritime Organization, Marine Environment Protection Committee, MEPC 65/INF.5. 8 February 2013. IMO, London

IMO (2013b) Information on the type approval of the Wärtsilä AQUARIUS® UV ballast water management system. Submitted by the Netherlands. International Maritime Organization, Marine Environment Protection Committee, MEPC 65/INF.11. 8 February 2013. IMO, London

IMO (2013c) Information on the type approval of the KBAL Ballast Water Management System. Submitted by Norway. International Maritime Organization, Marine Environment Protection Committee, MEPC 65/INF.12. 8 February 2013. IMO, London

IMO (2013d) Information on the type approval of the CrystalBallast® Ballast Water Management System. Submitted by Norway. International Maritime Organization, Marine Environment Protection Committee, MEPC 65/INF.13. 8 February 2013. IMO, London

IMO (2013e) Report of the twenty-fourth meeting of the GESAMP-Ballast Water Working Group. Note by the Secretariat. International Maritime Organization, Marine Environment Protection Committee, MEPC 65/2/9. 8 February 2013. IMO, London

IMO (2013f) Information on the renewal and update of the type approval for the Resource Ballast Technologies System (cavitation combined with ozone and sodium hypochlorite treatment). Submitted by South Africa. International Maritime Organization, Marine Environment Protection Committee, MEPC 65/INF.26. 8 March 2013. IMO, London

IMO (2013g) Report of the twenty-fifth meeting of the GESAMP-Ballast Water Working Group. Note by the Secretariat. International Maritime Organization, Marine Environment Protection Committee, MEPC 65/2/19. 22 March 2013. IMO, London

IMO (2013h) Report of the Marine Environment Protection Committee on its Sixty-Fifth Session. International Maritime Organization, Marine Environment Protection Committee, MEPC 65/22. 24 May 2013. IMO, London

IMO (2013i) Application for final approval of Ballast Water Management System with PERACLEAN® Ocean (SKY-SYSTEM®). Submitted by Japan. International Maritime Organization, Marine Environment Protection Committee, MEPC 66/2. 19 August 2013. IMO, London

IMO (2013j) Application for basic approval of the ECOLCELL BTs Ballast Water Management System. Submitted by Italy. International Maritime Organization, Marine Environment Protection Committee, MEPC 66/2/1. 10 September 2013. IMO, London

IMO (2013k) List of ballast water management guidelines, guidance documents and approved ballast water management systems. Note by the Secretariat. International Maritime Organization, Marine Environment Protection Committee, MEPC 66/INF2. 03 October 2013. IMO, London

IMO (2011za) Preview of global ballast water treatment markets. Submitted by the Institute of Marine Engineering, Science and Technology (IMarEST). International Maritime Organization, Marine Environment Protection Committee, MEPC 63/INF.11. 23 December 2011. IMO, London

IMO (2012za) Information on type approval of the OceanGuard™ Ballast Water Management System. Submitted by Norway. International Maritime Organization, Marine Environment Protection Committee, MEPC 65/INF.2. 29 October 2012. IMO, London

NSF International (2010) Generic protocol for the verification of ballast water treatment technology. For U.S. Environmental Protection Agency Environmental Technology Verification Program. In

cooperation with U.S. Coast Guard Environmental Standards Division, Washington, DC and U.S. Naval Research Laboratory, Center for Corrosion Science and Engineering Washington, DC
Köster E (2010) Sustainable Ballast Water Management Plant (BaWaPla), EU 6th Framework Programme. http://www.bawapla.com/. Accessed 05 June 2012
Liebich V, Stehouwer PP, Veldhuis M (2012) Re-growth of potential invasive phytoplankton following UV-based ballast water treatment. Aquat Invasions 7:29–36
Lloyd's Register (2011a) Ballast water treatment technology, Current status. June 2011. Lloyds Register Group
Lloyd's Register (2011b) Ballast water treatment technology, Update. September 2011. Lloyds Register Group
Martínez LF, Mahamud MM, Lavín AG, Bue JL (2012) Evolution of phytoplankton cultures after ultraviolet light treatment. Mar Pollut Bull 64:556–562
Mesbahi E (2004) Onboard treatment of ballast water (technologies development and applications) and application of low sulphur marine fuel (MARTOB), EU 5th Framework Programme. http://martob.ncl.ac.uk/Home.htm. Accessed 5 June 2012
Olivieri VP, Snead MC, Kruse CW, Kawata K (1986) Stability and effectiveness of chlorine disinfectants in water distribution systems. Environ Health Perspect 69:15–29
Shiferaw M (2012) EcoSeaSafe 7th Framework Programme. http://ecoseasafeproject.com/. Accessed 5 June 2012
STEP (2010) U.S. Coast Guard Shipboard Technology Evaluation Program. Ship Enrolment Application for Ballast Water Management System. Version 2.3

Risk Assessment in Ballast Water Management

Matej David, Stephan Gollasch, Erkki Leppäkoski, and Chad Hewitt

Abstract The risk assessment (RA) developed according to the BWM Convention is the most recently agreed global RA for bioinvasions. It was developed to enable a selective ballast water management (BWM) approach according to the BWM Convention and the G7 Guidelines. It describes three different BWM RA methods, "environmental matching", "species' biogeographical" and "species-specific" RA. The environmental matching RA between the areas of ballast water origin and discharge considers non-biological parameters as surrogates for the species survival potential in the new environment. The species' biogeographical RA identifies species with overlapping distribution in the donor and recipient ports and biogeographic regions which is taken as direct indications of the similarity of the environmental conditions and hence species survival in the new environment. The species-specific RA is focused on life history information and physiological tolerances to identify a species' physiological limits estimating its potential to survive or complete its life cycle in the new environment and considers target species. There are two fundamentally different RA approaches under the BWM Convention, the selective and the blanket approach. A blanket approach means that all ships intending to discharge ballast water in a port are required to conduct BWM. The selective approach means that appropriate BWM measures are required depending on different risk levels posed by the intended ballast water discharge. In one instance ships may be exempted from BWM requirements provided that the risk level of a ballast water discharge is acceptable. In another instance, if the risk is identified as (very) high,

M. David (✉)
Dr. Matej David Consult, Korte, Izola, Slovenia
e-mail: matej.david@siol.net

S. Gollasch
Gollasch Consulting (GoConsult), Hamburg, Germany
e-mail: sgollasch@aol.com

E. Leppäkoski
Department of Biosciences, Åbo Akademi University, Turku, Finland
e-mail: erkki.leppakoski@gmail.com

C. Hewitt
Faculty of Science and Engineering, University of Waikato, Hamilton, New Zealand
e-mail: chewitt@waikato.ac.nz

M. David, S. Gollasch (eds.), *Global Maritime Transport and Ballast Water Management*, Invading Nature - Springer Series in Invasion Ecology 8,
DOI 10.1007/978-94-017-9367-4_7

ships may be required to take additional measures based on the G13 Guidelines. The risk level is a RA result and input data reliability is of key importance. The chapter provides detailed step-by-step RA models for exemptions and for selective BWM measures, ready to be used by administrations.

Keywords Risk assessment • Selective ballast water management • Exemptions • Environmental matching method • Species specific method • Biogeographical method • Target species

Risk Assessment in Ballast Water Management

Risk Assessment and Its Key Principles

Risk is variously defined as the probability that an undesired event occurs in combination with the level of impact this event causes, frequently referred to as the consequence. Risk assessment (RA) is the process by which undesired events (hazards) are identified and the frequency and consequences of such undesired events are parameterized, typically including an expression of all uncertainties in the assessment process (e.g., Hewitt and Hayes 2002).

The RA may be defined by the following key principles (IMO 2007):

- **Effectiveness** – RA accurately measures the risks to the necessary extent to achieve an appropriate level of protection.
- **Transparency** – Reasoning and evidence supports the RA recommended action and uncertainty areas (as well as their possible consequences to those recommendations), are documented clearly and made available to decision-makers.
- **Consistency** – RA achieves a uniform high performance level, using a common process and methodology.
- **Comprehensiveness** – The full range of possibly affected values, including economic, environmental, social and cultural, will be considered when assessing risks and in the decision making process.
- **Risk Management** – Although risk scenarios exist, zero risk is not achievable, and therefore a risk should be managed by determining its acceptable level in each instance.
- **Precautionary** – RA incorporates a level of precaution when making assumptions and recommendations. This is to account for uncertainty, unreliability, and inadequacy of data. The absence of, or uncertainty regarding any data should therefore be considered as an indicator of potential risk.
- **Science based** – RA is to be based on the best available information that has been collected and analysed by scientific methods. Minimum data quality standards permitting a RA may be agreed.
- **Continuous improvement** – Any risk model should be reviewed and updated periodically to account for an improved understanding.

Risk Assessment of Harmful Species Introductions

Most RAs of marine biological invasions used in the past by different regulatory institutions are based on, or reflect the *Office Internationale des Epizooties* (OIE) framework (Hewitt and Hayes 2002). Here bioinvasions are understood as the culmination of a chain of events (see chapter "The Transfer of Harmful Aquatic Organisms and Pathogens with Ballast Water and Their Impacts"). A RA process to determine invasiveness requires an assessment of each event to attribute the degree of probability of successfully proceeding through that stage. The final RA of a 'successful' invasion is the result of the degree of probability attributed to each separately evaluated event. The OIE framework is efficient and simple to use for bioinvasions. Its efficiency may be improved further through the inclusion of quantitative RA fundamental principles. The quantitative RA includes five steps: (1) – hazard identification, (2) – frequency assessment, (3) – consequence assessment, (4) – risk estimation, and (5) – uncertainty analysis (e.g., Hayes 2000). The quantitative RA approach was developed for the application in complex industrial systems, but its constituent techniques and principles may also be adopted successfully within complex ecological systems.

An alternative approach bases the bioinvasion RA on environmental matching between the points of origin and destination (i.e., ballast water donor and recipient regions). One example of such an approach is that of the Queensland Ports Corporation, Australia which is based on a comparison of 40 environmental parameters (Hilliard and Raaymakers 1997). Other approaches have also addressed the issue of including environmental conditions including: a RA considering 34 parameters (GloBallast 2003), a German study based on climatic conditions and salinity (Gollasch 1996), a U.S. study considering salinity comparison alone (Carlton 1985), and a Slovenian study considered salinity as the only environmental parameter next to other species specific considerations (David 2007). In addition, an environmental match related RA was prepared for the Nordic Council of Ministers (Gollasch and Leppäkoski 1999) which was further developed for HELCOM (Leppäkoski and Gollasch 2006; Gollasch and Leppäkoski 2007).

Another approach is to consider target species, which was earlier adopted by the U.S. and Australia. This approach is based on a selection of species whose invasiveness in the examined area is likely and was confirmed in other areas. These RA activities resulted in two lists: 'America's Least Wanted' and the Australian 'Target Species List'.

These RA approaches may be supplemented by other elements. GloBallast's RA, further to the environmental matching method, includes some target species and additional risk quantifiers, such as voyage length and ballast tank size (GloBallast 2003). DNV's EMBLA also includes numerous parameters (Behrens et al. 2002; Endresen et al. 2004). Environmental matching combined with vessel voyage lengths and a target species list was also used in the Baltic to assess the risk of non-indigenous species introductions (Gollasch and Leppäkoski 1999, 2007). The Slovenian RA included ballast water sampling to confirm the presence of non-indigenous or other potentially harmful organisms in the ballast water which originated from the same biogeographic region (i.e., compatible environments) (David 2007). More recently a RA approach also in line with the IMO requirements was developed for the North and Baltic Seas (David and Gollasch 2010; David et al. 2013).

RA approaches can be differentiated in terms of data expressions, which can be qualitative, semi-quantitative, or quantitative (Norton et al. 1995). The qualitative approach aims to express the number of organisms or other parameters and uses descriptive values instead of figures (e.g., the quantity of organisms at origin: many, medium, few, the environmental match regarding salinity, e.g., high, medium, low). The quantitative approach is based on the quantification of all data in the RA system. Requirements on data intensity and the system complexity increase from the qualitative to the quantitative approaches. Different initiatives and approaches which were all developed prior the G7 Guidelines were adopted, and are presented in Table 1. Thereafter, to our knowledge, only one BWM related RA approach was yet prepared worldwide which strictly follows the G7 Guidelines and the precautionary approach (David et al. 2013). In Europe new approaches are currently being developed for the HELCOM/OSPAR area as regional activities, for the Baltic, North and western Mediterranean Seas during the VECTORS project,[1] and for the Adriatic Sea during the BALMAS project.[2]

Risk Assessment Process

The first RA steps are the introduction vector identification, followed by a hazard assessment relative to this vector and identified species. The RA approach should be selected depending on the objectives to be achieved and the data and resources availability. All these factors determine also the selection of the RA end-point.

Identification of the Vector of Transfer

More than a decade ago, 13 anthropogenic non-indigenous species transfer vectors were identified, addressing unintentional and intentional introductions (Gollasch and Leppäkoski 1999; Hewitt and Hayes 2002, see Table 2). In another summary more than 50 recognised vectors were listed (Minchin et al. 2005, 2009, see also chapter "The Transfer of Harmful Aquatic Organisms and Pathogens with Ballast Water and Their Impacts").

---

[1] Vectors of Change in Oceans and Seas Marine Life, Impact on Economic Sectors (VECTORS), http://www.marine-vectors.eu/

[2] Ballast Water Management System for Adriatic Sea Protection (BALMAS), http://www.balmas.eu/

Table 1 RA initiatives worldwide prior to the adoption of the G7 Guidelines

RA initiative	Goals	RA approach	Data expression approach	End-point
Australia (AQIS 1994)	Estimate costs of toxic dinoflagellate introductions in Australian waters	Species based tolerance, volume of ballast discharged and bloom dynamics	Quantitative	Estimate economic impact of toxic dinoflagellates on aquaculture, tourism, etc. (hazard assessment)
Germany (Gollasch 1996)	Risk identification for species invasions in German coastal waters	Environmental matching between donor and recipient localities	Qualitative	Species establishment potential assessment
Australian DSS[a] (Hayes and Hewitt 1998, 2000)	Identify low risk routes, vessels and tanks	Models four steps in the bio-invasion process: donor port infection, vessel infection, journey survival and survival in the recipient port	Quantitative	Target species life cycle completion in recipient port
Nordic countries (Gollasch and Leppäkoski 1999)	Risk identification for species invasions in Nordic ports and coastal waters	Environmental matching between donor and recipient localities	Qualitative	Species establishment potential assessment
DNV – EMBLA (Behrens et al. 2002, Endresen et al. 2004)	Identify low risk routes, vessels and tanks	Bioinvasion model based upon: donor port infection, vessel infection, journey survival and survival in recipient port	Quantitative	Target species life cycle completion in recipient port
GloBallast RA (e.g., Awad et al. 2004)	Enhance awareness and recommendations on BWM strategies	Environmental matching between donor and recipient localities, weighted by target species presence in the donor location and inoculation factors	Semi-quantitative	Identify and rank high and low risk ports
Slovenia 1 (David 2007)	Risk identification for species invasions in the Slovenian Sea, enhance awareness and recommend BWM strategies	Environmental matching between donor and recipient localities and identification of potentially harmful species in ballast water	Quantitative and qualitative	Identify and rank high and low risk donor ports

(continued)

Table 1 (continued)

RA initiative	Goals	RA approach	Data expression approach	End-point
Canada 1 (MacIsaac et al. 2002)	Estimate risk associated with NOBOB[b] vessels entering the Great Lakes	Species based tolerance, and taxa concentrations for NOBOB vessels	Quantitative	Journey survival of target species
Finland (Pienimäki and Leppäkoski 2004; Leppäkoski et al. 2005; Paavola et al. 2005)	Create baseline knowledge on the risks associated with NIS and shipping	Environmental matching between donor and recipient localities	Qualitative	Vector and establishment potential assessment
Slovenia 2 (David 2007)	Vessel-to-vessel (tank) assessment for BWM measures supported by DSS	Environmental matching between donor and recipient localities (bioregions) and species specific (inside bioregion)	Quantitative and qualitative	Identify and rank high and low risk donor ports and identify high risk species
EMBLA (for Croatia) (Dragsund 2005)	Recommend ballast water management plan for Croatia	Locality based and species tolerances	Qualitative	Hazard assessment
Canada 2 (MacIsaac et al. 2004)	Review and develop a ballast water RA framework	Target species (gravity model, i.e., rates and patterns of colonization)	Quantitative	Colonization prediction of target species

Enhanced after David (2007)
[a]*DSS* decision support system
[b]*NOBOB* no ballast on board

Table 2 Anthropogenic introduction vectors of aquatic organisms

Anthropogenic vectors	
Vessels	Accidental with vessel fouling (including boring into wooden hulls)
	Accidental with ballast water
	Accidental with solid ballast (e.g., rocks, sand)
	Accidental with anchor chains and in chain lockers
Fisheries	Deliberate translocations of fish and shellfish to establish or support aquaculture
	Accidental with deliberate translocations of fish and shellfish (e.g., epi- and endobionts as well as parasites and disease agents)
	Accidental with discharge of material from fish and shellfish processing plants
	Accidental with seaweed packing material for bait and fishery products
Plant introductions	Deliberate translocation of plant species (e.g., for erosion control)
	Accidental with deliberate plant translocations
Biocontrol	Deliberate translocation for biocontrol
	Accidental translocation with deliberate biocontrol release
Canals	Range expansion through man-made canals
Individual release	Deliberate and accidental release by individuals (e.g., from aquaria)
	Equipment used for recreation (e.g., diving bags, boats)
Scientific release	Deliberate and accidental release as a result of research activities

Enhanced after Hewitt and Hayes (2002). With kind permission of Springer Science+Business Media

In different world regions the importance of species introduction vectors varies. Nevertheless, in all regions considered the most important three vectors are (possibly in different order): ballast water, hull fouling, and aquaculture, so that shipping is considered to be the worldwide principal pathway by which species are spread (see chapter "The Transfer of Harmful Aquatic Organisms and Pathogens with Ballast Water and Their Impacts"). The vector identification for each species is extremely challenging as several species may be related to more than one vector. Vectors overlap which makes many of them indistinctive (Minchin 2007), as shown in Fig. 1.

All these overlapping vectors and multiple possibilities often create uncertainties regarding the vector identification and assignment. However, this information is very critical for vector management purposes. Different levels of certainty can be assigned to each vector (e.g., in the non-indigenous species database of the DAISIE[3] project three levels of certainty (i.e., direct evidence, likely, unspecified) are available for each transfer vector). This database is currently being updated and expanded during the EU-funded VECTORS project and it is expected that the new database, named AquaNIS,[4] will become publicly available in 2015. A vector identification is important to make vector management efficient, i.e., to regulate the most important species introduction vector first.

---

[3] Delivering Alien Invasive Species Inventories for Europe (DAISIE).

[4] http://www.corpi.ku.lt/databases/index.php/aquanis. last accessed December 2013.

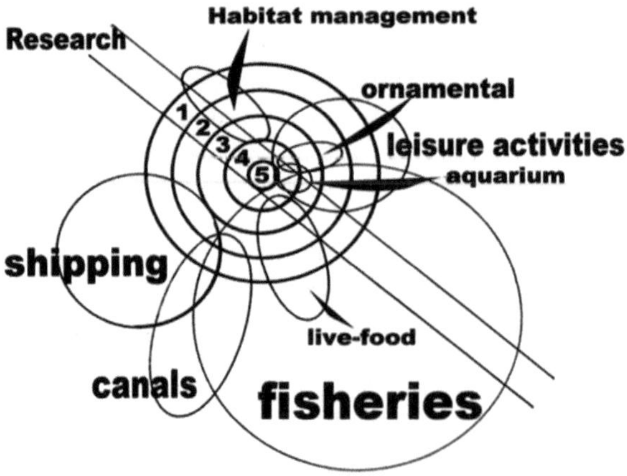

Fig. 1 Overlap of different species introduction vectors (Minchin 2007) (Reprinted from Minchin (2007), copyright 2007, with permission from Elsevier)

Identification of Hazards

Hazards may be defined as a situation to result in harm under certain circumstances, or, alternatively, as the likeliness of substances or activities to generate risk (Hewitt and Hayes 2002). In ecotoxicology a hazard is frequently considered merely as a function of the properties of a substance. However, a broader understanding would be more appropriate to include the fundamental properties of a substance as well as the circumstances. The implication inherent to the introductions of harmful aquatic organisms and pathogens (HAOP) RA is the assessment of the probability of the establishment of a species. This also depends on its potential invasiveness (i.e., its fundamental properties) and the recipient environment (i.e., circumstances).

The introduction of an organism and its possible invasiveness can be divided into several phases, or a chain of events (see above and chapter "The Transfer of Harmful Aquatic Organisms and Pathogens with Ballast Water and Their Impacts"): organism presence in the donor region, vector infection, transport survival, survival of the discharge process to the recipient environment, survival in the new environment, establishment in the new environment, and possibly spread and harm (invasiveness) in the new environment. The uncertainty relative to each step increases upon each following step, i.e., from the initial presence in the donor environment to the invasiveness in the recipient environment. In cases where the degree of uncertainty is high, quantitative methods for the definition of probability are inappropriate. Therefore, not all phases of the species invasion chain of events have to be quantified, but instead a combination of the empirical approach (based upon acceptable criteria) and the documented invasion history and adverse influences can be adopted.

IMO Risk Assessment Methods

The RA developed in the framework of the BWM Convention is the most recently agreed global RA framework for bioinvasions. It was developed to provide guidance how to implement a selective BWM approach according to the BWM

Convention and the G7 Guidelines. It describes three different BWM RA methods, "environmental matching", "species' biogeographical" and "species-specific" RA.

The environmental matching RA between the areas of ballast water origin and discharge considers non-biological parameters such as salinity and temperature as surrogates for the species survival potential in the new environment. The species' biogeographical RA seeks to identify species with overlapping distribution in the donor and recipient ports and biogeographic regions. These overlaps are taken as direct indications of the similarity of the environmental conditions and hence species survival in the new environment. The species-specific RA is focused on information on life history and physiological tolerances to identify a species' physiological limits and estimates its potential to survive or complete its life cycle in the new environment (IMO 2007).

Environmental Matching Related Risk Identification

The vector-related risk identification can be based on two fundamental elements:

- the likelihood of organism transfer (i.e., the quantity and origin of the discharged ballast water and abundance of propagules therein),
- the likelihood of organism survival in the recipient environment (match of selected environmental parameters of donor and recipient regions).

Different marine regions are typically defined as biogeographic regions, but all existing biogeographical schemes were developed for different purposes and not for biological invasions RA, e.g., Briggs (1974) and Springer (1982), IUCN bioregion system (Kelleher et al. 1995), Ekman (1953), Longhurst (1998) provinces, Spalding et al. 2007 and Briggs and Bowen (2012). IMO suggested to use the Large Marine Ecosystems (LME) approach (see Fig. 2) because at the time of drafting the G7 Guidelines this was considered the best available information, but local and regional adaptations may be necessary.

In the G7 Guidelines environmental matching was determined to assess the likelihood that species found in the ballast water donor region are able to survive in the recipient port. However, some uncertainty remains, namely the uncertainty to define the environmental conditions, which are predictive of the species to establish and cause harm in a new location. Another key point is the determination whether the risk of ballast water discharge is sufficiently low to be exempted from BWM requirements. Environmental matching RA is of limited use in cases where the differences between a donor region and a recipient port are small. In these cases, such as shipping within one biogeographic region, high similarity between donor and recipient areas is likely and indicates a high likelihood of successful species establishment. However, there are exemptions from this rule, e.g., areas with different water salinities in the same bioregion, which may be caused due to, e.g., run-offs of major rivers.

In addition to comparing the environmental conditions of biogeographic regions, this comparison should further be undertaken between the donor and recipient ports, i.e., in much smaller scale. Similarity of key environmental conditions between the

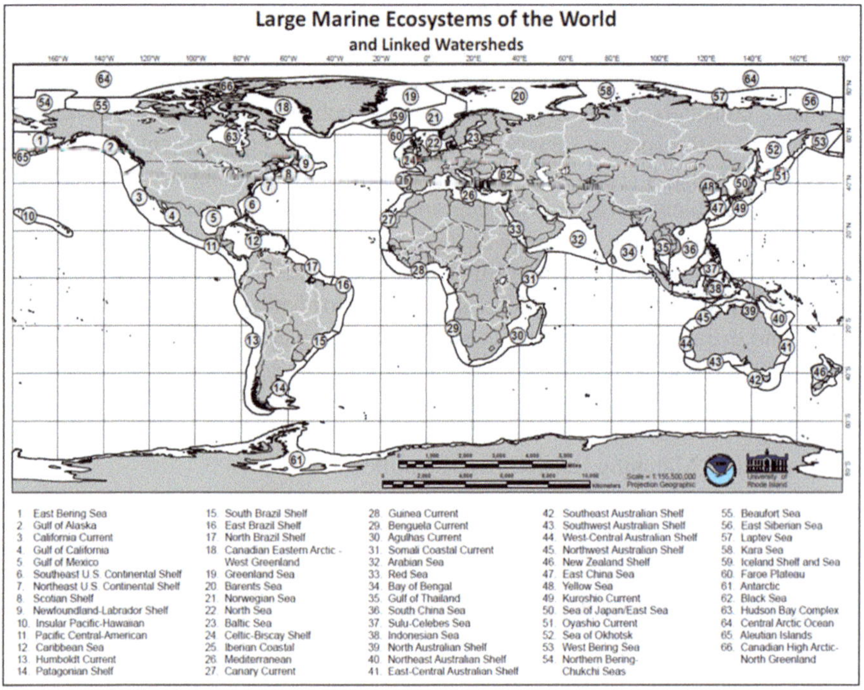

Fig. 2 Map of large marine ecosystems (Source NOAA, http://www.lme.noaa.gov/, last accessed in November 2013)

two ports to be assessed is a strong indication that species of the donor port will survive when released in the recipient port water.

The data needed to enable a RA using the environmental matching approach to determine the degree of environmental similarity between the donor and recipient environments (IMO 2007) include:

- the origin of the ballast water to be discharged in the recipient port,
- the biogeographic region of donor and recipient ports, and
- the average and range of environmental conditions, also considering seasonal differences, in particular salinity and temperature.

The analysis of the environmental similarity may be followed by an evaluation of species known to occur in the donor region, which tolerate extreme environmental differences. If such species are found, a species-specific approach should be used for RA associated with these species (IMO 2007). Such species include:

- species which migrate between fresh and marine environments to complete their life-cycle (anadromous species, such as salmon spend most of their life in the sea and return to fresh water to spawn, whereas the catadromous species, e.g., the Chinese mitten crab, do the opposite);
- species with a wide tolerance of temperature (eurythermal species) or salinity (euryhaline species).

Species' Biogeographical Risk Assessment

The species' biogeographical RA compares the distribution of non-indigenous, cryptogenic, and harmful native species presently occurring in the donor and recipient ports and biogeographic regions. Should species occurrences overlap in the donor and recipient ports and regions this is a direct indication of environmental similarity to enable a shared fauna and flora. The biogeographical approach may also be used to identify high risk species (see also the species-specific approach). As an example, harmful species in the ballast water donor biogeographic region which are known to have successfully invaded other (similar) biogeographic regions, but are not (yet) found in the recipient biogeographic region of the RA, could be considered as high risk species for the ballast water recipient region. As a general rule, the higher the number of biogeographic regions in which such species have invaded, the greater is the potential that those species would also be able to become established in the recipient port or biogeographic region. Another general risk indicator is given in case where the donor biogeographic region is a major source of species to other areas.

The data requirements (IMO 2007) to enable a species biogeographical approach RA include:

1. species invasion records in the donor and recipient biogeographic regions and ports;
2. records of native or non-indigenous species in the donor biogeographic region which may be transferred with ballast water and which have already invaded other biogeographic regions and the number and characteristics of these invaded biogeographic regions;
3. records of native species in the ballast water donor region which have the potential to affect human health or to cause substantial negative ecological or economic impacts after introduction to the ballast water recipient region.

The species' biogeographical RA may also be used to identify potential target species (see below) in the donor region(s). Criteria to identify such species include native species with a wide biogeographical or habitat distribution or species which are known as invaders in other biogeographic regions, which are similar to that of the ballast water recipient port.

Species-Specific Risk Identification

The identification of species-related risk focuses on the evaluation of the potential invasiveness of each selected species considering also the harm that it could cause in the new environment. Today we lack data and have insufficient knowledge concerning the invasiveness of organisms with some key questions remaining unanswered, e.g., What predicts invasiveness in a new environment? How does the degree of species tolerance regarding environmental conditions, food availability, reproduction behaviour and capabilities influence invasion success? How can we anticipate the harm that could be caused?

In many environments (or biogeographical regions) the knowledge on the taxonomy of indigenous organisms is deficient, while the identification of organisms originating from other parts of the world is even more demanding. Consequently, numerous organisms may remain unidentified.

For a target list of unwanted organisms, fundamental selection criteria must be defined. Based upon the IMO definition in the G7 Guidelines, at least all following factors need to be considered (IMO 2007) when identifying target species:

- evidence of prior introduction, i.e., thereby the species shows its capability to become introduced outside its native range;
- potential impact on environment, economy, human health, property or resources;
- strength and type of ecological interactions, i.e., severeness of its impact;
- current distribution within the biogeographic region and in other biogeographic regions; and
- relationship with ballast water as a vector, i.e., when the species was already found in a ballast tank or if the life cycle of the species include a larval phase which makes a ballast water transport likely.

Numerous attempts were undertaken to identify typical characteristics of an ‘ideal’ invasive species. It was discussed that species with high environmental tolerances and those with high reproduction rates may have a higher invasion potential (Safriel and Ritte 1980, 1983; Kareiva 1999; Hewitt 2003).

The objective of this approach is to consider species life history information and physiological tolerances to characterise physiological limits of a certain species which leads to its survival potential or potential to complete its life cycle in the recipient environment. In other words, the individual species characteristics need to be compared with the environmental conditions in the recipient port, which results in a determination of the likelihood of transfer and species survival.

A target species is not needed in all circumstances but may be useful to focus a surveillance action or may be necessary for legislative compliance. The species of concern (target species) need to be selected for a specific port, country, or biogeographical region. As a first step to generate a target species list, all species being potentially harmful and invasive (including cryptogenic and harmful native species) present in the donor port(s) should be listed and, secondly, target species are to be selected based on pre-defined criteria (see above).

A problem is subjectivity with the target species selection. It may occur that the assessment whether or not a species should become a target species will result with a degree of uncertainty associated with the approach. It is possible that species identified as harmful in some environments may not be harmful in others and vice versa.

In addition to the data referred above, the following information is needed to enable a RA using the species-specific approach (based on G7 Guidelines, IMO 2007):

- biogeographic region of donor and recipient port(s); the presence of all non-indigenous species (including cryptogenic species) and native species in the

donor port(s), port region and biogeographic region, not present in the recipient port, to allow identification of target species;
- the presence of all target species in the recipient port(s), port region, and biogeographic region;
- the difference between target species in the donor and recipient ports, port region, and biogeographic region;
- life history information on the target species and physiological tolerances, in particular salinity and temperature, of each life stage; and
- habitat type required by the target species and availability of habitat type in the recipient port.

Even when a target species has been reported, although its establishment status and abundance may be unknown, from the donor and recipient ports, its continued introduction into the recipient port(s) may increase the probability that it will become established and to cause negative impacts. This is especially the case when the target species occurs in higher abundance in the donor port compared to the recipient port.

As a starting point, a simple assessment may be conducted to evaluate whether a target species is present in the donor port, but not in the recipient port, and if it can be transported via ballast water. In a more comprehensive approach the following points may need to be evaluated (IMO 2007):

- Uptake – probability of viable stages entering the vessel's ballast water tanks during ballast water uptake operations;
- Transfer – probability of survival during the voyage;
- Discharge – probability of viable stages entering the recipient port through ballast water discharge on arrival; and
- Population establishment – probability of the species establishing a self-sustaining population in the recipient port.

An even more detailed scenario would be to determine the likelihood of a target species to survive each of the stages listed above. However, the required data may only be available in rare cases, especially when considering that all life stages of the target species need to be assessed also including seasonal variations in the target species presence in the donor port with seasonal conditions in the recipient port to meet the species abiotic tolerances (e.g., temperature and salinity). Consequently, the overall RA of unmanaged ballast water discharges should be determined based on the evaluation of all target species surviving all these stages.

To groundtruth the chosen species-specific RA approach, data may be gathered for already introduced species in the recipient port. This is to check whether or not the RA approach selected would have predicted this species to be able to survive in the ballast water recipient port. A failure to predict existing invaders correctly may indicate that the model under-predicts the risk, noting that species may have arrived by various vectors.

Risk Assessment End-Point

The risk of ballast water and sediment discharges may be defined as the likelihood of an undesired event to occur as a consequence of ballast discharge from a ship. The interpretation of this definition entirely depends on the assessment end-point. The end-point can be defined either as the discharge probability of potentially harmful organisms via ballast water, or their establishment in the new environment, or their invasiveness in and impact on the new environment.

When the identified end-point is the probability of impact,[5] a risk would need to be accurately defined through all RA stages from the bottom up (i.e., starting with the introduction and establishment probability of new organisms). The RA process was defined by the G7 Guidelines as "a logical process for objectively assigning the likelihood and consequences of specific events, such as the entry, establishment, or spread of harmful aquatic organisms and pathogens".[6]

The scenarios presented below describe the dependence of RA on the identified end-point under the assumption that the RA end-point is:

1. the discharge of HAOP via ballast water from a ship;
2. the establishment of HAOP in a novel environment;
3. the impact (invasiveness) of HAOP in a novel environment.

In scenario 1 the presence of HAOP in the discharged ballast water is understood as an undesired event. In scenario 2 an undesired event is defined as the establishment of a species, which means that the discharge of HAOP per se is not recorded as an undesired event in cases where they remain unestablished. In scenario 3 the undesired event is the impact while the discharge and establishment of a HAOP are not recorded as undesired events.

After the discharge of HAOP in a new environment many of the discharged individuals may not survive. Moreover, should they survive and establish themselves in the new environment, harm is not necessarily generated. However, considering the stochastic and complex array of factors which science is still unable to predict, one of the key points is that it is extremely difficult or practically impossible to conduct highly reliable assessments as to whether a new species introduced to a novel environment will cause harm or not. There are also cases of established of HAOP which have not caused harm for years but then, under certain circumstances, suddenly turned invasive. This lack of knowledge reveals that the conservativeness of the approach descends from the first to the third scenario presented above as does the degree of certainty of the identification of an event.

---

[5] i.e., various aspects of risk to human health, the natural environment, or the economy/resources.

[6] G7 guidelines, paragraph 5.1.

The decision as to the identification of the RA end-point is made by the risk assessor[7] and depends on the assessor's objectives, values, and abilities (Cothern 1996; Kirchsteiger et al. 1998). The perception of values can be highly diverse (Cothern 1996; Kirchsteiger et al. 1998; Souvorov 1999), e.g., the preservation of the native biological diversity in an environment will bear extraordinary value to a biologist whereas it might have a comparatively lower value to other stakeholders. A reverse relation would probably be observed when economic effects are considered (e.g., effects on fisheries and aquaculture). Therefore, we conclude that the perception of the degree of risk (within a broader circle of stakeholders in a state and usually in direct correlation to the country's level of development) exerts a significant influence on the acceptability degree of each risk.

Risk Assessment Errors

RA includes potential errors which can occur at any assessment step. The errors can be divided into two groups (Hayes 2000):

- Type I errors – to cause overestimates of the real risk situation;
- Type II errors – to cause underestimates of the real risk situation.

RA provides the basis for the implementation of preventive measures. Therefore, it can be assumed that a Type I error will result in higher protection from negative impacts yet concurrently laying the additional burden of preventive measures on the shipping industry. In contrast, a Type II error will result in a potentially lower degree of protection from negative impacts with consequently a lighter burden on the shipping industry.

The RA aims certainly to reflect the real situation as accurately as possible and implement appropriate measures in relation to the obtained results. However, given that the ballast water issue has not been extensively researched yet in this regard, the likelihood of error is high. In these cases the precautionary approach should be adopted, with primary emphasis laid on the avoidance of Type II errors through the entire RA and BWM process. In some cases Type II errors simply cannot be prevented (e.g., sampling on-board ships, data collection with ballast water reporting forms) and all possible measures aiming towards the error reduction have to be taken while the presence of the error has to be clearly recorded to allow for correct RA data interpretation also for the consideration of the error during the next step and the adoption of measures (Kirchsteiger et al. 1998; Hayes 2000).

[7] Given that the objective of RA is the prevention of undesired events via state regulation, the 'assessor' is to be understood as a state.

Application of Risk Assessment Under the Ballast Water Management Convention

There are two fundamentally different RA approaches under the BWM Convention, the selective and the blanket approach. The selective approach means that appropriate BWM measures are required depending on different risk levels posed by the intended ballast water discharge. This is further also depending on the BWM feasibility under certain circumstances. In one instance ships may be exempted from BWM requirements provided that the risk level of a ballast water discharge is acceptable based on the G7 Guidelines. In another instance, if the risk is identified as (very) high, ships may be required to take additional measures based on *Guidelines for Additional Measures Regarding Ballast Water Management Including Emergency Situations* (G13 Guidelines). The level of risk is a result of a RA. A blanket approach means that all ships intending to discharge ballast water in a port are required by the port State to conduct BWM.

Risk Assessment for Granting Exemptions from Ballast Water Management Requirements

Exemptions from BWM requirements may be given when a RA, prepared according to the G7 Guidelines, results in an acceptable low risk. This is specific for a ship, or different ships, sailing only between specified ports or locations. The exemptions may be granted for up to 5 years, but may also be withdrawn when the risk situation becomes unacceptable during this period (IMO 2007; David and Gollasch 2010). The RA developed under the BWM Convention is the newest and the only globally agreed RA framework for BWM purposes. This RA presented here was developed to enable a selective BWM approach (David 2007).

The need for a commonly agreed RA approach/model is outlined in section 6.5 Evaluation and decision-making of the G7 Guidelines. Paragraph 6.5.1 requires that port States considering to grant exemptions shall for both the evaluation and consultation processes especially consider Regulation A-4.3 which states that any exemption shall not negatively impact upon the environment, human health, property or resources of adjacent or other states. Any state potentially or adversely affected shall be consulted.

Furthermore, as stated in paragraph 7.4 of the procedures for granting exemptions, a RA model needs to be prepared, which is to be made available to exemptions applicants. It is also stipulated that if any Party (i.e., a country signatory to the BWM Convention) has decided that the shipowner or operator who applies for the exemption should conduct a RA, this Party should provide to that shipowner or operator all relevant information, including application requirements, the RA model to be used, the target species that should be considered and the required data reporting and collection standards. In turn, the shipowner or operator should make available all relevant information to this Party to enable a decision if an exemption can be granted (or not).

The RA itself could be conducted by any Party, or a Party may ask the applicant to prepare it. In both cases the Party which receives the application needs to have a common RA model available. Further this Party has to receive all necessary data and arrangements to conduct a RA with the aim to grant (or not) an exemption from BWM requirements. This is essentially needed as Parties are responsible to ensure that any action or decision taken may not cause harm to neighbouring or other states (see above). This process is globally applicable (Fig. 3).

Risk Assessment Framework

Data Reliability

The most critical point is to have reliable input data for the RA process as the decision taken by the Party has cost and legal consequences. However, there are known uncertainties, unpredictable stochastic events, as well as a lack of knowledge and

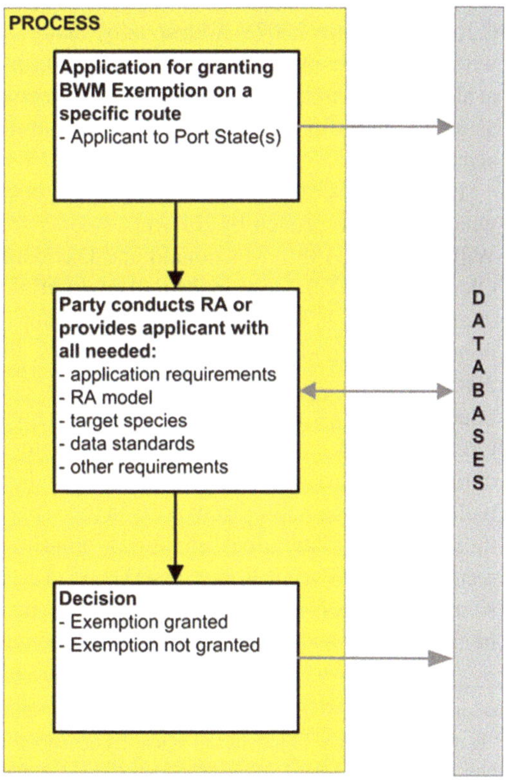

Fig. 3 General process and parties involved in the application for BWM exemptions (David et al. 2013) (Reprinted from David et al. (2013), copyright 2013, with permission from Elsevier)

data to characterise the introductions of harmful species via ballast water. Therefore, to keep the selective RA based BWM approach effective as much as possible, the precautionary principle[8] applies as a fundamental principle[9] in this RA process (EU Commission 2000; IMO 2007).

For the needs of environmental matching RA reliable environmental data need to be provided. For the needs of species-specific and species' biogeographical RA reliable biological data is needed. Critical issues identified regarding knowledge and data needs for RA include:

- the lack of data on harmful aquatic organisms and pathogens (HAOP) presence and abundance in ports (i.e., donor environment/port);
- the lack of knowledge regarding the survival of species during the voyage; and
- the lack of knowledge on their possible behaviour in the new environment.

Due to the poor general knowledge already mentioned, the weighting of importance of even a single parameter is difficult or may be impossible. Therefore we consider the risk parameters as of equal importance.

There have been relatively few comprehensive port baseline surveys conducted worldwide which have focused on collecting data regarding the presence of harmful species in ports and surrounding environments. In total, >100 port baseline surveys were conducted in more than 20 countries (Campbell et al. 2007; WGBOSV 2013; WGITMO 2013) which cover only ca. 1 % of the more than 9,400 ports in the world (Lloyd's Register 2007). Additionally, many of these studies are now out of date, with few continuous surveillance regimes in place (Hewitt et al. 2004a; Campbell et al. 2007). Consequently, the knowledge on cryptogenic and non-indigenous species as well as harmful native species in ports is limited, but essential for a comprehensive RA.

Introductions of new harmful aquatic species occur almost on a monthly basis, which has been proven by different studies around the world (e.g., Carlton 1985; Williams et al. 1988; Macdonald and Davidson 1997; Gollasch et al. 2000, 2002; Olenin et al. 2000; Carlton 2001; Hewitt et al. 2004b; David et al. 2007; Flagella et al. 2007). In ICES member countries a new species introduction forming a new population beyond its natural range occurs about every 9 weeks (Minchin et al. 2005). This includes the secondary spread of earlier introduced species in neighbouring areas (Minchin et al. 2005) (see chapter "The Transfer of Harmful Aquatic Organisms and Pathogens with Ballast Water and Their Impacts"). For instance, during the ballast water sampling study conducted in the Port of Koper (Slovenia), ballast water originating from ports in the same region (i.e., Mediterranean Sea, and mostly the Adriatic Sea) contained non-indigenous species that were not yet recorded in the Port of Koper area (David et al. 2007). This also leads to the conclusion that results from a port baseline survey by itself cannot last forever, but should be followed by a monitoring program to document possible new arrivals of harmful

[8] Communication from the Commission on the Precautionary Principle, Brussels, 02.02.2000.

[9] In the EU should be implemented when RA concerns environmental and human health protection and in the lack of robust scientific evidence (EU Commission 2000).

species. Further, the full comprehensive port baseline study may further need to be repeated to ensure most up-to-date information for RA (e.g., Hewitt and Martin 2001). In conclusion, biological data on a ballast water donor port can only be considered as reliable if a baseline survey for HAOP has been conducted and a regular monitoring program for HAOP is in place. The lowest frequency of surveys per time need to be decided depending on the target species group, e.g., harmful algae, indicator species for pathogens. Another way to determine the required frequency for sampling is proposed by Hewitt and Martin 2001, i.e., with a repeated survey one could then calculate the rate of arrival/establishment function which would then inform about a suitable re-survey frequency based on the acceptable level of protection/risk.

During the developing HELCOM/OSPAR RA port survey sampling data are regarded valid for granting an exemption for applicants for a period of in maximum 5 years. This means that the port survey data from the sampling in year one can also be taken up to 5 years later as a basis for granting an exemption, i.e., no new port baseline surveys are required (HELCOM/OSPAR 2013). We feel that a 5 year period is rather long considering that approximately two new primary introductions of non-indigenous species were found in this region per year over the last decade. In consequence, should this species introduction trend continue, this approach may overlook up to ten non-indigenous species thereby accepting the risk that such species are transported, which could have been avoided.

It should further be noted that introduced and cryptogenic species are registered only occasionally in continuous biological monitoring programs in Europe. The dominating first records of such species were made in projects and individual studies not part of regular monitoring programs. In some sampling studies the working standards are unclear, i.e., the data reliability is uncertain. In Europe only very few regular monitoring programs specifically target aquatic non-indigenous and cryptogenic species (e.g., in Estonia and Germany (WGITMO 2013)). However, reliable data are a crucial component for a proper RA (Lodge et al. 2006; David 2007). Further, introduced and cryptogenic species are also seldom targeted in port area monitoring programs in most European countries. In less than 10 European ports out of the more than 1,200 ports of all 22 coastal Member states[10] preliminary port baseline surveys were conducted to document the presence and abundance of non-indigenous and cryptogenic species. These port studies should be considered as preliminary because not all habitats were surveyed. Other continents are more advanced as, e.g., in North America, Australia and New Zealand the share of surveyed ports is much higher compared to Europe (Campbell et al. 2007).

Introductions of harmful species may occur every day also between ports within the same bioregion by secondary introductions and natural spread (e.g., Olenin et al. 2000; David et al. 2007; McCollin et al. 2008; Darling et al. 2012). As a result a one-time port baseline survey alone cannot be sufficient as a long-term basis for RA, but should be followed by a regular monitoring program for new (harmful) species (e.g., Hewitt and Martin 2001) and this should be done by experts in this field

[10] European Sea Ports Organisation (ESPO), http://www.espo.be/, last accessed November 2013.

to ensure reliable data quality. This is to avoid that exemptions are wrongly ongoing in cases of new species arrivals. We recommend that such monitoring (surveys) needs to be established regularly (e.g., every 6 or 12 months) to deliver reliable and current information.

Applying the precautionary principle, in cases where reliable data are lacking, no RA-based exemption can be granted. This is especially important where a RA relates to environmental and human health protection (EU Commission 2000; IMO 2007).

Risk Assessment Methods Applied

Environmental Matching Method

The environmental matching RA method uses environmental parameters as surrogates for species. Of the two most frequently used RA parameters, water temperature and salinity, the salinity variability is the only parameter common to all past RAs. Furthermore, the more variables a RA includes, the lesser transparent becomes the decision process. We believe that water salinity is the most "straight forward" concept, hence the RA presented here uses salinity as the only meaningful environmental parameter. Water temperature was also considered as a RA quantifying factor in the environmental match approach. However, we believe this is of lesser reliability to identify low risk scenarios because we assume that organisms are more flexible regarding temperature tolerances compared to salinity in temperate and polar regions. One reason for this assumption is the greater temperature difference compared to salinity difference over the annual seasons which the species need to tolerate. In the tropics this may be different as the temperature may be more similar throughout the year and here the rainy seasons may result in a stronger organism tolerance towards salinity. However, also the use of salinity shows its weakness. In cases when two ports may have totally different salinity ranges the RA result will be low risk. However, species salinity tolerance may cover both environments so that a high risk should have been the result (Hewitt and Hayes 2002; Hayes and Sliwa 2003). As a compromise, this RA uses salinity as the only environmental parameter. The difference between the ballast water donor and recipient ports as freshwater and marine ports respectively is the suggested acceptable salinity difference offering acceptable precaution levels to trigger a low risk result because the number of species being able to tolerate such a large salinity difference is comparably low (but not zero!).

In a two-step approach we considered that the minimum salinity difference to assume a low risk for a successful species transfer. A low risk was assumed when ballast water is moved between freshwater (<0.5 psu) and fully marine conditions (>30 psu). However, such conditions are rarely applicable in coastal shipping, but may occur in areas with larger estuaries, run-off of major rivers, when a port is situated on a river more inland etc. To cope with that situation other possibilities were considered. What could be acceptable, but at the price of a slightly higher risk, is when ballast water is transported between freshwater ports and higher saline brackish ports with salinities >18 psu. In these cases a species-specific method would be required in addition to the environmental match taking into account the species salinity tolerance ranges, especially considering species which have a known salinity tolerance higher than <0.5 psu and >18 psu.

The salinity limit of 18 psu is based upon the work of Remane (1934) and Remane and Schlieper (1958). They compared the diversity of freshwater, brackish and marine species along salinity gradients and showed that for many groups of species the minimum species diversity was found in low salinity conditions. A borderline used in their studies is at approximately 18 psu. It is interesting to note that the Venice salinity system (Venice System 1959) draws the line between polyhaline and mesohaline also at this psu level and it is found that this relates to a change in species diversity (den Hartog 1964). Paavola et al. (2005) more recently found the same trend for native and non-indigenous species. In European brackish seas, most non-indigenous species are well adapted to salinities with the lowest native species diversity. Also the non-indigenous species diversity maximum is frequently observed in the salinity ranges where the native species diversity reached a minimum. Bleich (2006) compared the macrozoobenthos diversity at different Baltic Sea sampling stations with different salinities. He found that the species diversity changed by more than 80 % at ca. 18 psu and concluded that this may be a salinity-related distribution limit. We therefore assume that the 18 psu salinity limit chosen is well enough justified.

Species-Specific Method

The identification of species-related risk takes into account the potential invasiveness of each selected species and the potential harm that it could cause in a new environment. The selection of target species was based on the IMO definition in the G7 Guidelines using the following criteria: (a) evidence of a prior introduction; i.e., where a species has become introduced outside its native range; (b) potential impact on the environment, economy, human health, property or resources; (c) strength and type of ecological interactions, i.e., severeness of its impact; (d) current distribution within the biogeographic region and in other biogeographic regions; and (e) relationship with ballast water as a vector, i.e., it has been shown to be carried in ballast water or it has a life-history stage that might be carried in ballast water.

The target species selection process should consider all harmful native, non-indigenous and potentially harmful cryptogenic species present within the donor and recipient ports and their surrounding areas. For a species-specific RA, an assessment results in an unacceptable risk if it identifies at least one target species that satisfies all following criteria: the target species is (a) likely to cause an unacceptable level of harm; (b) present in the donor port, but not in the recipient port; (c) likely to be transferred to the recipient port with ballast water; and (d) likely to survive in the recipient port.

Further, should both the ballast water donor and recipient regions have the identical target species, but these occur in very different abundances, each species case needs to be examined separately to qualify the level of risk. This is because a target species may occur at a low level of abundance in a recipient port not with a fully self-sustaining population, but further releases from a donor port, where abundance of this species is higher, may lead to a self-sustaining population in the recipient port. As a result it is unacceptable to transfer unmanaged ballast water in cases a target species occurs in much higher abundance in any of the donor ports compared with a recipient port.

In addition to human-assisted movements, aquatic native and non-indigenous biota have the potential to spread naturally from a donor to a recipient port without being moved by a vector. The ability to spread naturally is species-specific, and in the RA this is acceptable only in situations where all target species of concern could easily spread naturally from a donor to a recipient port.

A coastal state may introduce a control or eradication program for the most unwanted species already introduced into their waters; this has RA implications. A control or eradication program would only be undertaken to manage high impact species. Should these species be potentially carried in ballast water, then their inoculation in a recipient area would not be acceptable. Therefore any such control or eradication program conducted in the donor port indicates a high risk.

Combined Environmental Matching and Species-Specific Method

In this RA we considered that it may be still acceptable that ballast water is moved between freshwater ports and brackish ports with salinities >18 psu, in which case a species-specific method would additionally be required. This would especially consider the species with known higher salinity tolerances than <0.5 psu and >18 psu. The presence of one such species in only one of the donor ports considered results in the situation that a low risk cannot be assumed.

Species' Biogeographical Method

The study focus was laid on species movements within the same biogeographical region. The species' biogeographical method is considered here through the target species selected (see section "Species-specific method").

Shipping Vector Factors

Species Survival of the Voyage

Prerequisites for a species to be successfully transported from a donor to a recipient port with ballast water include that it first needs to enter the vessel during the ballasting process, survive the physical stress during ballasting, survive the likely unfavourable conditions inside the tank during a voyage, become discharged from the vessel and to survive the deballasting process. This "chain of events" needs to coincide with opportunities in the recipient area that they find suitable environmental conditions and food sources so that they can survive and reproduce. The latter requirements are termed 'invasion windows'. The survival of species during vessel voyages has been studied earlier. It was assumed that longer containment inside ballast tanks negatively affects species survival.

In contrast it was found that species survive several months in ballast tanks. Resting stages may even be viable for many years (e.g., Hallegraeff and Bolch 1992; Gollasch et al. 2000; Olenin et al. 2000; David et al. 2007; McCollin et al. 2008). Further, a RA model as the chain of events was prepared (Hayes 2000; Bailey et al. 2011). However, the high diversity of potential species in transit with their stochastic behaviors, e.g., some species have even been found to reproduce in ballast tanks (Gollasch et al. 2000), it can be assumed that some species will survive a vessel voyage (see chapter "The Transfer of Harmful Aquatic Organisms and Pathogens with Ballast Water and Their Impacts") so that survival en-route is considered as not robust or reliable enough to be used as a risk quantifying factor.

Noting the above and applying the precautionary principle, this RA model assumes that all species present in a ballast water donor port which can theoretically be transported with ballast water will become discharged alive in a recipient port. However, it is impossible to predict at which point in time this might happen. This means that, *a priori*, ballast water discharges from a donor port with a harmful species is an undesirable (unacceptable) event.

Quantity and Frequency of Ballast Water Discharges

Other shipping factors such as the quantity and frequency of ballast water discharges also relate to the risk level (Bailey et al. 2011; Chan et al. 2013). We assume that the higher the number of introduced organisms is and also the higher the introduction frequency is, the greater is the expected probability of a successful species introduction. However, this is species-specific and certainly depends on many conditions in each new environment where the species is introduced (Briski et al. 2012).

We found that the total number of ballast water discharge events and their temporal distribution in the recipient environment are insufficiently studied regarding their possible risk level impact and influence, and were therefore not considered in this RA model. Ruiz et al. (2013) concluded recently that there was no relationship between the quantity and frequency of ballast water discharges of foreign vessels with the number of introduced ballast water mediated species in 16 large bays in the United States. Furthermore, to our knowledge there is not even a single study to quantify the minimum number of organisms (propagule pressure) which would need to be discharged with ballast water to enable a species establishment with a self-sustaining population which may subsequently become invasive in a new environment.

In conclusion we consider that even small quantities of harmful organisms present in discharged ballast water may result in a successful transfer of a species which in turn may have negative consequences. As a result the RA described here does not consider the ballast water volume discharged in a recipient port and neither the discharge frequency as a risk level indication.

Definition of Potential Impacts

Studies have proven that organisms even after entering a new environment may not survive, reproduce or cause harm. However, other species introductions resulted in drastic negative impacts on various stakeholders (see chapter "The Transfer of Harmful Aquatic Organisms and Pathogens with Ballast Water and Their Impacts" for examples). In many cases it was shown that the process of introduction and species adaptation to the new environment, before they cause harm, may last for years. If a newly arrived species is not being studied in depth case by case (i.e., for each recipient environment) it is very difficult, if not almost impossible, to predict the species behaviour in the new environment(s) with an acceptable reliability. Hence, a prediction of these stochastic events seems impractical and almost impossible.

As a result, the precautionary approach for the RA decision process considers all aquatic non-indigenous organisms as harmful, and assumes that all harmful species present in the ballast water donor port, if discharged, will cause harm in the recipient environment. In conclusion this means that the discharge of ballast water from a donor port that contains harmful species is already an undesirable event.

The Main Risk Assessment Model Premises

As outlined above, the RA model in the decision making process considers different premises, which are based on best available scientific knowledge covering the expertise from different fields (e.g., invasion biology, maritime transport, BWM, RA, regulatory affairs, environment and human health protection, etc.). In summary, the premises on which this RA model (see Fig. 4) is based are:

- The input environmental (i.e., salinity) and biological data for the RA must be reliable.
- Biological data may be considered as reliable if a port baseline survey for HAOP has been conducted, and a regular monitoring program for HAOP is in place.
- If salinity based RA results in acceptable low risk, no biological data is needed.
- If a species is present in the ballast water donor port it will be discharged alive with ballast water in the recipient port.
- The voyage length, quantity of ballast water discharged and the frequency of discharges as RA factors are difficult to be defined to a reliable level to change the RA result.

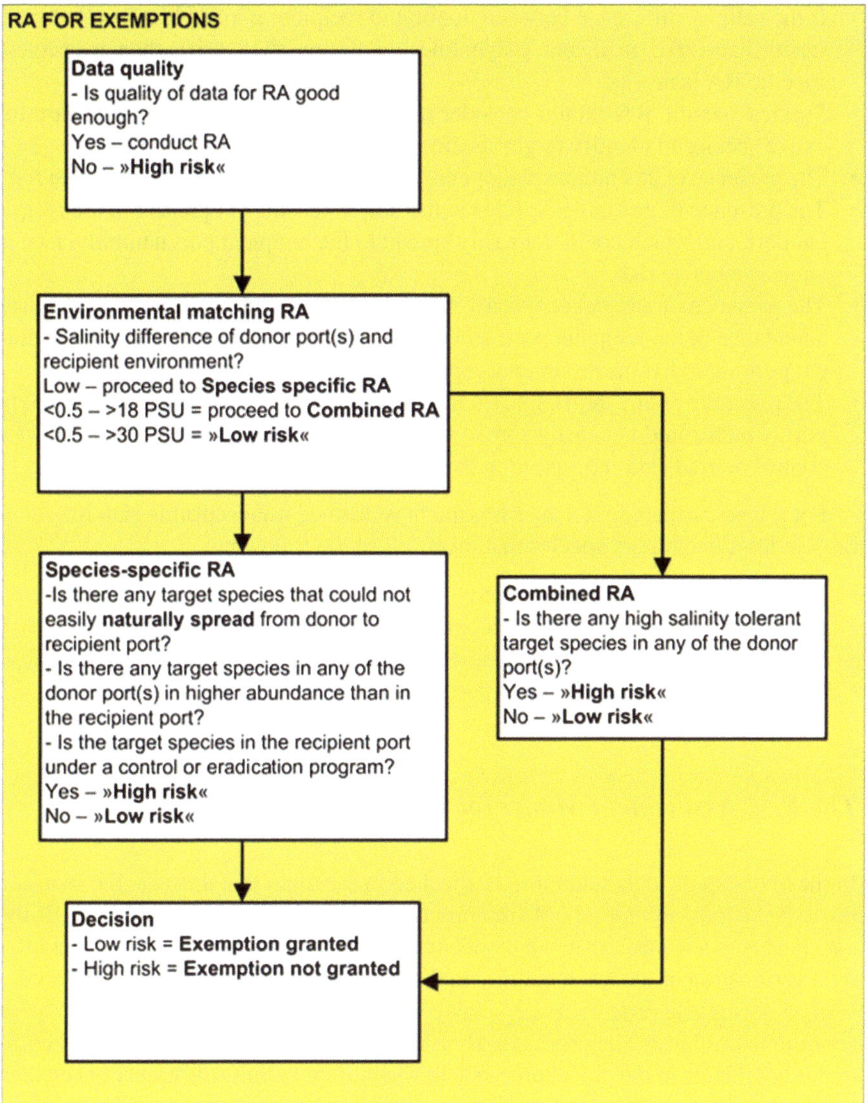

Fig. 4 Basic principles for the RA for exemptions

- Salinity is the only enough reliable parameter for the environmental matching RA.
- RA would result in acceptable low risk only if the donor and recipient ports are located one in freshwater (<0.5 psu) and the other in fully marine conditions (>30 psu).
- If the donor and recipient ports are located one in freshwater (<0.5 psu) and the other in polyhaline conditions (>18 psu), than a combined approach with species-specific RA is needed to consider high salinity tolerant species.

- If the salinity difference between donor and recipient ports is less than between freshwater (<0.5 psu) and polyhaline conditions (>18 psu), than a species-specific RA is needed.
- Species-specific RA should consider non-indigenous, cryptogenic and harmful native species to identify target species, and human pathogens.
- The presence of any human pathogens in the donor port means unacceptable risk.
- The presence of any target species in the donor port not yet present in the recipient port, and which could not easily spread to the recipient port naturally, means an unacceptable risk.
- The presence of any target species in the donor port and its occurrence in lower abundance in the recipient port, and which could not easily spread to the recipient port naturally, means an unacceptable risk.
- The presence of any target species in the donor port also present in the recipient port, which could not easily spread to the recipient port naturally, but is under a control or eradication program in the recipient port, means unacceptable risk.

For a species-specific RA, an assessment is deemed **unacceptable risk** if it identifies at least one **target species** that meets all of the following:

- likely to cause unacceptable harm;
- present in the donor port or biogeographic region, but not in the recipient port;
- likely to be transferred to the recipient port through ballast water; and
- likely to survive in the recipient port.

The Risk Assessment Model for Granting Exemptions

In the first step the data reliability is checked to ascertain that this is at the required level. If the data are not reliable the process ends with an unacceptable risk. If the data quality is adequate, then the model proceeds to the environmental matching RA with verification of the water salinity in the donor and recipient ports. If the salinity is of an acceptable difference, i.e., between freshwater (<0.5 psu) and fully marine conditions (>30 psu), the process ends with an acceptable risk result. If this condition is not met, than the model proceeds to verify if the salinity difference is between freshwater (<0.5 psu) and euryhaline conditions (>18 psu). If this condition is met then the model proceeds with a species-specific approach, but considering human pathogens and only high salinity tolerant target species. While if none of the environmental (miss)-matching conditions are met, then the process proceeds with a complete species-specific approach, i.e., considering all target species and human pathogens. The model in the next steps checks if species could spread naturally to the recipient port, if these are already present in the recipient port and in which abundance, and if these are under any control or eradication program. The RA result depends on answers to all these questions.

Human pathogens were here defined as microbes or microorganisms (virus, bacterium, prion, or fungus) that cause a disease in humans. It should be noted that many human pathogens are difficult to identify in water. Therefore IMO suggested to use "indicator microbes" such as *Escherichia coli* and Enterococci and to limit their acceptable numbers in ballast water discharges. Although these indicator microbes themselves are usually harmless, natural mutations may result in human diseases, as recently shown by a strain of bacteria known as enterohaemorrhagic *E. coli* (EHEC), a natural mutation of *E. coli* (Carter et al. 2012). Further, the presence of elevated numbers of human faecal bacteria like *E. coli* and Enterococci in water indicates an improper wastewater treatment system and the water may consequently also include other more problematic species, such as disease agents. IMO further includes the toxic strains of *Vibrio cholerae*, the agent of the Cholera disease, in this standard (D-2 standard).

In the context of this model less abundant target species in the recipient port means a considerable difference in species abundance, e.g., if in the donor port a species occurs with 100 ind/m^2 and in the recipient port with 10 organisms, the recipient port clearly inhabits a less abundant target species population. However, should the target species occur in the donor port with 2,000 ind/m^2 and in the recipient port with 1,500 ind/m^2 this can be considered as a comparable abundance. These numbers should give an indication only, but need to be reconsidered as per the species concerned.

The BWM RA model in the form of a flow chart is presented in Fig. 5.

Risk Assessment for Selective Ballast Water Management Measures

Risk Assessment Framework – Background, Principles, Assumptions and End-Point

The precautionary principle[11] is applied as a fundamental principle (EU Commission 2000) in this RA process which considers all aquatic non-indigenous organisms as being harmful, and assumes that all HAOP, if present in the ballast water donor port, if discharged, will cause harm in the recipient environment. This sets the RA end-point "at discharge" and means that already the discharge of ballast water from a donor port with HAOP is an undesirable event (see above).

The quantity of discharged ballast water is also one of the factors possibly related to the risk level. However, RA here does not relate the risk level to the quantity of discharged ballast water as also a small quantity of harmful organisms present in the discharged ballast water may result in critical consequences in the recipient environment.

[11] Communication from the Commission on the Precautionary Principle, Brussels, 02.02.2000.

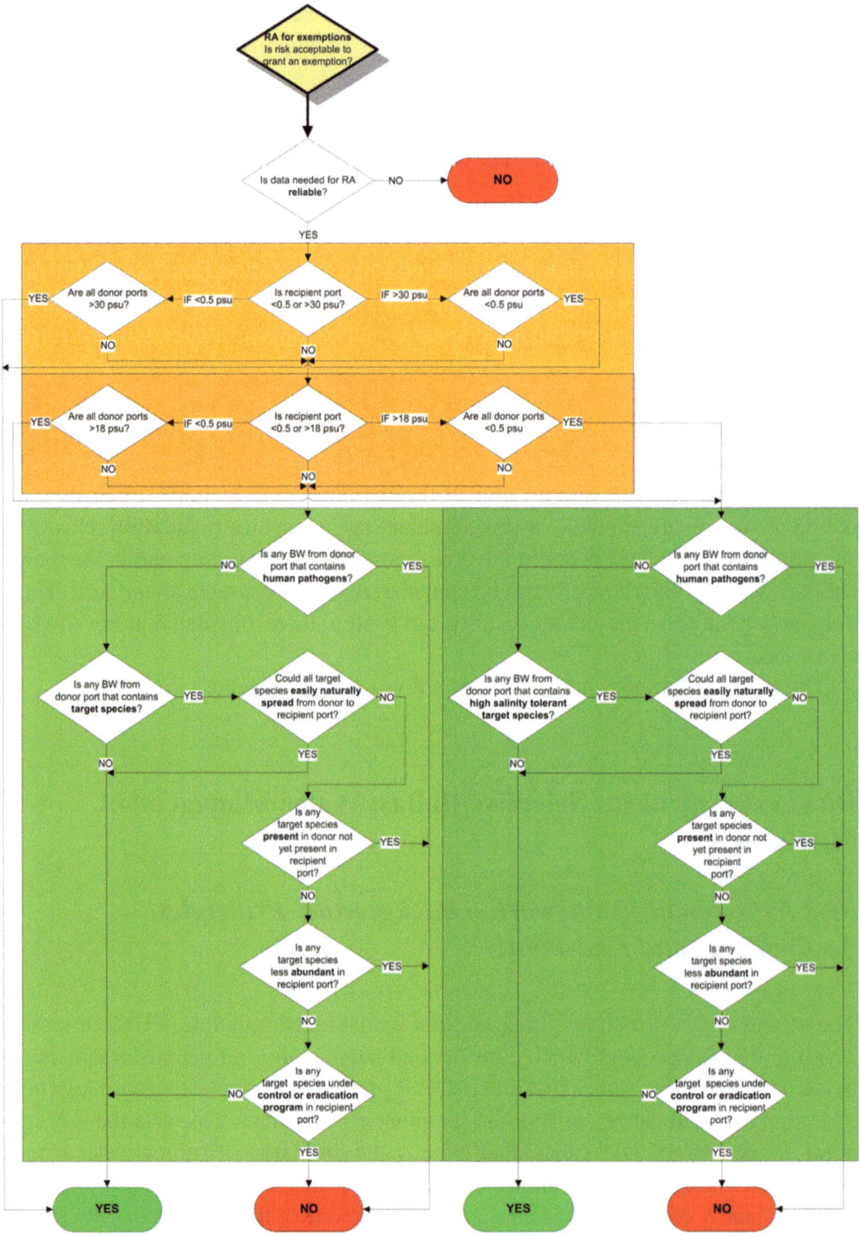

Fig. 5 The RA model for granting exemptions from BWM requirements. The *orange* box area is the environmental matching RA process, in the *green* box area is the species-specific RA process, in the shaded dark orange and green area is the combined RA approach. Reprinted from David et al. (2013), copyright 2013, with permission from Elsevier (This figure can be downloaded from http://extras.springer.com/)

The level of risk is assigned based on the different approaches as described above: environmental matching, species specific and considering biogeographical approach with target species.

In line with the G7 Guidelines on RA (IMO 2007), the Large Marine Ecosystems (LME)[12] approach was chosen as units for regions. For the RA and according to the LME philosophy this means that ports inside each LME have higher biological similarity and environmental compatibility. In cases when the ballast water donor port is in a different region (LME) from the recipient port, this means that species living in that region are by default considered non-indigenous to the recipient environment. However, there also may be an overlap of species between bioregions as, e.g., the Baltic and North Seas have many species in common, but are two separate LMEs. The more distant the LMEs are located, the more different seems the species assemblage.

The number of different risk rankings is directly related or actually dependent on BWM needs, i.e., how many different BWM responses are needed. This RA has a four level approach that was chosen as appropriate and detailed enough for BWM responses with different needs. Nevertheless, this can be easily adapted to more or fewer levels if there are different needs.

The selected risk levels are:

- low risk,
- intermediate risk,
- high risk, and
- extreme risk,

each of them resulting from a different ballast water source situation, and in the following steps triggering different BWM requirements.

The environmental matching RA is based on salinity. The input environmental (i.e., salinity) and biological data for the RA must be reliable. Biological data may be considered as reliable if a port baseline survey for HAOP has been conducted, and a regular monitoring program for HAOP is in place. If salinity based RA results in acceptable low risk, no biological data is needed.

The species specific RA is included with the questions on the presence of different species in the donor port that are associated with different levels of risk posed. The presence of HAOP in the donor port triggers different levels of risk, depending on their presence and abundance in the recipient port and whether they were included in a control program.

The logic behind this is:

- if a HAOP is not yet present in the recipient port, its introduction poses a high risk;
- if a HAOP is present also in the recipient port and was not included in any control program, the perception of it's harmfulness from the recipient port State is

[12] http://woodsmoke.edc.uri.edu/Portal/

uncritical, and hence the level of risk is lower than intermediate, but still not acceptable for unmanaged ballast water discharges, however,

- if a HAOP is present also in the recipient port and was included in a control program, this means that it was perceived and selected by the port State as critical. Therefore, the level of associated risk is extreme.

If a port State has selected target species which it does not want to become discharged in its jurisdictional waters, then these by default trigger the extreme risk should these species occur in the donor port or region. Target species are selected based on selection criteria (see section "Species-specific method"). The species' biogeographical method is considered through the target species selection.

When considering human pathogens, these are certainly one of the most unwanted species, and therefore have also been selected to trigger the same level of extreme risk. In the case of toxic algae, the approach is split in two levels. In many cases, these are present in ports as resting stages in sediments and may not cause blooms. However, these can be loaded on board ships with ballast water. This may occur when sediments are stirred-up in the water column so that some resting stages of toxic algae may also be present in the ballast water, and therefore have been selected as posing a high risk. In case these algae are in the bloom state, these will certainly be loaded on board the vessel within the ballast water in millions and possibly form resting stages in the ballast tank to survive the voyage. Hence, they represent a serious threat to the ballast water recipient environment and have also been selected to trigger extreme risk. After a vessel has loaded ballast in an algal bloom state, it may be expected that water and/or sediments inside a ballast tank will have a great potential to contain harmful algae, which may last for a longer time, i.e., also multiple ballasting operations in their next ports of call may not remove those organisms completely. Therefore, the cleaning of tanks and notifications issued by port State authorities to vessels in case of harmful algal blooms is critical.

Risk Assessment Model for Selective Ballast Water Management Measures

The discharge of ballast water will be deemed as posing a **low risk** in conditions when:

- the ballast water is moved between ports with freshwater (<0.5 psu) and fully marine conditions (>30 psu), independent of whether the donor and recipient ports are in the same region; or
- the ballast water is from a donor port that does not contain HAOP and is from the same region as the recipient port.

The discharge of ballast water will be deemed as posing an **intermediate risk** in conditions when:

- the ballast water is from a donor port that contains HAO that are already present in the recipient port and also occur in a similar abundance, where these are not under any control program.

The ballast water will be deemed as posing a **high risk** in conditions when:

- there is no reliable data about environmental (i.e., salinity) or biological conditions in the donor port; or
- the ballast water is from a donor port that contains HAO (i.e., non-indigenous species and toxic algae (not in the blooming state), which are not present in the recipient port).

The RA will result in an **extreme risk** in conditions when:

- the ballast water is from a donor port that contains target species, especially when those occur in much higher abundance as in the recipient port;
- the ballast water is from a donor port that contains toxic algae that are in a bloom state;
- the ballast water is from a donor port that contains human pathogens; or
- the ballast water is from a donor port that contains HAO that are already present in the recipient port, where these are under any control program in the donor port.

The BWRA model to assess the level of risk posed by ballast water to the recipient port is shown in Fig. 6. According to each level of risk identified different BWM measures may be applied (see chapter "Ballast Water Management Decision Support System").

Implementation of Selective Ballast Water Management Based on Risk Assessment

The advantages of the blanket approach include low data and skill requirements and it is simple for port State implementation. However, the main disadvantages are that more burden is placed on ship crews with "unnecessary" BWM requirements (in case of low risk), which will result in more costs for the shipping industry. Depending on the BWM method used also more pressures may be placed on the environment (e.g., in case chemical treatment of ballast water is required which may result in residual toxic components in discharged ballast water or in the addition of neutralization agents before ballast water discharge).

The selective approach places less "unnecessary" BWM burden on vessels, but it requires more extensive data gathering for port States as well as more data and reporting requirements for vessels. It may require higher skills and knowledge for port State personnel; however with an appropriate decision support system (DSS) this can be overcome.

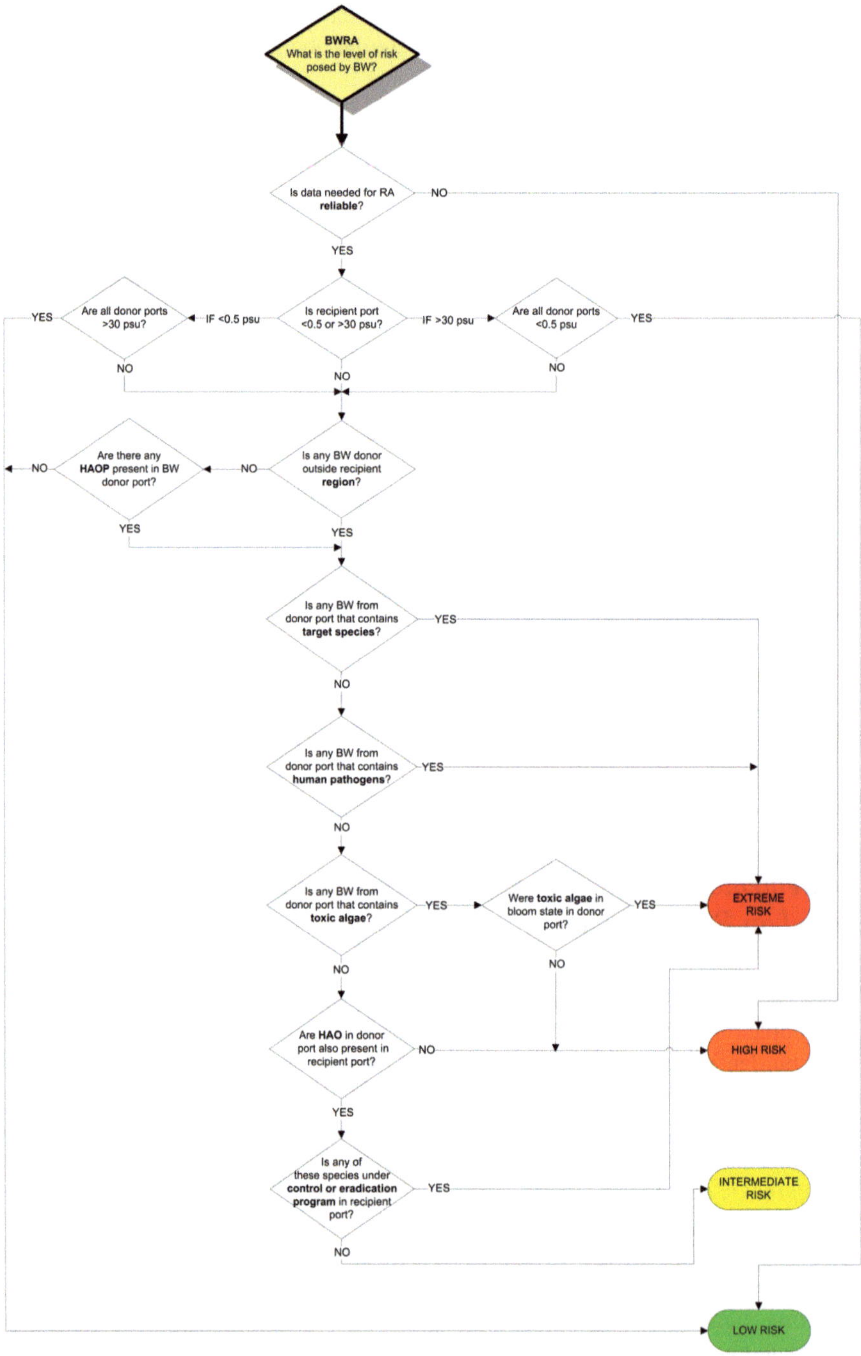

Fig. 6 RA model resulting in four different risk levels (Enhanced after David 2007) (This figure can be downloaded from http://extras.springer.com/)

With too many limiting factors for vessels discharging ballast water in a port, the blanket approach becomes ineffective. Further to the feasibility, a decision on the appropriate (blanket or selective) approach can be taken considering their advantages and disadvantages which we summarize here:

The advantages of the blanket approach include:

- low data requirements for the port State;
- low skill requirements for the port State personnel to come to a RA result; and
- simple implementation for the port State.

The disadvantages of the blanket approach include:

- all vessels conduct BWM, even those that do not carry harmful organisms and pathogens;[13]
- more burden on vessels crew by requiring "unnecessary" BWM measures;
- more costs with BWM; and
- depending on the BWM method used also some additional environment pollution or pressures.[14]

The advantages of the selective approach include:

- less "unnecessary" BWM burden for vessels;
- lower costs for the shipping industry; and
- less unnecessary environment pollution or pressures.[15]

The disadvantages of the selective approach include:

- more extensive data requirements[16] for port State;
- more data and reporting requirements for vessels;
- more complex BWM approach requiring the use of a RA system;
- more complex BWM system requiring DSS;
- higher skill and knowledge requirements for port State personnel; and
- in cases of a lack of data or false data, the risk may be underestimated and consequently "high risk" ballast water may be discharged.

As stated above, the implementation of the BWM Convention under the blanket approach is clearly simpler. However, there are many factors arising from unique situations/conditions worldwide that may limit the possibility of its implementation, which, at the same time, favours the selective approach. On the other side, the selective approach is without doubt more demanding, which would appear to limit its application. Hence, appropriateness should be studied and decisions taken on a case by case – port by port basis.

---

[13] Source ports may be in the same region and not infected by harmful organisms and/or pathogens.

[14] e.g., more oil consumption and gas emissions for creating more power supply necessary for ballast water treatment or exchange, chemicals (active substances) used for treatment.

[15] e.g., more oil consumption and gas emissions for creating more power supply necessary for ballast water treatment or exchange, chemicals (active substances) used for treatment.

[16] i.e., quantitative and especially qualitative.

Acknowledgements The research leading to part of these results has received funding from the European Community's Seventh Framework Programme (FP7/2007–2013) under Grant Agreement No. [266445] for the project Vectors of Change in Oceans and Seas Marine Life, Impact on Economic Sectors (VECTORS). Part of this publication has been produced with the financial assistance of the IPA Adriatic Cross-Border Cooperation Programme – strategic project Ballast Water Management System for Adriatic Sea Protection (BALMAS). The contents of this publication are the sole responsibility of authors and can under no circumstances be regarded as reflecting the position of the IPA Adriatic Cross-Border Cooperation Programme Authorities.

References

AQIS (1994) Bio-economic risk assessment of the potential introduction of exotic organisms through ship's ballast water. Australian Quarantine and Inspection Service (AQIS), Report No. 6 of the Ballast Water Research Series. Australian Government Publishing Service, Canberra

Awad A, Clarke C, Greyling L, Hilliard R, Polglaze T, Raaymakers S (2004) Ballast water risk assessment, port of Saldanha Bay, Republic of South Africa, November 2003: final report, GloBallast monograph series No. 13. IMO, London

Bailey SA, Deneau MG, Jean L, Wiley CJ, Leung B, MacIsaac HJ (2011) Evaluating efficacy of an environmental policy to prevent biological invasions. Environ Sci Technol 45:2554–2561

Behrens HL, Haugom GP, Mordal Bakke S (2002) EMBLA concept; risk assessment methodology and impact and consequence methodology scheme. DNV Rep. No. 2002–1269, Det Norske Veritas, Høvik, Norway

Bleich S (2006) Messung der β-Diversität entlang eines Salzgehaltsgradienten anhand von Makrozoobenthoszönosen der Ostsee. Master of Science thesis. Universität Rostock, Rostock

Briggs JC (1974) In: Beach RE, Laufer RS, Willey WJ (eds) Marine zoogeography. McGraw-Hill, New York

Briggs JC, Bowen BW (2012) A realignment of marine biogeographic provinces with particular reference to fish distributions. J Biogeogr 39:12–30

Briski E, Bailey SA, Casas-Monroy O, DiBaccio C, Kzaczmarska I, Levings C, MacGillvary ML, McKindsey CW, Nasmith LE, Parenteau M, Piercey GE, Rochon A, Roy S, Simard N, Villac MC, Weise AM, MacIsaac HJ (2012) Relationship between propagule pressure and colonization pressure in invasion ecology: a test with ships' ballast. Proc R Soc B 279:2990–2997

Campbell ML, Gould B, Hewitt CL (2007) Survey evaluations to assess marine bioinvasions. Mar Pollut Bull 55:360–378

Carlton JT (1985) Transoceanic and interoceanic dispersal of coastal marine organisms: the biology of ballast water. Annu Rev Oceanogr Mar Biol 23:313–374

Carlton JT (2001) Introduced species in U.S. coastal waters: environmental impacts and management priorities. Pew Oceans Commission, Arlington

Carter MQ, Louie JW, Fagerquist CK, Sultan O, Miller WG, Mandrell RE (2012) Evolutionary silence of the acid chaperone protein HdeB in enterohemorrhagic *Escherichia coli* O157:H7. Appl Environ Microbiol 78:1004–1014

Chan FT, Bailey SA, Wiley CJ, MacIsaac HJ (2013) Predictive relative risk assessment for ballast-mediated invasions at Canadian Arctic Ports. Biol Invasion 15(2):295–308

Cothern CR (1996) Handbook for environmental risk decision making: values, perceptions and ethics. Lewis Publishers/CRC Press LLC, Boca Raton

Darling JA, Herborg L-M, Davidson IC (2012) Intracoastal shipping drives patterns of regional population expansion by an invasive marine invertebrate. Ecol Evol 2(10):2552–2561

David M (2007) A decision support system model for ballast water management of vessels: doctoral dissertation. University of Ljubljana, Portorož

David M, Gollasch S (2010) Ballast water risk assessment for intra North Sea shipping. Final report. Prepared for Interreg IVB North Sea Ballast Water Opportunity project. NIOZ, Texel

David M, Gollasch S, Cabrini M, Perkovič M, Bošnjak D, Virgilio D (2007) Results from the first ballast water sampling study in the Mediterranean Sea – the Port of Koper Study. Mar Pollut Bull 54:53–65

David M, Gollasch S, Leppäkoski E (2013) Risk assessment for exemptions from ballast water management – the Baltic Sea case study. Mar Pollut Bull 75:205–217

den Hartog C (1964) Typologie des Brackwassers. Helgol Wiss Meeresunters 10:377–390

Dragsund E (2005) Ballast water risk assessment – the DNV approach used for Croatia based on the DNV EMBLA method. In: ICES. 2005. Report of ICES/IOC/IMO Working Group on Ballast and Other Ship Vectors (WGBOSV), Arendal ACME:04

Ekman S (1953) Zoogeography of the sea. Sidgwick and Jackson, London

Endresen Ø, Behrens HL, Brynestad S, Andersen AB, Skjong R (2004) Challenges in global ballast water management. Mar Pollut Bull 48:615–623

EU Commission (2000) Communication from the Commission on the Precautionary Principle. Com.(2000)1. Brussels

Flagella MM, Verlaque M, Soria A, Buia MC (2007) Macroalgal survival in ballast water tanks. Mar Pollut Bull 54:1395–1401

GloBallast (2003) Ballast water risk assessment, user guide for the ballast water risk assessment database/GIS system. GEF/UNDP/IMO, Global Ballast Water Management Program. URS Meridian-GIS. International Maritime Organization, London

Gollasch S (1996) Untersuchungen des Arteintrages durch den internationalen Schiffsverkehr unter besonderer Berücksichtigung nichtheimischer Arten. Diss., Univ. Hamburg, Verlag Dr. Kovac, Hamburg

Gollasch S, Leppäkoski E (1999) Initial risk assessment of alien species in Nordic coastal waters. In: Gollasch S, Leppäkoski E (eds) Initial risk assessment of alien species in Nordic coastal waters. Nord 1999: 8. Nordic Council of Ministers, Copenhagen, pp 1–124

Gollasch S, Leppäkoski E (2007) Risk assessment and management scenarios for ballast water mediated species introductions into the Baltic Sea. Aquat Invasive 2(4):313–340

Gollasch S, Lenz J, Dammer M, Andres HG (2000) Survival of tropical ballast water organisms during a cruise from the Indian Ocean to the North Sea. J Plankton Res 22(5):923–937

Gollasch S, Macdonald E, Belson S, Botnen H, Christensen J, Hamer J, Houvenaghel G, Jelmert A, Lucas I, Masson D, McCollin T, Olenin S, Persson A, Wallentinus I, Wetsteyn B, Wittling T (2002) Life in ballast tanks. In: Leppäkoski E, Gollasch S, Olenin S (eds) Invasive aquatic species of Europe: distribution, impacts and management. Kluwer Academic Publishers, Dordrecht, pp 217–231

Hallegraeff GM, Bolch CJ (1992) Transport of diatom and dinoflagellate resting spores in ships' ballast water: implications for plankton biogeography and aquaculture. J Plankton Res 14(8):1067–1084

Hayes KR (2000) Quantitative ecological risk assessment, a ballast water case study. Doctoral dissertation, Heriot-Watt University, Edinburgh

Hayes KR, Hewitt CL (1998) A risk assessment framework for ballast water introductions. CRIMP technical report 14, Division of Marine Research. CSIRO, Hobart

Hayes KR, Hewitt CL (2000) A risk assessment framework for ballast water introductions – volume II, CRIMP technical report 21. Division of Marine Research, CSIRO, Hobart

Hayes KR, Sliwa C (2003) Identifying potential marine pests – a deductive approach applied to Australia. Mar Pollut Bull 46:91–98

HELCOM/OSPAR (2013) Joint HELCOM/OSPAR Guidelines on the granting of exemptions under the International Convention for the Control and Management of Ships' Ballast Water and Sediments, Regulation A-4. This document is a part of the 2013 HELCOM Ministerial Declaration and was adopted by the 2013 HELCOM Ministerial Meeting. Helsinki, Finland

Hewitt CL (2003) Marine biosecurity issues in the world oceans: global activities and Australian directions. Ocean Yearb 17:193–212

Hewitt CL, Hayes KR (2002) Risk assessment of marine biological invasions. In: Leppäkoski E, Gollasch S, Olenin S (eds) Invasive aquatic species of Europe: distribution, impact and management. Kluwer Academic Publishers, Dordrecht, pp 456–466

Hewitt CL, Martin RB (2001) Revised protocols for baseline port surveys for introduced marine species – design considerations, sampling protocols and taxonomic sufficiency. CRIMP technical report number 22. CSIRO Marine Research, Hobart

Hewitt CL, Willing J, Bauckham A, Cassidy AM, Cox CMS, Jones L, Wotton DM (2004a) New Zealand marine biosecurity: delivering outcomes in a fluid environment. N Z J Mar Freshw Res 38:429–438

Hewitt CL, Campbell ML, Thresher RE, Martin RB, Boyd S, Cohen BF, Currie DR, Gomon MF, Keough MJ, Lewis JA, Lockett MM, Mays N, McArthur MA, O'Hara TD, Poore GCB, Ross DJ, Storey MJ, Watson JE, Wilson RS (2004b) Introduced and cryptogenic species in Port Phillip Bay, Victoria, Australia. Mar Biol 144:183–202

Hilliard RW, Raaymakers S (1997) ballast water risk assessment for 12 ports in Queensland, Australia. In: Proceedings of Pacific Coasts & Ports' 97 conference. Centre for Advanced Engineering, Christchurch

IMO (2007) Guidelines for risk assessment under regulation A-4 of the BWM Convention (G7). Marine Environment Protection Committee, Resolution MEPC.162(56), 13 July 2007. International Maritime Organization, London

Kareiva P (1999) Developing a predictive ecology for non-indigenous species and ecological invasions. Ecology 77:1651–1652

Kelleher G, Bleakeley C, Wells S (1995) A global representative system of marine protected areas, vol 1–4. The Great Barrier Reef Marine Park Authority, The World Bank, The World Conservation Union (IUCN), Washington, DC

Kirchsteiger C, Christou MD, Papadakis GA (1998) Risk assessment and management in the context of the Seveso II Directive. Elsevier, Science & Technology, Elsevier, Amsterdam

Leppäkoski E, Gollasch S (2006) Risk assessment of ballast water mediated species introductions – a Baltic Sea approach. Report prepared for HELCOM, Helsinki

Leppäkoski E, Laine AO, Paavola M, Pertola S (2005) BITIS-project ("is the biological integrity of the Baltic Sea threatened by invasive non-native species?" funded by the Finnish Academy of Sciences for 2003–2005 within the Baltic Sea Research Programme BIREME). ICES 2005. Report of ICES/IOC/IMO Working Group on Ballast and Other Ship Vectors (WGBOSV). ACME:04, Arendal

Lloyd's Register (2007) Lloyd's Register – Fairplay. http://www.lrfairplay.com/default.html. Accessed 01 April 2007

Lodge DM, Williams S, MacIsaac HJ, Hayes KR, Leung B, Reichard S, Mack RN, Moyle PB, Smith M, Andow DA, Carlton JT, McMichael A (2006) Biological invasions: recommendations for U.S. Policy and management. Ecol Appl 16(6):2035–2054

Longhurst A (1998) Ecological geography of the sea. Academic, San Diego

Macdonald EM, Davidson RD (1997) Ballast water project, Final report. FRS Marine Laboratory, Aberdeen

MacIsaac HJ, Robbins TC, Lewis MA (2002) Modeling ships' ballast water as invasion threats to the Great Lakes. Can J Fish Aquat Sci 59:1245–1256

MacIsaac HJ, Borbely JVM, Muirhead JR, Graniero P (2004) Backcasting and forecasting biological invasions of inland lakes. Ecol Appl 14(3):773–783

McCollin TA, Shanks AM, Dunn J (2008) Changes in zooplankton abundance and diversity after ballast water exchange in regional seas. Mar Pollut Bull 56:834–844

Minchin D (2007) Aquaculture and transport in a changing environment: overlap and links in the spread of alien biota. Mar Pollut Bull 55:302–313

Minchin D, Gollasch S, Wallentinus I (2005) Vector pathways and the spread of exotic species in the sea, vol 271. ICES Cooperative Research Report, Copenhagen

Minchin D, Gollasch S, Cohen AN, Hewitt CL, Olenin S (2009) Characterizing vectors of marine invasion. In: Rilov G, Crooks JA (eds) Biological invasions in marine ecosystems, vol 204, Ecological Studies. Springer, Berlin, pp 109–116

Norton TM, Beer T, Dovers SR (eds) (1995) Risk and uncertainty in environmental management. In: 1995 Australian Academy of Science Fenner Conference on the Environment. Centre for resource and Environmental Studies, The Australian National University, Canberra, Australia

Olenin S, Gollasch S, Jonusas S, Rimkute I (2000) En-route investigations of plankton in ballast water on a ship's voyage from the Baltic to the open Atlantic coast of Europe. Int Rev Hydrobiol 85(5–6):577–596

Paavola M, Olenin S, Leppäkoski E (2005) Are invasive species most successful in habitats of low native species richness across European brackish water seas? Estuar Coast Shelf Sci 64:738–775

Pienimäki M, Leppäkoski E (2004) Invasion pressure on the Finnish Lake District: invasion corridors and barriers. Biol Invasion 6:331–346

Remane A (1934) Die Brackwasserfauna. Verh Deut Zool Ges 36:34–74

Remane A, Schlieper C (eds) (1958) Die Biologie des Brackwassers – Die Binnengewässer 12. Schweizerbart'sche Verlagsbuchhandlung, Stuttgart

Ruiz GM, Fofonnoff PW, Ashton G, Minton MS, Miller AW (2013) Geographic variation in marine invasions among large estuaries: effects of ships and time. Ecol Appl 23(2):311–320

Safriel UN, Ritte U (1980) Criteria for the identification of potential colonizers. Biol J Linn Soc 13:287–297

Safriel UN, Ritte U (1983) Universal correlates of colonizing ability. In: Swingland IR, Greenwood PJ (eds) The ecology of animal movement. Clarendon, Oxford, pp 215–239

Souvorov AV (1999) Marine ecologonomics. The ecology and economics of marine natural resources management. Elsevier, Amsterdam

Spalding MD, Fox HE, Allen GR, Davidson N, Ferdaña ZA, Finlayson M, Halpern BS, Jorge MA, Lombana A, Lourie SA, Martin KD, McManus E, Molnar J, Recchia CA, Robertson J (2007) Marine ecoregions of the world: a bioregionalization of coast and shelf areas. Bioscience 57:573–583

Springer VG (1982) Pacific Plate biogeography with special reference to shore fishes. Smithson Contrib Zool 376:1–182

Venice System (1959) The final resolution of the symposium on the classification of brackish waters. Arch Oceanogr Limnol 11(suppl):243–248

WGBOSV (2013) Report of the ICES/IOC/IMO Working Group on Ballast and Other Ship Vectors (WGBOSV) in Montreal, Canada, 18–20 March 2013. International Council for the Exploration of the Sea, Copenhagen, Denmark. ICES CM 2013/ACOM:29

WGITMO (2013) Report of the ICES Working Group on Introduction and Transfers of Marine Organisms (WGITMO) in Montreal, Canada, 20–22 March 2013. International Council for the Exploration of the Sea, Copenhagen, Denmark. ICES CM 2013/ACOM:30

Williams RJ, Griffiths FB, Van der Wal EJ, Kelly J (1988) Cargo vessel ballast water as a vector for the transport of non-indigenous marine species. Estuar Coast Shelf Sci 26:409–420

Ballast Water Sampling and Sample Analysis for Compliance Control

Stephan Gollasch and Matej David

Abstract In the past, the purpose of ballast water sampling studies was limited to general scientific interest, awareness raising or the determination of organism numbers per water volume. In this chapter we focus on compliance control sampling with BWM requirements as set out in the BWM Convention. Key aspects described are sampling methods and approaches to take a representative ballast water sample and the need for a harmonised sampling approach, to avoid that the ballast water of a vessel is proven compliant in one port, but would not be proven compliant in another port just because of different sampling methods or approaches used. In this chapter we describe suitable compliance control sampling methods and approaches and address both indicative and detailed sampling. Details on possible sampling access points, equipment and other details recommended for in-tank and in-line sampling are given. Further, recommendations are given how samples should be handled, including suitable sample transport and storage conditions. Another subject of this chapter addresses organism detection technologies for indicative and detailed sample analysis for compliance control with BWM standards. Suitable organism detection technologies are recommended in the end of the chapter.

Keywords Representative ballast water sampling • Indicative and detailed sampling • Sampling gear • Sampling access points • Compliance monitoring and enforcement • Sample handling • Sample transport • Sample storage • Sample analysis • Indicative analysis • Detailed analysis • Organism detection technology

---

S. Gollasch (✉)
Gollasch Consulting (GoConsult), Hamburg, Germany
e-mail: sgollasch@aol.com

M. David
Dr. Matej David Consult, Korte, Izola, Slovenia
e-mail: matej.david@siol.net

M. David, S. Gollasch (eds.), *Global Maritime Transport and Ballast Water Management*, Invading Nature - Springer Series in Invasion Ecology 8,
DOI 10.1007/978-94-017-9367-4_8

Introduction

The purpose of a ballast water sampling studies was so far limited to general scientific interest, awareness raising or the determination of organism numbers per water volume. Each study objectives triggers the use of different sampling gear and strategies. Further the high variability of organism type, size and behaviour observed in ballast water samples combined with the complexity of physical and chemical characteristics of ballast water implies to use a variety of sampling methods to catch all organisms present in the water (e.g., Medcof 1975; Hallegraeff and Bolch 1991; McCarthy and Khambaty 1994; Gollasch 1996; MacDonald and Davidson 1997; Hay et al. 1997; Sutton et al. 1998; Oemcke and van Leeuwen 1998; Lenz et al. 2000; Ruiz et al. 2000; Murphy et al. 2002; Gollasch et al. 2003; David and Perkovic 2004; David et al. 2007; IMO 2010a).

In this chapter we focus on compliance control sampling with the standards as set forth in the IMO BWM Convention. However, before we describe suitable compliance control sampling methods and approaches, we briefly address how different sampling results are when different sampling methods are used. This is addressed here to amplify the need for a harmonised sampling approach to avoid that the ballast water of a vessel is proven compliant in one port, but would not be proven compliant in another port just because of different sampling methods used.

During a comparison of sampling technologies previously used for ballast water studies it was observed how different the performance of the individual methods was. In this experiment it became clear that some methods over- and other methods undersample the organism concentration. Therefore, by the selection of the wrong sampling method samples may be wrongly identified as compliant with BWM standards although the organism concentration could have been above the compliance threshold.

Noting the influence on results by the use of different sampling methods and approaches a clear need for an international harmonised ballast water compliance control sampling approach was stated. Consequently the ballast water working group of IMO developed a sampling guideline, i.e., *Guidelines for Ballast Water Sampling* (G2 Guidelines), adopted in 2008. The G2 Guidelines and all additional BWS related developments at IMO prepared after the adoption of G2 were reviewed and taken into account during the preparation of this chapter. This refers especially to the:

- ***aide-memoire*** for the sample analysis developed at BLG13 (March 2009). BLG agreed that a detailed guidance document on sampling procedures could not yet be developed because of the lack of results of relevant sampling studies (IMO 2009). This *aide-memoire* is limited to different types of sample analysis, including recommendations for indicative and detailed sample processing.
- preparational documents of a BWS circular (e.g., IMO 2010a, b, c, d, e). This IMO Circular was aimed to provide guidance on BWS and sample analysis, including sample representativeness and sample analysis protocols.
- report of the Ballast Water and Biofouling Working Group from the last BLG meeting (BLG17, February 2013). It contains the draft BWS circular (i.e.,

Guidance to Ballast Water Sampling and Analysis for Trial Use in Accordance with the BWM Convention and Guidelines G2). The draft BWS circular was adopted at MEPC65, May 2013 (IMO 2013). It includes a list of sample analysis protocols, methodologies and approaches for D-1 and D-2 standards compliance tests as well as recommendations for a trial period during which sampling experience may be gained. It was agreed that this experience will be used to update the BWS circular accordingly. However, it is stated in the document that representative sampling methods are still developing (IMO 2013) so that no detailed requirement regarding the number of samples to take, or on their volume could be included.

Our sampling experiences contributed substantially to the BWS methods suggested here and were gained from the author's involvement in various national and international research studies, expert, scientific and/or governmental working groups or organisations (i.e., ICES/IOC/IMO Working Group on Ballast and Other Ship Vectors, Ballast Water Management Sub Commission for the Adriatic Sea, IMO/MEPC Ballast Water Working Group, Global Ballast Water Management Programme of GEF-UNDP-IMO, the European Maritime Safety Agency and the relevant national authorities). In these activities different aspects (i.e., biological, nautical, technical, logistical) of ballast water sampling were addressed.

We further like to highlight our findings and experiences from three studies on representative BWS for compliance monitoring, which were conducted for the Federal Maritime and Hydrographic Agency, Hamburg, Germany in 2009 and 2012, and for the European Maritime Safety Agency, Lisbon, Portugal in 2010 (Gollasch and David 2009, 2010a, b, 2013). This was in addition complimented by our sampling experience gained on more than 80 shipboard tests for the type approval of 18 different BWMS which were conducted since 2004. A comprehensive report, prepared by David (2013) for WWF International, provides recommendations for representative ballast water sampling for compliance control with the BWM Convention and was considered in detail when drafting this book chapter.

Sampling for Compliance Control

After the BWM Convention enters into force, IMO Member states will be required to check compliance of vessels with the standards of the BWM Convention, and one way of doing this is sampling ballast water. As per Article 9.1 *Inspection of Ships*, it is stated that ships to which the BWM Convention applies may be subject to inspections for the purpose of revealing possible violations of the provisions of the BWM Convention. These inspections shall:

- Verify that a valid Ballast Water Management Certificate is carried on board;
- Verify that a Ballast Water Management Plan is on board which is specific to the ship and is also approved by the Flag state;
- Undertake a review of the on board Ballast Water Record Book.

As part of the Port State Control efforts to demonstrate compliance with the BWM Convention standards, port authorities may consider sampling of ballast water for subsequent analyses. The ballast water sampling guidance approved by IMO as the Guidelines G2 is mainly limited to general information. In this book chapter we focus on the selection of appropriate sampling methodologies to assess compliance with the BWM Convention standards, i.e., the ballast water exchange (Regulation D-1) and performance (Regulation D-2) standards.

If the sampling event has to demonstrate D-2 standard compliance, then a numerical documentation of viable organisms greater than or equal to 50 μm in minimum dimension is especially challenging because only less than 10 viable organisms per 1 m^3 of water are acceptable in the discharged ballast water. It becomes obvious that various difficulties can occur including to collect more than 1,000 l of water to proof compliance, and several replicates may have to be sampled to meet general scientific standards and accuracy levels. Further, the accuracy of the sampling technique used must be validated because inefficient sampling techniques may result in inappropriate results.

Different vessel specifics, considering vessel types, sizes and cargo profiles, result in very different ballast water discharge profiles and times. Ballast water may be discharged "at once" or "in sequences" which may last from approximately 1 h, e.g., emptying of two tanks in parallel on a container vessel, up to several days for larger bulk carriers or tankers. The tank discharge duration is also depending on the length of the cargo operation, e.g., tankers, bulk carriers, very large container vessels, and sometimes also general cargo vessels, load cargo over up to several days duration. Therefore, the ballast water operation is frequently conducted in sequences over time until the cargo operation is completed. This factor is important to be considered when planning a sampling event as it is difficult to assume that the PSC officer and/or sampling team will stay on board a vessel for several days.

Sampling Methods According to the Sampling Access Point

Ballast water sampling access points may be divided into in-tank and in-line (at discharge) sampling points. In-tank sampling points enable ballast water access directly from a tank and this may be achieved either via opened ballast tank manholes, sounding or air pipes. In-line sampling points are located in the ship's pipe work, preferably after the ships ballast water pumps.

For D-1 standard compliance monitoring in-tank or in-line samples may be taken to either proof the presence of coastal biota or for water salinity checks. This may be done by utilizing all possible sampling access points including sounding pipe, manhole and the vessels main ballast water line. However, the latter is not recommended to be used because a discharge to sea may most likely occur in this sampling approach and in case of non-compliance the non-complaint water would be pumped into the recipient environment during the sampling event. As the D-1 standard is not a numerical organism standard quantitative biological sampling is not needed.

In contrast, compliance control for the D-2 standard, which is a numerical, biological, discharge standard, samples should be taken from the ballast water discharge line (but see below). Here, a quantitative biological approach is needed as the numerical standard refers to viable organisms above 10 μm in minimum dimension no matter what type they are. In contrast, for the indicator microbes as stated in the D-2 standard, both qualitative and quantitative approaches are needed so that the concentration of colony forming units of certain indicator microbes can be documented. Although it seems that in-line sampling is the most appropriate way to assess compliance with the D-2 standard, this view changes in cases when the ballast water originates from a high risk area, i.e., an area with a known occurrence of target species. In these situations samples may preferably be taken from the ballast tank prior discharge, which would make non-compliance actions possible before the water is discharged into the recipient environment. The in-tank approach is also advisable for D-2 standard compliance checks for those tanks which have direct discharge to sea, e.g., top-side tanks on some bulk carriers.

Tank Selection – Which Tank to Sample (First)?

Vessels may have ballast water on board to be discharged which originates from different sources and also with different uptake dates (holding time on board). Ballast water from all different sources might need to be tested. If this is impossible or in cases of a need to have results as soon as possible, possibly even prior any discharge, tank(s) to be sampled first should be selected based on a risk assessment approach. This risk assessment will focus to identify which ballast water may contain potentially harmful species for the recipient port.

Such a risk assessment may consider the following elements, but may not be limited to:

- the environmental compatibility of both, the ballast water source area and the ballast water recipient area;
- the presence of harmful aquatic organisms and pathogens (HAOP) in the area of ballast water origin;
- if appropriate, the presence of target species in the area of ballast water origin; and
- the duration of the in-tank holding time.

The tank(s) with higher environmental compatibility of the origin and discharge area, tank(s) filled in a ballast water origin area where HAOP or target species are present, and tank(s) with shorter in-tank holding time should be given priority for ballast water sampling because these would likely pose the highest risk to introduce HAOP.

When in-tank sampling methods are applied, the ease of the sampling access point may be used as an additional criterion to identify the tank(s) to be sampled, also considering that some tanks may not be accessible at all for in-tank sampling (David and Perkovič 2004).

For in-line sampling the tank(s) which is(are) currently discharged when PSC comes on board may be prioritized for sampling, to avoid that PSC has to wait many hours until the "targeted tank" is ready to be discharged.

In general, tanks currently being discharged or those to be discharged first may be sampled first as this would give the opportunity to PSC officers to decide for appropriate management measures in cases of non-compliance, or in cases where an indication of possible non-compliance is identified.

Sampling for Compliance Control with the D-1 Standard

Salinity measurements of ballast water may be used to verify if the water was exchanged according to the BWM Convention requirements (D-1 standard). Should the measured ballast water salinity be low, e.g., below 30 psu, it can with a high level of confidence be assumed that the ballast water originates from coastal areas with freshwater influence. This means the water was not exchanged with ocean water as required, i.e., outside 50 or 200 nautical miles from nearest land and at water depths higher than 200 m, because this water would clearly have a higher salinity. Consequently, with this D-1 standard compliance control option non-compliance would be assumed when the inspected vessel has loaded ballast in a lower salinity or freshwater port.

The D-1 standard requires that at least 95 % of the water needs to be exchanged. Therefore up to 5 % of water may remain in the ballast tank unexchanged. When a vessel has taken up ballast water in a freshwater port (100 % tank volume) and 95 % are exchanged in mid-ocean, the possibly remaining 5 % freshwater in the tank will dilute the salinity of the ocean water taken up during the exchange. As a consequence this salinity dilution may result in a false non-compliant indication in cases when the remaining freshwater from the previous tank filling would be ignored. From our on board studies we know that sometimes more than 5 % of water remain as unpumpable ballast inside ballast tanks. This depends, e.g., on the vessels trim and tank design. We observed salinity differences of ca. 4 psu when a ballast tank was filled in Hamburg (freshwater) and the water from this tank was exchanged with marine water according to the depth and distance requirements as stated above, i.e. the freshwater from the previous tank filling "diluted" the marine water during the water exchange.

When seasonally averaged the lowest ocean salinity is ca. 30 psu (see Fig. 1). Consequently, ballast water salinities below 30 psu likely indicate that the exchange occurred less than 50 nautical miles from the nearest land with influence of freshwater from nearby rivers or estuaries, because otherwise the salinity should be higher. Therefore a low salinity measurement indicates that the ballast water was exchanged closer to land than required.

In an experiment we have shown that the salinity of the ballast water in a ballast tank was not homogenous when the salinity was measured over different depths of a sounding pipe. The deeper the salinity sensor was lowered in the sounding pipe

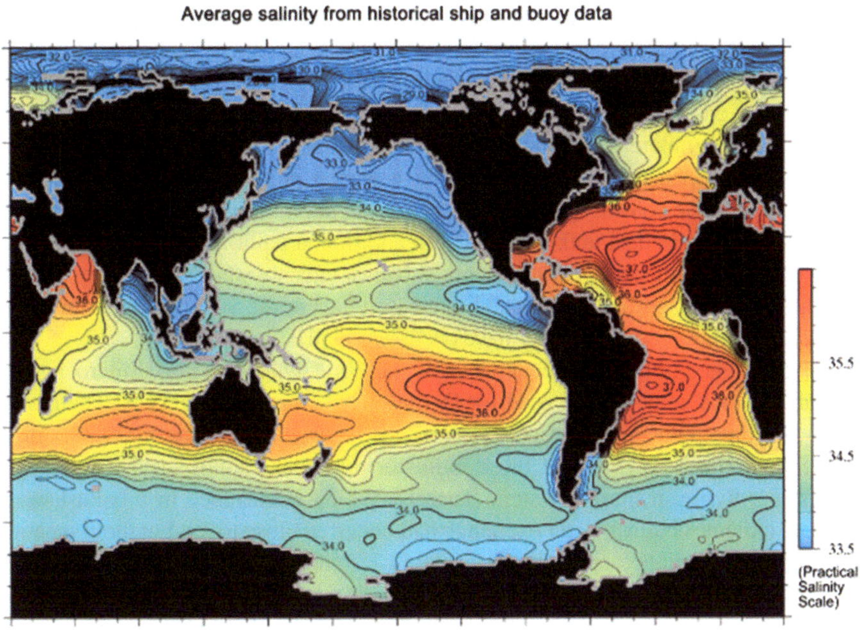

Fig. 1 World ocean's surface salinity (Source: Antonov et al. 2010, NOAA National Oceanographic Data Center)

the higher was the salinity. Further, during ballast water discharge of this vessel the salinity was measured over time. Over time, i.e. towards the end of the tank emptying process, the salinity value increased (Gollasch et al. 2012).

This is a clear indication that salinity sampling for compliance control is not trivial and that more than one sample may need to be taken to cover potentially varying salinities of ballast water in the same tank. This may either be done by lowering a salinity meter to different depths in a ballast tank or by taking multiple samples from the ships ballast water discharge pipe during the water discharge of the tank.

Documenting the presence of coastal biota as a D-1 standard compliance control test is also of a limited value as very few organisms are restricted in their occurrence to coastal waters alone. Candidate species for this approach include, e.g., harpacticoid copepods and barnacles. Most barnacles, but not all, are fouling species found along the tidal zone of hard-bottom shores. Therefore their presence in a ballast water sample could indicate coastal origin of the water. However, barnacles are also frequently found in the biofouling of vessels. It is theoretically possible that two vessels follow each other closely and the barnacles in the hull fouling of the first vessel may release their larvae into the water and these larvae may be pumped into a ballast tank of the second vessel during a ballast water exchange operation. In this scenario the exchanged ballast water, although exchanged in mid-ocean, may contain barnacle larvae from the hull fouling of the first vessel thereby indicating coastal origin. Adult barnacles also occur inside ballast tanks in the fouling on the tank walls (Gollasch and David, own observation), they may reproduce inside the tanks and release their larvae into the ballast water. The presence of barnacle larvae originating from in-tank reproduction may then wrongly be assumed as originating from a ballast water exchange in coastal areas. Although this scenario may be of very low probability, but it cannot be completely excluded. The other organism group mentioned here, adult harpacticoid copepods, are benthic species and their presence in ballast water may therefore indicate coastal water origin . However, adult harpacticoids were also frequently found in ballast tank sediments. In addition, a ballast water sampling study with daily sampling events of the identical ballast tank showed that the harpacticoid copepod numbers in that ballast tank increased during the voyage which indicated that an in-tank reproduction may have occurred (Gollasch et al. 2000).

The presence of human faecal bacteria, such as *Escherichia coli*, Enterococci or *Vibrio cholerae* may also be used for D-1 standard compliance checks. Their presence in water indicates improper waste water management along the coasts in or close to urban areas. Most of these indicator microbes cannot survive for longer times free living in marine waters. Therefore, their occurrence in a ballast water sample indicates ballast water exchange had occurred close to land without meeting the depth and distance requirement of the D-1 standard. These indicator microbes are unlikely to survive a (longer) vessel voyage in a ballast tank outside their human (or other) "hosts" and therefore this method seems less reliable for longer voyages.

As shown above, the analysis of the biota for D-1 standard compliance control delivers results only with a limited level of certainty. Therefore, in cases when non-compliance is indicated by these methods, we believe that these data are not robust enough to justify a non-compliance action with all its logistical, costs and legal implications.

Another option for D-1 standard compliance control may be to document tracers of human influence on the sea. It seems to be logical that human influence is greater in near shore regions compared to the high seas. Candidate methods include to document Nitrogen or Phosphorous levels of which high concentrations may result from river run-offs in areas with human settlements, but this method may deliver regionally very different results. In several oceanic regions, especially in oceanic island states or in coastal environments with low human populations, the Nitrogen or Phosphorous levels may be very low even close to shore. Although these shortcomings are known, it was concluded that the absence of trace elements may be used to identify the oceanic origin of ballast water thereby evaluating whether or not ballast water was exchanged at sea (Murphy et al. 2002, 2004; Hunt et al. 2007). Consequently a tracer detection tool was developed for compliance control (Murphy et al. 2008).

Murphy et al. (2006) suggested also that fluorescence may be used to verify ballast water exchange for most samples of high salinity ballast water, but water contamination with, e.g., fuel oil, may influence the measurements.

Other instruments may be used to measure the characteristics and concentration of chromophoric dissolved organic matter (CDOM) in water (Murphy et al. 2008). CDOM is a result of the decay process of (terrestrial) plants and it is believed that higher CDOM concentrations indicate near shore ballast water exchanges.

Because of all limitation as stated above, biological sampling is not recommended for D-1 standard compliance control. The non-biological methods also seem to generate results with low confidence levels so that only the measurement of ballast water salinity seems to be a pragmatic option to check the compliance with the D-1 standard. However, as explained above, all vessel which load ballast in a marine port would always be in compliance when using this method. Therefore, salinity D-1 standard compliance checks should be done together with checking the vessels logbook and the salinity in the ballast water uptake port(s).

In-Tank D-1 Standard Sampling

Selection of Ballast Water Sampling Equipment and Methods for In-Tank D-1 Standard Sampling

As suggested here, the in-tank D-1 standard compliance control should be limited to a non-biological analysis of the water and salinity measurements seem to be a good parameter for this analysis. Already a very small volume of water is sufficient for this measurement.

The water used for salinity measurements may be collected by a water column sampler or a pump. Alternatively the sensor of a salinity meter may also be lowered into the tank or sounding pipe to enable direct measurements (see Table 1). A detailed description of the sampling equipment and sampling arrangements is given in section "Sampling Equipment and Sampling Point Arrangements".

Measuring the water salinity from ballast tanks is not as trivial one may assume. Experiments have shown that salinity measurements via sounding pipes, with separate readings in different heights, resulted in different salinity values over the length of the sounding pipe. The deeper the salinity meter was lowered down the sounding pipe the higher was the salinity value (Gollasch et al. 2012). This is a reasoning for our recommendation that more than one sample should be taken to cover potentially varying water salinities in the same ballast tank. This may either be done by lowering the salinity sensor to different water depths or by operating a water column sampler to sample different water depths.

Table 1 Sampling access points, equipment and other details recommended for compliance control sampling with the D-1 standard (Modified after David 2013)

Sampling point	Equipment	Water volume	Number of samples
Sounding pipe, manhole or air vent	Water column sampler or pump	ca. 50 ml	1 integrated sample from possibly whole water column
Sounding pipe, manhole or air vent	Point-source sampler or pump	ca. 50 ml	1 integrated sample from 3 different depths

Description of Sampling Equipment and Methods for In-Tank D-1 Standard Sampling

Water Column Sampler

In order to obtain an integrated sample from the whole water column, a water column sampler may be lowered at best to the bottom of the tank or to the deepest point accessible. During lowering the water column sampler, water will enter the sampler from an opening at the top. The water will be proportionally sampled from the entire water column provided the sampler is lowered with a constant speed through the water column. As a relatively low water volume needs to be sampled, i.e., ca. 50 ml, the water column sampler may need to be lowered down only once.

Point-Source Sampler

To take salinity samples from different depths, a point-source sampler has to be lowered three times to different depths, i.e., the surface, somewhere in the middle of the water column, and possibly close to the tank bottom or to the deepest accessible point. Each time the valve of the sampler is opened, which is done by pulling the rope connected to a valve at the bottom of the sampler, the water can enter the sampler. The valve can be closed again when the rope is relaxed, and then the sampler is pulled up. We recommend that the three samples from different depths are mixed and one salinity value is measured. As relatively low water volumes have to be sampled, i.e., ca. 50 ml, the point-source sampler may need to be lowered only once per each desired sampling depth, all together three times.

Pump

A pump can be used to receive an integrated sample from three different water depths or even from the whole water column. The pump itself, or the suction opening of the hose connected to the pump, may be lowered to three desired water depths, i.e., the surface, somewhere in the middle of the water column, and possibly close to the tank bottom or to the deepest point accessible. From each depth water can be pumped up. Alternatively, when lowering the pump or the suction opening of the hose, water may be pumped up constantly from the top surface water to the deepest point accessible. In this approach the limiting factor to be considered to retrieve a sample is the pumping head. As relatively low water volumes are to be sampled, i.e., ca. 50 ml, only a very short pumping time is needed to get a sample from each of the three desired water depths. This is the same in case the water is pumped up is constantly when lowering the pump to the deepest point accessible in the tank.

In-Line D-1 Standard Sampling

Selection of Ballast Water Sampling Equipment and Methods for In-Line D-1 Standard Sampling

In-line ballast water sampling for the D-1 standard seems unlikely to happen as vessels without BWMS installed lack in-line sampling points. Nevertheless, in case a sampling point is available somewhere in the line or a tap is available at the ballast pump, this could be the chosen sampling approach, especially in case when the ballast water discharge was already started and is ongoing.

However, D-1 standard sampling may also be used for an early indication of potentially non-compliant ballast water with the aim to apply appropriate management measures, but one key problem remains. This is that compliance or non-compliance can in this way only be proven while the ballast water is pumped overboard. Therefore, in cases when a risk assessment identifies ballast water as of high risk, in-line sampling during discharge should be avoided, but in-tank sampling should be undertaken for a compliance check (see above).

The suggested sampling method and equipment is described in Table 2 and for a detailed description of the sampling equipment and sampling arrangements please refer to section "Sampling Equipment and Sampling Point Arrangements".

As some salinity stratification may occur in the tank, multiple measurements, e.g., one in the beginning, one in the middle, and one in the end of the discharge, could identify such differences. However, such a measurement approach could be impractical especially when ballast water is discharged over longer times. Secondly, this approach would also be inappropriate in the interest to have the result as soon as possible before all possibly non-compliant ballast water is discharged from the vessel.

Description of Sampling Equipment and Methods for In-Line D-1 Standard Sampling

For in-line D-1 standard compliance tests a small sample bottle is sufficient. In cases a conductivity meter will be used, it is recommended to chose sample collection bottles with a wider opening that the conductivity sensor can be inserted to the sample right away.

Table 2 In-line sampling equipment and other details recommended for compliance control sampling with the D-1 standard (David 2013)

Sampling point	Equipment	Water volume	Number of samples
In-line	Sampling bottle	ca. 50 ml	1 sample as soon as possible during the discharge

Sampling for Compliance Control with the D-2 Standard

Compliance control with the D-2.1 standard is purely quantitative, thereby ignoring the types of organisms with the exception of the indicator microbes (D-2.2 standard). The numbers of viable organisms per size class document (non-)compliance. As the BWM Convention in Regulation D-2 reads "Ships conducting Ballast Water Management in accordance with this regulation shall discharge less than 10 viable organisms per m^3 ..." the D-2 standard is clearly understood as a discharge standard. This indicates that the most appropriate sampling point to proof D-2 standard compliance has to be installed in the discharge line of the vessels ballast water system. This is also recommended in the Guidelines G2, i.e., "samples should be taken from the discharge line, as near to the point of discharge as practicable, during ballast water discharge whenever possible."

However, in-tank sampling should also be considered as a valid option for D-2 standard compliance control. This is because some vessels, e.g., certain bulk carriers and tankers, may have upper side wing tanks which are emptied via direct overboard discharge valves and not by using ballast pumps and pipework (see Fig. 2). In such cases, the Guidelines G2 indicate that in-tank sampling may be an appropriate approach.

Fig. 2 Ballast water discharge above pier level from the upper wing tanks of a bulk carrier (Photo: Jure Barovič, Port of Koper)

Indicative Sampling for Compliance Control with the D-2 Standard

The indicative sample analysis is addressed in the Guidelines G2. Paragraph 6.3 reads: "Prior to testing for compliance with the D-2 standard, it is recommended that, as a first step, an indicative analysis of ballast water discharge may be undertaken to establish whether a ship is potentially compliant or non-compliant. Such a test could help the Party identify immediate mitigation measures, within their existing powers, to avoid any additional impact from a possible non-compliant ballast water discharge from the ship."

For ballast water sample analyses, certainly, as a very first step, a sampling event needs to be performed, but Guidelines G2 do not address explicitly how an indicative sampling event would need to be conducted. Implicitly, an indicative analyses could be performed with a sample, or a part of it, which was taken during the detailed D-2 standard compliance control sampling process, or just on any standalone sample.

One important point to note is that an indicative sampling event may be targeted towards only one group of organisms addressed by the D-2 standard. The results from each of these organism groups alone may already be taken as an indication that a BWMS is not performing properly. From the author's experience of on-board type approval sampling of BWMS, it is likely that, indicator microbes and organisms less than 50 μm in minimum dimension and greater than or equal to 10 μm in minimum dimension meet the D-2 standard, but it was observed that at the same time organisms greater than or equal to 50 μm in minimum dimension may be found in too high concentrations so that the acceptable organism numbers in the D-2 standard are exceeded for this organism group.

Different groups of organisms in general require different sampling approaches. In general organisms greater than or equal to 50 μm in minimum dimension require larger water volumes to be sampled to collect them compared to organisms less than 50 μm in minimum dimension and greater than or equal to 10 μm in minimum dimension. This is because there are usually lower concentrations of larger organisms in the water as for smaller organisms. Consequently, indicative sampling methods may be very different for each organism group, and may differ in, e.g., sampling duration, timing, volume, and further in the recommendation which sampling point is to be used.

Without a known performance history of a certain BWMS it is very difficult to predict in advance which group of organisms should be considered to identify possible non-compliance with the D-2 standard. Not knowing this in advance it would be most effective to use a sampling method which enables an analyses of all organism groups. This would also enable a step-by-step process of consequtive analysis methods. One analysis method for one organism group may be applied first, and in cases this shows an indication or even does not give an indication of non-compliance, the second organism group may be tested next with another sample analysis method.

Another issue to consider are the consequences which may arise from an indicative sample analysis. Does an indicative D-2 standard test result trigger the need for further tests, i.e., a detailed D-2 standard compliance test? Or should a vessel be banned from discharging ballast water solely based on the indicative test result? Based upon paragraph 6.3 of the G2 Guidelines, it is understood that an indicative analysis was included to give a Party an opportunity to identify potential non-compliant ballast water in an early stage, i.e., the detailed compliance test may show results only after all ballast water was already discharged, the indicative analysis was implemented to avoid any impact from a possibly non-compliant ballast water discharges from a ship.

Indicative In-Tank Sampling for Compliance Control with the D-2 Standard

Sampling will likely be conducted on a number of different ship types in a port. After ship selection according to the sampling program, i.e., targeting of vessels based on PSC, the ballast tanks for sampling are to be selected. Hereby the sampling access plays a crucial role and determines if the ballast water is available for sampling at all. Therefore, a flexible approach with sampling equipment suitable to be used via various sampling points, is in most cases crucial to obtain a sample.

Another aspect is the need for in-tank sampling. On certain occasions it may be appropriate to avoid taking a sample from the ballast water discharge line as G2 recommends, i.e., during the discharge overboard. This refers to cases when it is known that a vessel carries ballast water from areas with documented outbreaks, infestations, or populations of HAOP, e.g., toxic algal blooms. Here the sampling during the overboard discharge should be avoided and a risk assessment should be used to identify high risk ballast water. Such ballast water discharges overboard would pose a risk to the environment, human health, property or resources. Instead it is recommended that in these cases it is preferred to take an indicative ballast water sample directly from the tank prior any ballast water discharge. Although such a sampling event may not be representative of the whole discharge, it enables an indicative compliance control test without taking the risk to discharge this high risk ballast water into the environment.

Selection of Ballast Water Sampling Equipment and Methods for Indicative In-Tank D-2 Sampling

Studies have shown that sampling for zooplankton via the sounding pipes does not result in a representative sample. Comparisons of sounding pipe and manholes samples taken simultaneously from the same tank found that net samples contained a higher biological diversity. Sounding pipe samples contained only 0–60 % of the organisms of a net sample, which indicates the need to sample ballast tanks via opened manholes. It was an interesting observation that pumps operated via open

Table 3 Possible sampling access points, equipment and other details recommended for indicative compliance control in-tank sampling with the D-2 standard (David 2013)

Organism group	Sampling point	Equipment	Water volume [litre]	Number of samples
>50 μm	Manhole	Plankton net	300–500	1 integrated sample from possibly the whole water column
	Manhole, sounding pipe or air vent	Pump	100	1 integrated sample from possibly the whole water column or from 3 different depths
<50 and >10 μm	Manhole, sounding pipe or air vent	Pump or water column sampler	5–6	1 integrated sample from possibly the whole water column or from 3 different depths
	Manhole, sounding pipe or air vent	Pump or point-source sampler	5–6	1 integrated sample from 3 different depths
Indicator microbes	Manhole, sounding pipe or air vent	Pump or water column sampler	1	1 integrated sample from possibly the whole water column
	Manhole, sounding pipe or air vent	Pump or point-source sampler	1	1 integrated sample from 3 different depths

manholes delivered more diverse samples compared to net samples. Therefore pumps may also be considered for manhole sampling. As a result, future indicative compliance control in-tank studies should note that sampling via sounding pipes is inferior when selecting appropriate sampling techniques. However, in many occasions manholes cannot be opened for sampling due to, e.g., overlaying cargo or on-going cargo operations in the area where the manhole is located. In these instances sounding pipe sampling might be the only solution to sampling.

The suggested in-tank sampling methods and equipment are outlined in Table 3. For a detailed description of the sampling equipment and sampling arrangements note section "Sampling Equipment and Sampling Point Arrangements".

Description of Sampling Equipment and Methods for Indicative In-Tank D-2 Sampling

For organisms greater than or equal to 50 μm in minimum dimension it is suggested to use a plankton net because of the ease to filter the suggested 300–500 l water volume. As a second choice a pump may be used to pump up ballast water and to filter 100 l. In case a tank would only be accessible through a sounding or air pipe, a pump would probably be the only appropriate choice to get a sample for the purpose of D-2 standard compliance sampling.

For compliance checks with organisms less than 50 μm in minimum dimension and greater than or equal to 10 μm in minimum dimension, and for indicator microbes only small water quantities need to be sampled. This can be achieved by using a pump or different water samplers. In case the pump is already chosen to sample for organisms greater than or equal to 50 μm in minimum dimension, where 100 l may be filtered, at the same time additional 5–6 l of a continuous drip sample (see section "Recommended Sample Quantity") may be collected in a bucket for checking organisms less than 50 μm in minimum dimension and greater than or equal to 10 μm in minimum dimension and for indicator microbes. The recommended indicator microbes sample of 1 l may be subsampled from this bucket.

Plankton Net for In-Tank Sampling

For in-tank sampling D-2 standard compliance control sampling it is recommended to lower the net to the maximum accessible tank depth, wait a minute and retrieve the net by hand with an approximate speed of 0.5 m per second. The plankton net cod-end should be emptied into a sample bottle and the process be repeated to meet the desired water volume for sampling.

Pumps for In-Tank Sampling

A pump can be used to obtain an integrated sample from three different depths or alternatively from the entire water column. The pump should be lowered to three desired depths, i.e., the surface, somewhere in the middle of the water column, and possibly close to the tank bottom or to the deepest point accessible. From each depth water needs to be pumped up. Alternatively, when lowering the pump, water may also be pumped out constantly from the surface water to the deepest point accessible. The limiting factor to be considered is the pumping head, which may not enable to pump up water from greater depths. From each of the three desired depths or constant pumping may be used when lowering the pump to the deepest point accessible in the tank.

Water-Column Sampler for In-Tank Sampling

To obtain an integrated sample from the whole water column, the water column sampler is lowered to the bottom of the tank or to the deepest point accessible. When lowering the water column sampler, water will start entering the sampler from its top opening. The water will be proportionally sampled from the entire water column provided (a) the sampler is lowered with a constant speed, and (b) the time used to lower the sampler from the surface to the deepest tank point accessible is the same as the time needed to fill the sampler with water. The water column sampler may need to be lowered down multiple times until the desired water volume is met.

Point-Source Sampler for In-Tank Sampling

To obtain three samples from different depths, the point-source sampler is lowered three times to the desired depths, i.e., the surface, somewhere in the middle of the water column, and possibly close to the tank bottom or to the deepest point accessible. Each time the valve of the sampler is opened by pulling the rope which is connected to the valve at the bottom of the sampler allowing the water to enter the sampler. The valve is closed again when the rope is relaxed, and then the sampler is pulled up. The three samples are integrated (mixed together) and one subsample is taken for subsequent analysis. As a relatively low water volume is to be sampled, i.e., a few litres, the point-source sampler may need to be lowered only a few times per the desired depth.

Indicative In-Line Sampling for Compliance Control with the D-2 Standard

Selection of Ballast Water Sampling Equipment and Methods for Indicative In-Line D-2 Sampling

Considering the above and especially as in-line sampling may develop also for detailed D-2 standard compliance checks, we recommended for an in-line indicative ballast water sampling event that one sequential sample is taken using the same sampling methodology as for a detailed D-2 standard compliance test (see sections "Detailed In-Line D-2 Standard Sampling" and "Recommendations for a Ballast Water Sampling Protocol that Is Representative of the Whole Discharge").

For taking one sequential sample a relatively short sampling time is sufficient. The sample analysis may be conducted with a variety of different methods (see below). The results generated this way may also represent very solid grounds for different non-compliance actions which PSC may have available. These include:

- the requirement that, when in doubt, more comprehensive tests are needed and to proceed to a detailed compliance D-2 standard check,
- redirecting the vessel to a designated ballast water discharge area,
- to require ballast water discharges to a port reception facility, or even
- to ban the vessel from further ballast water discharges.

Which (non-)compliance action to take depends on the sampling results obtained. For instance, should the organism concentration identified just be above the D-2 standard, this may possibly indicate that further checks are required. In contrast, should the organism concentration be much higher than the D-2 standard, i.e., gross exceedence, non-compliance actions as listed above may instantly apply.

The suggested sampling methods and equipment are outlined in Table 4.

Table 4 Possible sampling access points, equipment and other details recommended for indicative in-line compliance control sampling with the D-2 standard (David 2013)

Organism group	Sampling point	Equipment	Water volume [litre]	Number of samples
>50 µm	In-line	Plankton net	300–500 in each sequence	1 sequential sample of ca. 10 min duration, avoiding the very beginning and very end of the tank discharge event
<50 and >10 µm	In-line	Bucket	5–6	1 continuous drip sequential sample, may be simultaneously collected during sampling of organism group > 50 µm
Indicator microbes	In-line	Bucket	1	1 continuous drip sequential samples, may be sub-sampled from the bucket

Description of Sampling Methods for Indicative In-Line D-2 Sampling

We recommend that for an in-line indicative ballast water sampling event, one sequential sample is taken using the same methodology as for a detailed D-2 standard compliance test (see sections "Detailed In-Line D-2 Standard Sampling" and "Recommendations for a Ballast Water Sampling Protocol that Is Representative of the Whole Discharge").

Detailed Sampling for Compliance Control with the D-2 Standard

Detailed In-Tank D-2 Standard Sampling

Because D-2 is a discharge standard, in-tank sampling for a detailed analysis is of limited value. However, in cases when in-tank sampling reveals very high organism numbers, non-compliance may be assumed also when this ballast water is discharged. To illustrate this, should a sample from the tank contain 1,000 viable organisms greater than or equal to 50 µm in minimum dimension and the tank capacity is 100 t, the organism concentration would exceed the D-2 standard when the water is discharged.

Detailed In-Line D-2 Standard Sampling

Selection of Sampling Equipment and Methods for Detailed In-Line D-2 Standard Sampling

Our previous on board ballast water sampling studies on commercial vessels have shown that different sampling approaches, i.e., short/long sampling times, result in different concentrations of viable organism (Gollasch and David 2009, 2010a, b, 2013).

Consequently the selection of inappropriate sampling approaches will influence the compliance control result. In consequence, the concentration of viable organisms in the ballast water discharge may be underestimated, so that ballast water managed with an inefficient BWMS could be recognised as compliant. On the other hand, concentrations of viable organisms may also be overestimated, and ballast water complying with the D-2 standard may fail in compliance tests.

Sequential ballast water sampling trials documented different organism numbers in each sequence of one test of the identical ballast water tank which indicates the patchy organism distribution inside the tank. This was observed during all sampling events conducted by Gollasch and David (2009, 2010a, b, 2013) and for both organism groups studied. It was therefore concluded that sampling during ballast water discharge is biased by tank patchiness of organisms.

For organisms greater than or equal to 50 μm in minimum dimension our previously undertaken studies have shown that the samples taken over the entire discharge time of a tank contained much lower concentrations of viable organism compared to the organism count in the sequences. It was therefore concluded that sequential sampling may deliver more representative results. Comparative studies have further shown that sequential samples of approximately 10 min duration are suitable for in-line D-2 standard compliance tests. In most of our tests the highest count of viable zooplankton organisms was found in the last sequence so that sampling at this time may "oversample" the real organism concentration. It was also observed that sequential samples taken in the very beginning and very end during a ballast tank is emptied are unlikely to provide representative results of the viable organism concentrations because in these samples the organism count showed very high variations. This could result in an under- or oversampling the organism concentration. Even when avoiding these time windows for sampling the concentration of viable organisms still seems to be patchy and we therefore recommend to take at least two sequential samples but excluding the very beginning and the very end of the pumping times when a ballast tank is emptied. The mean value of the viable organism concentration in these two sequential samples may be taken to assume the real organism concentration.

In the group of organisms less than 50 μm in minimum dimension and greater than or equal to 10 μm in minimum dimension counts of viable individuals in sequences when compared to the samples taken over the entire time of a tank discharge showed lower viable organism concentrations in the sequential samples, which is in contrast to the larger organism results. Comparisons of the smaller organism concentration between the different sequences of all tests showed that no clear trend can be identified during which time window a more representative sample may be taken. It is therefore recommended to take at least two sequential samples during the discharge of a ballast water tank but avoiding sampling times during the very beginning and end of the discharge of a tank or tanks. The mean organism count in these two or more sequential samples may be seen as the real viable organism concentration in the ballast water.

Suggested methods and equipment are outlined in Table 5 and for a detailed description of the sampling equipment and sampling arrangements see section "Sampling Equipment and Sampling Point Arrangements".

Table 5 Possible sampling access points, equipment and other details recommended for detailed in-line compliance control sampling with the D-2 standard (David 2013)

Organism group	Sampling point	Equipment	Water volume [litre]	Number of samples
>50 μm	In-line	Plankton net	300–500 in each sequence	2 (or more) sequential samples of ca. 10 min duration each, avoiding the very beginning and very end of the tank discharge event
<50 and >10 μm	In-line	Bucket	5–6 in each sequence	2 (or more) continuous drip sequential samples collected at the same time as for organism group >50 μm
Indicator microbes	In-line	Bucket, sampling bottle	1 in each sequence	2 (or more) continuous drip sequential samples sub-sampled from the bucket

Recommendations for a Ballast Water Sampling Protocol that Is Representative of the Whole Discharge

As stated above, the results from our previous ballast water sampling studies showed that different sampling approaches influence the results regarding organism concentrations (Gollasch and David 2009, 2010a, b, 2013). The organisms are potentially affected by the approach chosen, so that the selection of an inappropriate sampling approach may have an influence on the compliance control sampling result. Consequently, the organism concentrations in the ballast water discharge may be underestimated so that ballast water managed with an underperforming BWMS could falsely become recognised as compliant. In contrast organism concentrations may also be overestimated and a BWMS who's application results in ballast water to comply with the D-2 standard may fail a compliance test.

We observed that a certain level of pragmatism is required during on board ballast water compliance control sampling because the work is not undertaken under controlled laboratory conditions. In any case, all attempts should be made to avoid negative impacts of organism survival during the sampling process. This is especially relevant for organisms greater than or equal to 50 μm in minimum dimension. PSC are unlikely to have available larger water collecting tanks, e.g., >500 l, during the sampling event and will therefore likely need to work with nets to concentrate the ballast water sample during the sampling procedure. Guidelines G2 also address these aspects: "sampling should be undertaken in a safe and practical manner; and samples should be concentrated to a manageable size".

We observed that the main factors to influence viable organism concentrations results include sampling duration, i.e., length of the sampling process, the timing, i.e., the point in time during the ballast water discharge when the sampling is conducted, the number of samples and the water quantity sampled (Gollasch and David 2009, 2010a, b, 2013).

Recommended Sampling Duration

The organisms greater than or equal to 50 μm in minimum dimension are negatively affected by longer sampling times as sampling study results have documented. The findings of our studies also showed that shorter sampling times result in representative samples, so that we recommend a sampling time for a sequential sample of approximately 10 min. It was concluded that longer sampling times will likely result in an underestimation of the viable organism concentration in the ballast water discharged. This is especially the case for organisms greater than or equal to 50 μm in minimum dimension.

Recommended Sampling Timing

We documented that organism concentrations vary considerably if the sampling event is conducted at the very beginning or at the very end of the ballast water discharge process because at these times the patchy organism distribution inside ballast water tanks was the greatest indicating that organism are not homogenously distributed inside the tank. Therefore, it is recommended to avoid taking a sample at the first 5 min or at the last 5 min of the ballast water discharge event because an under- or overestimation of organism concentrations may have to be expected. It is therefore recommended that the sampling is conducted with random sequence(s) of approximately 10 min duration anytime in the middle of the ballast water discharge from a tank, starting not before 5 min from the start of discharge and ending not after 5 min before the end of the ballast water discharge event from a tank.

Recommended Number of Samples

It was previously documented that the organism concentration of all organism groups addressed by the D-2 standard varies in ballast water samples due to their patchy distribution inside the tanks. Due to this variation a single 10 min sequential sample may under- or overestimate the real concentration of organism discharged. It was also observed that the average organism concentrations of two random sequential samples provide a very similar result to the average of three random samples. Therefore, we recommend that compliance control sampling is carried out by undertaking at least two random samples, and that the samples are analysed immediately after each sampling event has ended. For the final result the organism concentrations of the two sequences sampled should be averaged.

Recommended Sample Quantity

During our earlier studies (Gollasch and David 2009, 2010a, b, 2013) we conducted sequential ballast water sampling over different time durations of the sequences, i.e., 5, 10 and 15 min, with the average water flow rate ranging from 30 to 50 l per minute. To obtain most representative results we recommended that:

- for the organisms greater than or equal to 50 μm in minimum dimension between 300 and 500 l should be filtered and concentrated at the sampling point;
- for the organisms less than 50 μm in minimum dimension and greater than or equal to 10 μm in minimum dimension a "continuous drip" sample with a total volume not less than 5 l should be taken. To achieve this we recommend to collect about 0.5 l of sample water every minute during the entire sequential sampling time duration. Alternatively 0.5 l of sample water may be collected every 30–45 l of the ballast water sampled. The resulting 5 l of collected sample water should be mixed and sub-sampled in two sets. One set of samples should be kept alive and another preserved. We recommend approximately 60–100 ml as sub-sample volume;
- for indicator microbes samples, a sample of approximately 1 l should be collected as a sub-sample after mixing from the 5 l continuous drip sample (see bullet point above).

Other Recommendations

It is further suggested that the flow rates during sampling may have an additional influence on the viability of organisms. In case lower flow rates are obtained by partially closing valves at the sampling point this may result in sheer forces at the valve which likely will damage (especially larger) organisms during the sampling process. A similar negative viability effect may be caused by very strong flow rates, which may affect mainly the organisms greater than or equal to 50 μm in minimum dimension. Hence, the flow rate or "valve" effect, may result in an underestimation of viable organisms as organisms may have died during the sampling process. To avoid this unwanted effect it is recommended that the valve at the sampling point is opened as much as possible. However, the flow rate should not exceed 50 l/min so that the water pressure in the sampling net is not too high during sample concentration because this may also negatively affect organism survival.

Sampling Logistics Feasibility

Vessels of different types, sizes and cargo profiles have very different ballast water discharge patterns (see chapter "Vessels and Ballast Water"). The ballast water discharge may be carried out as a one time event "at once" or sequentially over longer time durations. The ballast water discharge may last for approximately 1 h, e.g., a fast discharge of two tanks in parallel on a container vessel. Longer discharge durations may stretch up to several days according to the length of the cargo operation. This may be the case on tankers, bulk carriers and sometimes general cargo vessels which load cargo over several days duration. Hence, the ballast water operation is frequently conducted in sequences over the time of the entire cargo operation.

The time factor is important to be taken into account as it is difficult to assume that a PSC officer and/or sampling team member(s) would stay on board the vessel for longer time periods of up to even several days.

Another aspect is the daytime of the sampling event, i.e., cargo operations and the corresponding ballast water operations, are regularly conducted also in night shifts, but PSC officers and/or sampling team and/or the laboratory for analyses may only be available at daytime.

Another challenge is the need to obtain a representative sample of the whole discharge of a vessel when the vessel will be discharging ballast water which originates from different ballast water source areas. In such cases it is recommended, if possible, that at least one sequential sample per ballast water source area is taken. However, if a single tank was filled with ballast water from multiple source areas this does not trigger the necessity for two or more samples to be taken.

Sampling Equipment and Sampling Point Arrangements

Here, sampling equipment is meant to include all equipment a PSC officer or sampling team needs to bring on board a vessel to conduct a compliance control sampling event. In addition sampling arrangements listed here include all the arrangements which would need to be setup on vessels enabling sampling for compliance monitoring.

Sampling Equipment

The use of light-weight and robust equipment of compact design is recommended to ease the transport and its use on board a vessel. The sampling equipment presented here was tested and used in on board sampling studies of commercial vessels and also in BWMS type approval tests, and is included here only to give examples.

Plankton Net for In-Line Sampling

When nets are used for in-line sampling events, the diameter of the net should not be larger than 50 cm and it should be shorter than 100 cm in length because such a net design eases the handling at the sampling point.

For the net mesh size it is recommended to use a mesh of 36 μm in square dimension, which results in a diagonal dimension of 50 μm. This is in line with Guidelines G8, i.e., "If samples are concentrated for enumeration the samples should be concentrated using a sieve no greater than 50 μm mesh in diagonal dimension."

The plankton net should at best be equipped with a removable cod-end, preferably with filtering panels so that the sample can be concentrated effectively during the sampling process. A valve at the bottom of the cod-end is beneficial as it eases the extraction of the concentrated sample. Should multiple samples be taken, it is recommended that the filtering sieve of the cod-end can be replaced between the sampling events so that no organisms become stuck from one sampling event and could be erroneously added to another sample. An example of such a net with a removable cod-end is given as Fig. 3.

Wash Bottle

For cleaning of the plankton net an unbreakable wash bottle may be used. Such a wash bottle may also be used when emptying the cod-end content, i.e., to concentrate the sample into an (unbreakable) sample bottle to ensure that all organisms caught are transferred into the sample bottle. An example is given in Fig. 4.

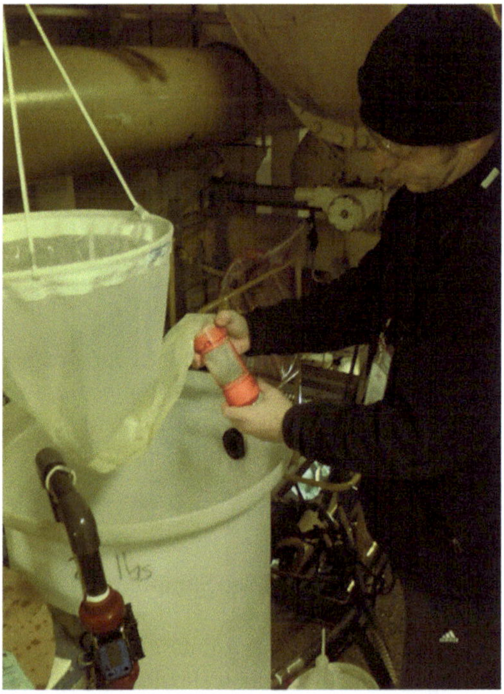

Fig. 3 Plankton net for in-line sampling with a removable cod-end with filtering panels (David 2013)

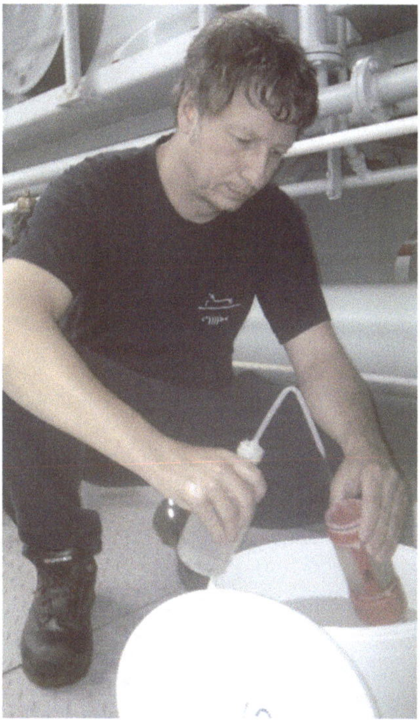

Fig. 4 Wash bottle used to drain all organisms caught in the cod-end

Plankton Net for In-Tank Sampling

For in-tank sampling events via an opened manhole a short plankton net with a cone-shaped opening and with a small diameter is beneficial as it can easily be lowered through a manhole. Studies have shown that the conical net top increases the sampling performance. At the same time this net design reduces the risk that the net becomes stuck inside the ballast water tank. A removable cod-end helps to clean the net between sampling events of different tanks so that an "organism contamination" from sample to sample can be avoided. An example is given as Fig. 5.

The sampled water volume can be calculated by considering the net opening dimension and the distance of the vertical net haul. It is recommended to use a metered rope when lowering the net that the depth from which the net is pulled up is known.

Flow Meter

During in-line sampling events a calibrated flow meter should be used to enable an accurate measurement of the water volume filtered through the plankton net. It is further recommended that the flow meter should also show the sampling flow rate, which is important for appropriate sampling planning and setup. Two flow meter examples are shown in Fig. 6.

Fig. 5 Plankton net for in-tank sampling (David 2013)

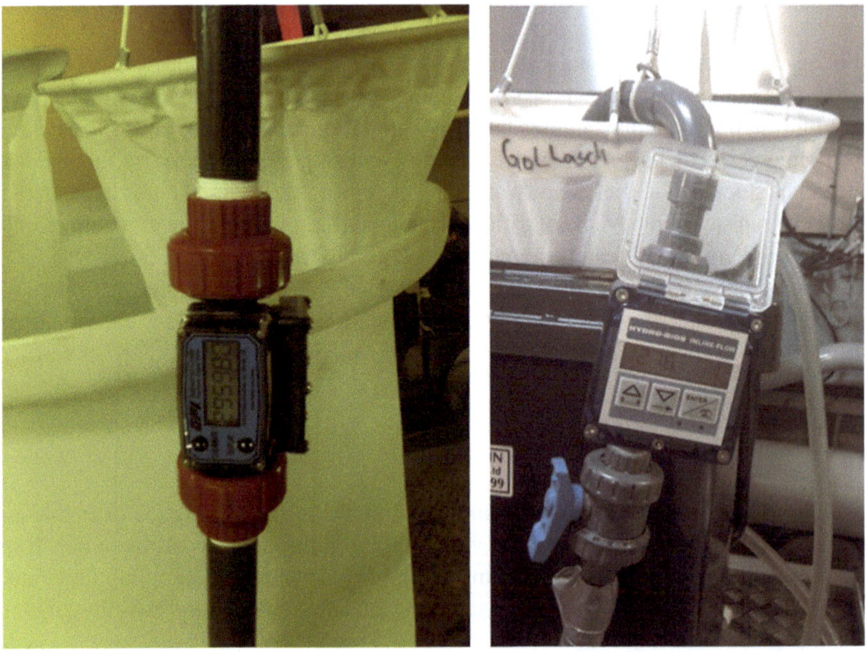

Fig. 6 Both flow meters are battery powered, the left one is intrinsically safe for use on tankers

Fig. 7 Water column sampler suitable for ballast water sampling via sounding pipes

Water Column Sampler

Water column samplers, one example is shown in Fig. 7, should be of dimensions which allow entering the ballast tanks via sounding pipes, but they may also be used via manholes.

The sampler may be lowered into the tank via a sounding pipe or manholes until the bottom is reached and than pulled back up. The water enters the sampler through a 6 mm opening at the top. The time to fill this sampler is approximately 10 s. When lowering the sampler, water will be proportionally sampled from the entire water column provided the sampler is lowered through the entire water column with a constant speed. A maximum sample volume of 0.2 l of ballast water may be sampled per one pull with this water column sampler. To increase the volume of water sampled multiple replicates may be applied.

Point-Source Sampler

The point-source sampler is of dimensions which allow it to enter ballast tanks through sounding pipes, but it can also be used via manholes. The sampler is lowered down the sounding pipe to the desired water depth for sampling and then the valve of the sampler is opened by pulling a rope which is connected to this valve. The point-source sampler can be used also to sample the ballast water and sediments at the ballast tank bottom, simply by lowering it to the bottom and here the valve, when touching the tank bottom, opens automatically thereby allowing water and sediment to flow into the sampler at its bottom.

Fig. 8 Point-source sampler used for sampling ballast water through sounding pipes

The valve of this sampler has a 3 mm diameter. The time needed to completely fill the sampler is approximately 1 min and its capacity is 0.225 l which may be sampled by one pull. Multiple replicates could be used to increase the water volume sampled. An example is given as Fig. 8.

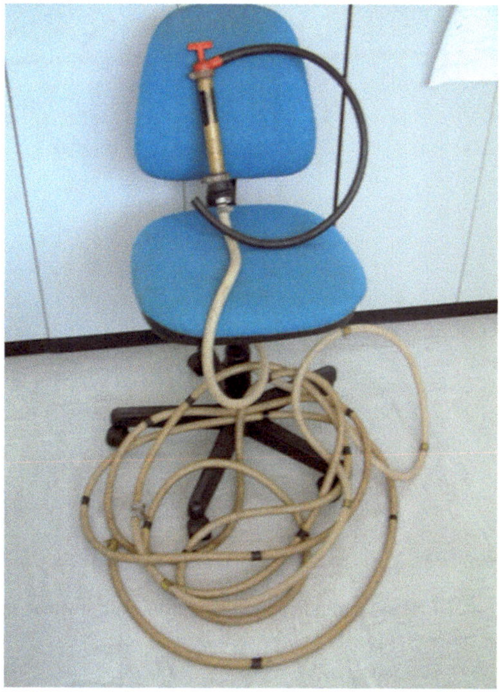

Fig. 9 Hand pump used for ballast water sampling

Pumps

Hand Pump

This hand pump is light-weight and of compact design (ca. 30 cm long, diameter ca. 5 cm). The hand pump may be used, without priming with water, until a maximum pumping head of ca. 9 m. The hose to be used on the suction side should have extra supported hose walls to resist under pressure. To sample, the hose is lowered through a manhole or sounding pipe to the desired water depth and the water is pumped up into a bucket or directly through a filtering device (e.g., plankton net). A hand pump example is given in Figs. 9 and 10.

Air-Driven Well Pump

The air-driven well pump was especially designed to be used for ballast water sampling via sounding pipes or manholes. It samples (a) at a desired water depth through a sounding pipe or a manhole, (b) the water column and/or (c) ballast water and related sediments at the tank bottom. The pump is operated by pressurized air as supplied on a vessel (5–7 bars) to pump up water from the desired water depth. The operation of the air-driven well pump is depth independent and it can be used to pump up ballast water from greater depths, e.g., >30 m. The flow rate of this pump is between 1.3 and 2.0 l/min. An example is given as Fig. 11.

Fig. 10 Hand pump in operation during ballast water sampling

Fig. 11 Air-driven well pump used for sampling via sounding pipe or manhole/tank hatch

Bucket

Experience has shown that a bucket of 10 l capacity is suitable for the sampling events as it is a compromise considering the minimum volume of samples needed and its portability on board. To avoid objects and dust etc. to be blown into the bucket during the sampling event and also to ease the transport of the sample a bucket with a (water tight) lid is required. Further, a volume scale on the bucket is helpful to allow readings of the water volume collected.

In-Tank Sampling Arrangements on Vessels

Today ships lack in-tank sampling points for compliance control tests. However, ballast water may under certain conditions be accessed via manholes, sounding pipes and air vents. The availability and accessibility of these in-tank "sampling points" is critical.

The availability and accessibility of sampling point is specific on vessels depending on ship type, design, age, dimensions and also on current ship operations. Three different general patterns were identified:

- ships which do not carry cargo on the weather deck generally provide easier access to sampling points located on that deck. This is critical especially for the access to ballast tank manholes;
- larger vessels in general have more suitable and accessible sampling points, e.g., due to more space to install and operate the sampling equipment. Further sounding pipes may be wider on larger vessels which eases the operation of the sampling gear;
- newer ships and those being better maintained show easier sampling points access because no or less rusty screws and nuts on manholes or venting pipes need to be removed. Further, no or less rust may occur inside sounding pipes etc.

As a rule, manholes are available on all vessels to access all ballast tanks. However, the experience from Gollasch (1996) and David and Perkovič (2004) has shown that only 20 % could be opened for a sampling event. The manhole opening limitations observed included very rusty screws and nuts (which would have to be cut-off to open the manhole cover), some tank covers were cemented and could therefore not be opened at all, and sometimes the manhole was covered with cargo. In other cases the access was limited because of ongoing cargo operations (which occurred in 80 % of the sampling attempts). Air venting pipes are also available on all ballast tanks and were mostly accessible. However, most of their covers are fixed with rusty nuts thereby limiting the sampling access. The most frequent and easiest accessible sampling points on all inspected vessels were sounding pipes. Another benefit of using sounding pipes is that no crew member needs to be involved to get access to the ballast water.

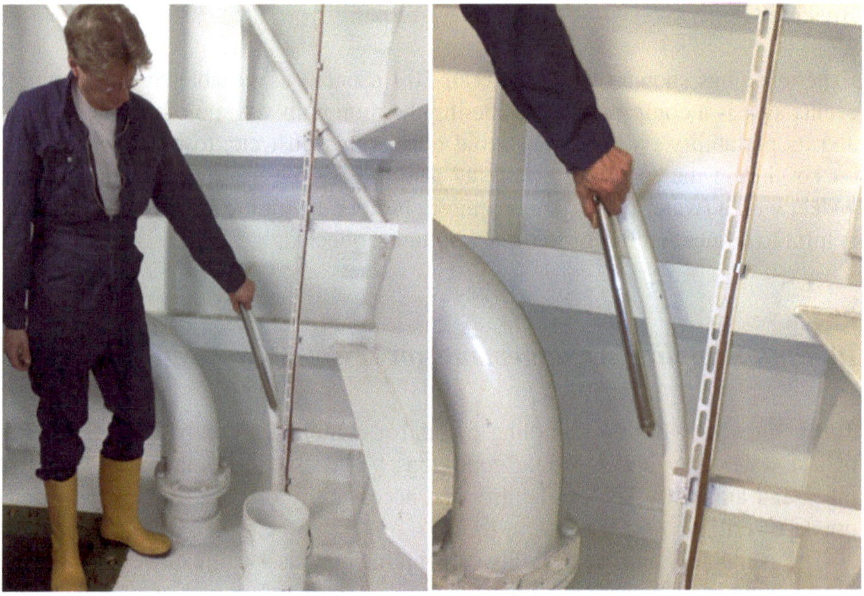

Fig. 12 Slightly bended sounding pipe on a car carrier which would not enable all sampling equipment described above to be lowered to the tank bottom

The sounding pipe requirements (rules) of some members of the International Association of Classification Societies, London (IACS) regarding the construction of the pipes have been analysed. The minimum sounding pipe requirements include (David and Perkovič 2004):

- all ballast water tanks should have sounding pipes, which need to be as straight as practicable (see Fig. 12),
- sounding pipes should not be less than 32 mm of internal diameter, and
- they must always be accessible.

Certain technical limitations were also identified when analysing the sampling accessibility of ballast water through sounding pipes. These are the pipe diameter and the distance of the water depth level inside the tank from the sampling point on deck. David and Perkovič (2004) further noticed that most sounding pipes have a welding under their cover, which narrows the access into the pipe by a reduced pipe diameter. Due to the limited sounding pipe diameter sampling equipment such as plankton nets, buckets etc. cannot be used for sampling. Further, suction pumps are practically excluded from an application if the pumping head (distance from the sampling point to the water level inside the tank) is more than 9 m. Therefore, sounding pipe sampling at greater depths to, e.g., double bottom tanks (some ships may only have double bottom tanks, or ballast water may just be carried in double bottom tanks) will require a well pump of smaller diameter to be lowered down the pipe.

Should a pump be used, this should have a capacity to pump up water from greater depths and at the same time its application should not cause damage to organisms. Several pumps are available (see examples given above), but all were not specifically designed for shipboard ballast water sampling. Pumps which require (external) power supply, impose a limiting factor especially for their use on board vessels that transport oil and oil products or different dangerous cargoes. To overcome this limitation pumps which are driven by compressed air may be used as compressed air is already available on almost all ships. The current sounding pipes design allows the water and associated sediment only to enter from the bottom end of the pipe, which is also recognized as a possible factor impacting the representativeness of sounding pipe samples.

In-Line Sampling Arrangements on Vessels

To enable sampling from a vessels' ballast water discharge line, appropriate permanent sampling arrangements need to be installed on the vessel in an area with sufficiently enough space to safely conduct a sampling event. The permanently installed sampling arrangements would include a:

- sampling point installed in the ballast water discharge line (see below);
- isokinetic sampling facility (see below),
- discharge point for the discharge of the sampled water (after filtration) which may be installed in the ballast water discharge line after the sampling point,
- space sufficient to place a sampling bin,
- hook or other installation that the plankton net can be hang ca. 100 cm directly over the middle over the sampling bin, and
- discharge pump of adequate capacity to empty the sampling bin during sampling. The pump to pump out the exciding water from the sampling bin, after it was filtered through the sampling net, should be of a capacity to withstand the head pressure in the ballast discharge line. It is also important to install a valve which allows for the regulation of the discharge flow from the sampling bin to provide for an adequate level of water in the bin during sampling, i.e., after having achieved an adequate level of water in the sampling bin to enable the sampling plankton net sitting in the water as much as possible a simultaneous water discharge from the bin is needed in the same dimension as the sampling water inflow to the bin to avoid an overflow and sample water spillage at the sampling point.

The equipment for a temporary use at the sampling point, but which should be stored on the vessel include:

- a sampling bin used to place the plankton net in water during sampling,
- a valve at the discharge point of the sampling bin and/or a pump to manipulate the water level in the sampling bin,

- hoses to connect from the sampling point to the flow meter and plankton net as well as from the sampling bin to the discharge point of the sampled water.

This equipment should be stored on the vessel that the sampling team would not need to bring it on board which eases the logistics.

Thc sampling bin should be sufficiently big that the plankton net can be placed into the bin at best completely. A sampling bin with dimensions of ca. 100 cm in height and approximately 50 cm in diameter has proven suitable. This size allows enough space to ensure that the sampling net is placed in water in the bin when filtering ballast water so that the maximum of its filtering surface is permanently submerged. This is very important to avoid a negative impact of the filtered organisms because of exposure to different stresses during filtration (Figs. 13 and 14).

As an alternative it may further be considered to permanently install a sampling arrangement on board so that the PSC only needs to bring the "consumables" on board, including a new plankton net, sample bottles, buckets etc. Such sampling skids are developing and tested with first tests being conducted on commercial vessels (IMO 2008; Lemieux et al. 2010; Schillack 2013; Wier 2013). However, rigorous on board validation tests are currently lacking.

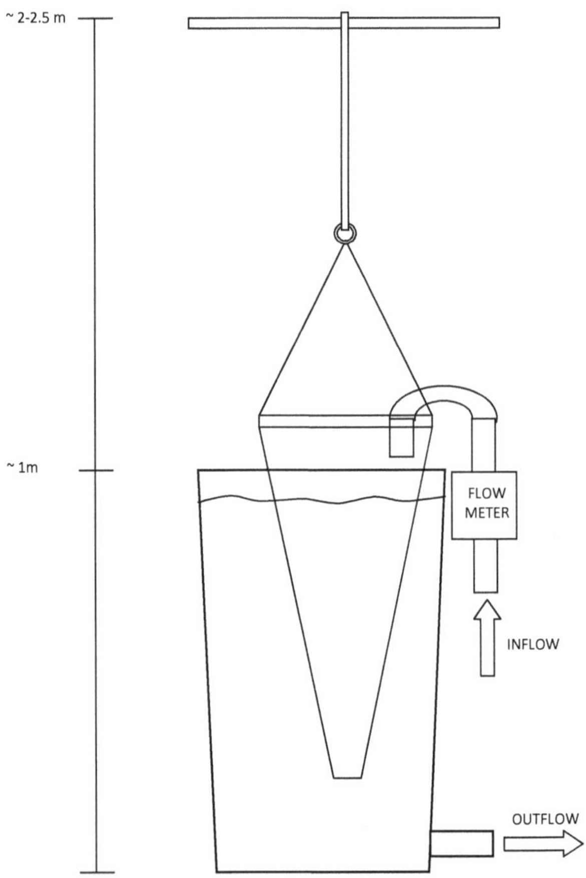

Fig. 13 Elements of a sampling arrangement for in-line sampling on a vessel (David 2013)

Fig. 14 Sampling arrangements for in-line sampling on vessels, on the *left* in a more open space, and on the *right* in more confined space of the engine room

Sampling Point

Vessels are of very different sizes, design and arrangements, mainly depending on their purpose and age. Consequently, also the ballast water systems are very different in capacities and designs. Regarding appropriate sampling points the G2 Guidelines recommend the installation of an isokinetic sampling point/facility, which diameter is related to the diameter of the ballast discharge line where it is installed. Therefore it is expected that a range of different sampling points may be found on different vessels. This poses a real challenge to PSC to have the adequate sampling equipment for all different sampling points available. All hoses and connections would then also need to be of the same size/diameter as the sampling point.

As per the authors sampling experience for type approval tests of BWMS most vessels have installed 1 in. sampling points, which has shown to deliver enough water flow for the sampling purpose.

Guidelines G2 defines the "sampling facilities" as the equipment installed to take the sample and the "sampling point" as that place in the ballast water piping where the sample is taken. This means that the sampling point is part of the vessel's main pipe where the sampling facility is installed. Guideline G2 provides further details for isokinetic sampling facilities:

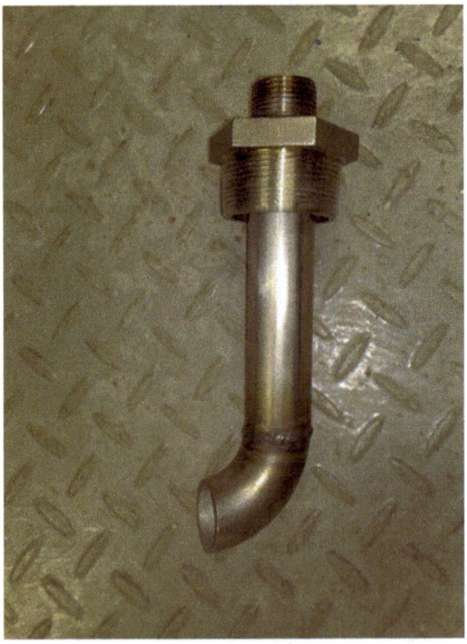

Fig. 15 Elbow-shaped sampling facility to be installed within a straight stretch of the ships' ballast water line

In order to undertake an accurate measurement on the organism concentration in the ballast water, it is recommended to install an "isokinetic" sampling facility. Isokinetic sampling is intended for the sampling of water mixtures with secondary immiscible phases (i.e., sand or oil) in which there are substantial density differentials. In such conditions, convergence and divergence from sampling ports is of significant concern. Since most organisms are relatively neutrally buoyant, true isokinetic sampling is unnecessary. However, the mathematics related to isokinetic sampling are deemed to be useful as a basis for describing and specifying sampling geometries. Isokinetic sampling is necessary to ensure that a sample contains the same proportions of the various flowing constituents as the flow stream being sampled. During isokinetic sampling the sampling device does not alter the profile or velocity of the flowing stream at the moment or point at which the sample is separated from the main flow stream. Under isokinetic conditions, the velocities of both the sample and the main flow are equal at the point at which the sample is separated from the main flow. To achieve isokinetic sampling conditions, a sampler is designed to separate a subsection of the total flow-stream in a manner that does not encourage or discourage water entry other than that which is otherwise in the cross-section of the sampler opening. In other words, flow streams in the main flow of the pipe should not diverge or converge as they approach the opening of the sampler.

Sampling facility pipes should be elbow-shaped with its opening located in the middle of the ships' main ballast water pipe and directed towards the water flow in the line (see Figs. 15 and 16).

Sample Handling

As addressed in the G2 Guidelines, all samples should be transported, handled and stored under proper conditions and we recommend below how this can be achieved.

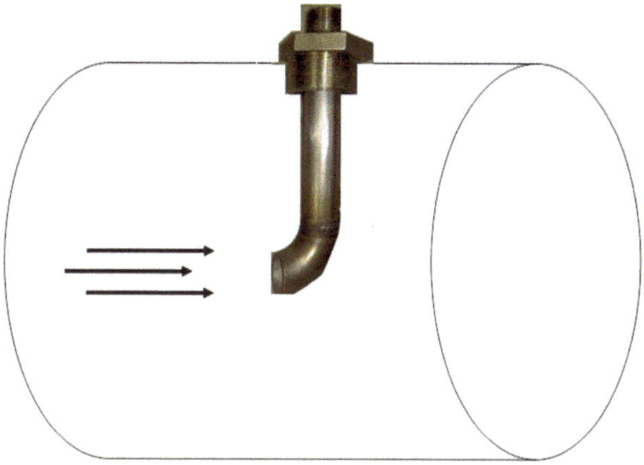

Fig. 16 Elbow-shaped sampling facility as it should be installed within a straight stretch of ships' ballast water line. *Arrows* indicate direction of ballast water flow

Sample Labelling

Guidelines G2 recommend that each sample container should be labelled by, e.g., using a waterproof permanent marker and additional paper which may also be deposited inside the sample container. Alternatively we recommend to use self-adhesive labels and pencil for writing. To ensure that samples are not mixed the sample bottle is to be labelled (not the lid which may be confused in cases several bottles are opened at the same time). As an additional sample integrity assurance the stickers are fixed to the sample bottle with transparent tape that they cannot fell off.

The information recorded should include specific information, e.g., the date, ship name, sample identification code. Should the sample be concentrated the original volume should be added to the label. It is recommended that each sample will clearly be numbered. Additional information should be recorded on paper as a sample collection data form as suggested in Guidelines G2 and the form and sample bottle should be stored together in a sealed plastic bag or storage box.

Sample Transport

In accordance with Guidelines G8 "The samples should be analysed as soon as possible after sampling, and analysed live within 6 hours or treated in such a way so as to ensure that proper analysis can be performed." This is to avoid a die off or other negative impacts of the organisms. Should samples need to be transported to a laboratory, e.g. for a detailed compliance check with the D-2 standard, leak-proof sample bottles should be used to avoid water leakage during transport.

All samples should be transported in, e.g., Styrofoam boxes to avoid (rapid) temperature changes. In this approach a question arose how the storage time and conditions would influence organism survival. During our studies several experiments to address this point were conducted and it became clear that the samples should be stored in the dark in a slightly cooler environment as the original sample temperature to slow down the metabolism of the organisms but not to expose the samples to too cool or warm conditions to avoid a temperature shock of the organisms. A temperature difference between the ambient sampling conditions and during storage of 10–15 °C, seems appropriate to avoid such a temperature shock and cold-induced stress. However, a temperature drop below ca. 5 °C should be avoided (Gollasch and Kacan 2014).

Recommendations for organisms greater than or equal to 50 µm in minimum dimension:

- The organisms sampled should not be concentrated below 1 l of water to enable appropriate sample transport to avoid negative effects resulting in organism mortality.
- After sampling the sample needs to be transferred into the transport box as soon as possible and the lid placed back on the box to avoid light penetration or heating up.

Recommendations for organisms less than 50 µm in minimum dimension and greater than or equal to 10 µm in minimum dimension:

- The sample should not be concentrated. Subsamples should be taken of not less than 100 ml each to enable appropriate sample transport.
- After sampling the sample needs to be transferred into the transport box as soon as possible and the lid placed back on the box to avoid light penetration or heating up.

Recommendations for indicator microbes

- The water should not be concentrated. Subsamples should be taken of not less than 1 l to enable appropriate sample transport.
- The 1 l sample taken should after sampling be transferred into the transport box as soon as possible and the lid placed back on the box to avoid light penetration or heating up.

Chain of Custody

A chain of custody procedure should be implemented to document sample handling and transport. The chain-of-custody record should be kept with the samples and should identify

- all names and parties being involved in the sample transfer and handling,
- type of sample,
- location of the sampling event,
- date and time of the sampling event,
- number of samples,

- vessel details,
- type of analysis, and
- date of transfer and signature of each involved party representatives.

Sample Analysis

There are two fundamentally different approaches to analyse ballast water samples to proof compliance with BWM requirements, i.e., the samples may be analysed indicatively or in detail. Comprehensive organism detection method reports were prepared by Gollasch and David (2010b) for EMSA and by Gollasch et al. (2012) in the framework of the Interreg IVB Project Ballast Water Opportunity. Both reports were considered in detail when drafting this part of this book chapter. It was concluded that organism detection technologies that enable both an indicative and detailed inspection of ballast water samples are available today. This conclusion was also supported by our tests conducted on board of commercial vessels to evaluate the suitability of such technologies for practical work by PSC.

In general, an indicative sample analysis is meant to identify potentially non-compliant ballast water in an early stage to avoid such ballast water discharges. Should an indicative sample analysis result in doubts whether or not the BWM requirements were met, a detailed compliance control sample analysis may follow. However, a detailed sample analysis may also be conducted without a preceding indicative sample analysis.

Criteria for the Selection of Organism Detection Technologies for Ballast Water Compliance Control

Should sample processing be needed to prove compliance with the standards D-1 and/or D-2, the methods to be used for both indicative and a detailed sample analysis may have to be selected according to the following criteria.

Accuracy/Precision

The accuracy and precision of the sample processing method is critical as inappropriate sample processing techniques may result in a wrong compliance determination due to, e.g., missing organisms. Therefore only tested methods or at best standard methods, such as ISO methods, should be used for the purpose of compliance control tests. For the analysis of bacteria such methods exist, but for phyto- and zooplankton organisms new methods had to be developed or existing methods had to be adapted. One of the reasons is that the D-2 standard refers to viable organisms, i.e. a viability test of the organisms sampled needs to be undertaken which was not addressed in most standard methods in the past. This implies that selected methods had to be tested and validated rigorously.

The viability assessment of zooplankton organisms may be done by using a stereomicroscope and the analyst gently poking complete organisms to initiate a response. Moving organisms are to be counted as viable. For phytoplankton, stains may be used to assess viability and the most promising staining methods are (Chloromethyl-)fluorescein Diacetate ((CM)FDA) and Sytox. (CM)FDA is a stain which stains living cells, whereas Sytox stains dead cells. For a viability assessment (CM)FDA may be applied to a sample and the cells which take up the stain may be counted as viable cells. Using Sytox, which stains dead cells, the principle is reversed. As a first step the stain is applied and all cells which take up the stain are counted, thereby resulting in the total number of dead cells with living cells uncounted. As a second step the sample is treated that all viable cells are killed and the stain is applied again. The two counts delivered are put in relation so that the number of viable cells can be calculated, i.e. the result of step number two (all cells dead) minus the result of step number one (only dead cells stained, living cells unstained) gives the number of viable cells in the sample.

A particular problem is to identify whether or not resting stages and cysts are viable because they are not taking up stains and a microscope inspection would not enable a viability assessment. One of the most promising ways to proof viability would be to conduct hatching experiments of cyst and resting stages. However, this is difficult as the hatching conditions of many organisms are unclear so that some inaccuracy exists as the cyst and resting stages may not have hatched due to wrong culturing conditions. In addition, the hatching experiments are time consuming efforts. Noting all these difficulties, resting stages are so far recommended to be excluded from viability tests.

In addition to the viability of organisms, their size needs to be documented in minimum dimension. The minimum dimension measurement should be based upon an investigation of the organism "body", thereby ignoring sizes of thin spines, antenna etc. In e.g. flat worms or diatoms the minimum dimension should be the smallest part of their "body", i.e. the dimension between the body surfaces when looked at the individual from the side. In ball shaped organisms the minimum dimension should be the spherical diameter. This approach is in-line with the views expressed at the relevant IMO discussions. In summary, for the measurement of the minimum dimension the smallest visible axis of an organism should be chosen and the smallest point on this axis be measured. One way to interpret the minimum dimension of organisms is shown by the examples given in Fig. 17. However, the minimum dimension measurement leaves room for interpretation and can therefore not be understood as uniformly applied world-wide.

The minimum dimension measurement makes it extremely challenging also for counting machines such as flow cameras (e.g., Fig. 18) or flow cytometers (e.g., Fig. 19) to estimate the size of an organism. These machines "calculate" the size, which is especially in non-spherical objects, very difficult and may lead to a wrong size categorisation of organisms. For these machines it would possibly be easier to measure the maximum dimension of an organism.

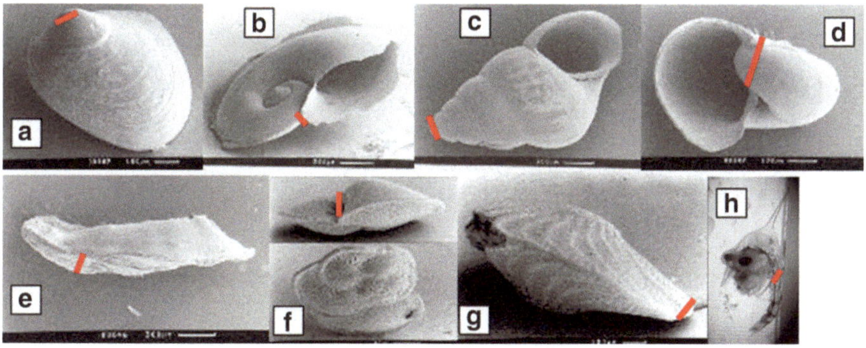

Fig. 17 One way to interpret the minimum dimension measurements (*red line*) for selected organism types: *A* mussel larvae, *B–D* gastropod larvae, *E* young oyster, *F* Foraminifera (phytoplankton), *top* shows individual from the side and bottom the same individual from the top, *G* Foraminifera and *H* decapod larvae. All organisms shown here are well above 50 μm in minimum dimension (Photos *A–G* Stephan Gollasch, *H* Matej David)

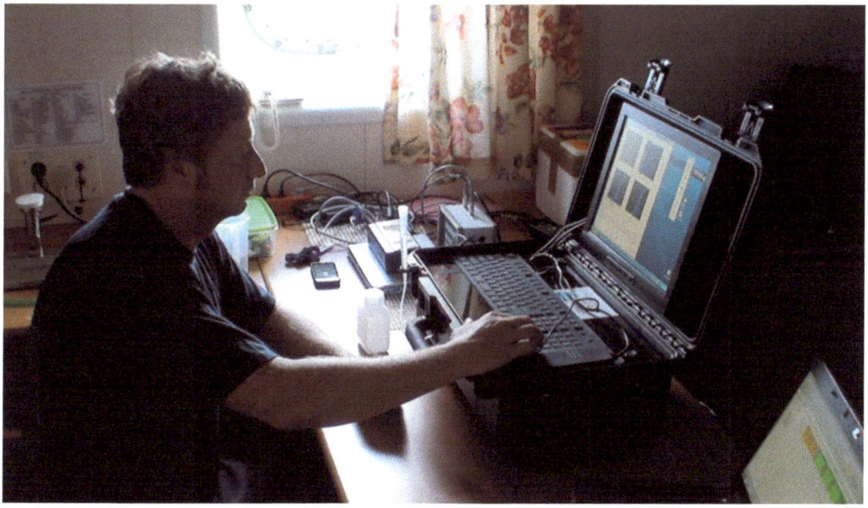

Fig. 18 On board test of a flow camera

Another issue are colony forming species. A question arose in which size category a colony falls when the single cell is below 50 μm in minimum dimension but the colony is above 50 μm in minimum dimension. A team of experts, i.e. the ICES/IOC/IMO Working Group on Ballast and Other Ship Vectors, believes that in those cases the individual specimen size should be measured. This group finding, which was also accepted by IMO, is based upon the D-2 standard as it refers to organisms and not to colonies. Further, viability assessments should address the smallest unit able to reproduce which is the individual and not the colony. Based on these conclusion the size of the individuals should be measured and not the colony. For an experienced analyst this is not a problem, but machine counts may result here again in a wrong size categorisation of organisms.

Fig. 19 On board test of a flow cytometer

Calibration

Scientific instruments need to be calibrated according to the manufacturers specifications to ensure correct results. After a calibration exercise possible adjustments of the technologies may be needed and much of this work cannot be done in the field, but requires a laboratory environment. An organism detection method which needs calibration after each individual use makes it therefore disadvantageous. Such a scenario would not permit to analyse several samples consecutively in a timely manner and the compliance inspection team would need to bring the instrument back to a laboratory for calibration after each use.

Time to a Result

At best the organism detection method should deliver prompt results. The sooner the results are obtained the earlier a potentially non-compliant ballast water discharge can be stopped or other non-compliance actions can be initiated. It is especially valid for an indicative sample analysis to reveal a prompt (non-)compliance indication. Therefore, both the indicative as well as the detailed sample analysis methods should be chosen in such a way that they deliver results promptly to enable timely non-compliance response action(s).

A rapid sample processing is also an advantage in cases where PSC is undertaking the sample analysis on board. Because of all other PSC duties a comprehensive and lengthy processing procedure of ballast water samples may result in overcommitted work.

Expertise

At best a PSC officer should be able to use the organism detection method(s) with some special training, but without a need for academic education in biology or chemistry.

Portability

Portable sample analysis equipment, which may be carried onboard a vessel by one person, is beneficial because the analysis can be undertaken "on the spot", i.e., on the vessel. Alternatively, the samples may be taken from the vessel to a laboratory or to a van on the pier equipped with necessary sample analysis equipment.

Costs

The methods selected should not be too expensive (capital costs) and should also be cost effective regarding running costs (e.g. operating costs, consumables). However, the accuracy and precision of methods is of prime importance. In general, an expensive, but prompt and accurate organism detection technology is to be preferred to a cheaper not as accurate system.

Recommended Technologies to Proof Compliance with the D-1 Standard

Sample analysis tools to proof compliance with the D-1 standard, such as tracers of human activities or a salinity meter are available and were already described above.

Recommended Technologies to Proof Compliance with the D-2 Standard

It is important to note that a ballast water sample analysis may be undertaken for all three groups of organisms of the D-2 standard. King and Tamburri (2010) as well as Gollasch and David (2010b) concluded that it will be very difficult to promptly assess whether or not the ballast water of a vessel meets all limits of all organism groups in the D-2 standard. However, should one organism group already show to be above the required concentration of viable organisms, non-compliance is indicated and the other two organism groups may not need to be tested.

Indicative Sample Analysis Methods

The easiest way for an indicative analysis of a sample is to show presence and absence of viable organisms in a sample. This may be done by DNA, RNA, ATP or Chl *a* detection. However, the presence/absence indication cannot assume organism numbers and some of the approaches cannot distinguish between viable and dead organisms as required by the D-2 standard. Efforts are underway to "translate" the reading of these detection technologies into numbers, but the validation of such an approach is not completed. We note that some of these detection technologies are semi-quantitative, so that a higher reading of the instrument means a higher viable biological content in a sample. From our experience so far, the best indicative sample analysis tool for phytoplankton analysis is Pulse-Amplitude Modulated (PAM) fluorometry. It provides a rapid measurement of photosynthetic activity which is considered as an indicator of viable cells, the method is portable and easy to apply (Gollasch and David 2010). The on board operation of two PAM instruments is shown in Figs. 20 and 21. For the measurement of the instrument shown in Fig. 20 the sample water can be analysed without additional processing. In contrast, the instrument pictured in Fig. 21 requires a sample filtration step.

For zooplankton, a stereomicroscope may be used (see Fig. 22). It would theoretically be possible to get a PSC trained to indicatively analyse a zooplankton sample through a stereomicroscope, but even the most compact designed stereomicroscopes are far from pocket-size dimensions. Further, PSC would need comprehensive training to distinguish between viable and dead organisms so that this method seems unlikely for routine PSC compliance control checks. Alternatively, a sample may be visually inspected for viable zooplankton organisms. It should be noted that without magnification a visual inspection will result only in bigger organisms being detected in a sample. It is assumed that organisms bigger than

Fig. 20 On board operation of a PAM instrument

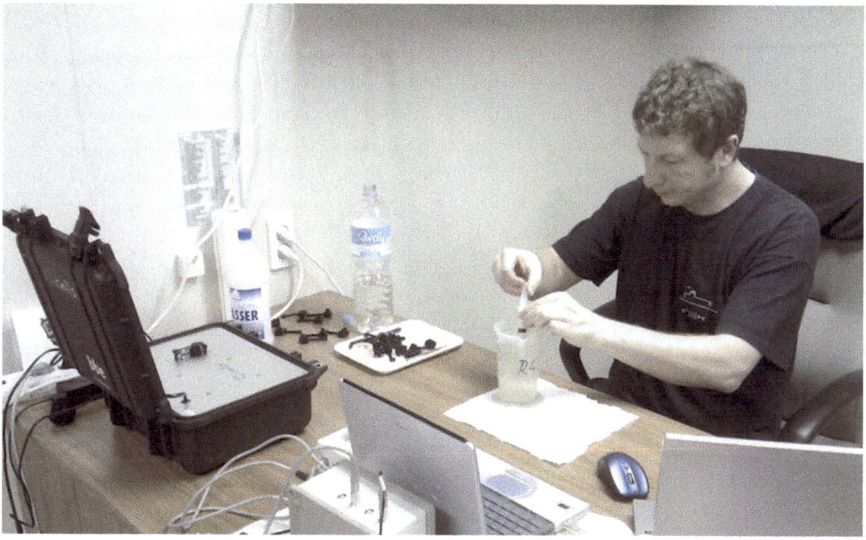

Fig. 21 On board operation of a PAM instrument with a filtration step

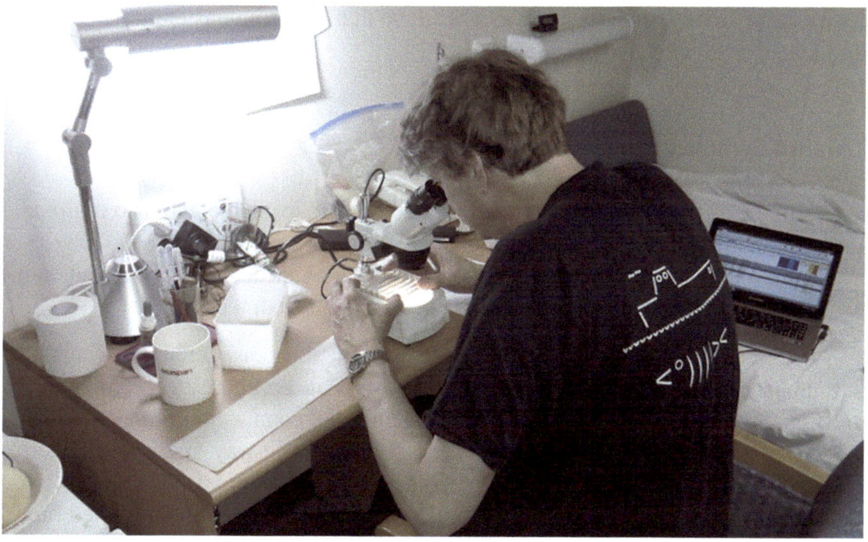

Fig. 22 On board operation of a stereomicroscope

1,000 μm in minimum dimension may be determined in such way and their movement in a sample could be taken as viability indication.

Presence/absence methods are available for the D-2 indicator microbes. Species specific or bacteria group specific enzyme detection methods were developed and these methods show in a short time whether or not the indicator microbes are present

Fig. 23 Hand-held fluorometer used on board

in a sample by e.g. hand-held fluorometers (Fig. 23). To document colony forming units and for an enumeration of these, samples need to be incubated and it takes at least 6 h incubation time to come to a result so that this is an unsuitable method for an indicative sample analysis.

Detailed Sample Processing Methods

A wealth of different sample processing methods are available and also regionally different methods are used for organism detection.

As time and portability of methods is here not as critical as for the indicative sample analysis, the following methods are recommended:

- Phytoplankton. Although machine counts may be considered, it seems that human counts using a counting chamber and an epifluorescence microscope (Fig. 24) deliver the highest accuracy in viable cell counts. Stains may be used to separate viable from dead cells and here, in the same way as for an indicative analysis, (CM)FDA and Sytox are recommended (see above).
- Zooplankton. An analysis with a stereomicroscope, operated by a trained biologist, will result in a viable organism count. Exposing the organisms to light or gentle poking may trigger movement and fully intact moving organisms should be considered viable.
- Indicator microbes. Selective media should be used to grow bacteria colonies which can subsequently be counted by an analyst or machine. The incubation time is at least 2 days and the documentation of *Escherichia coli* and Cholera bacteria may require the use of more than one medium during the incubation effort.

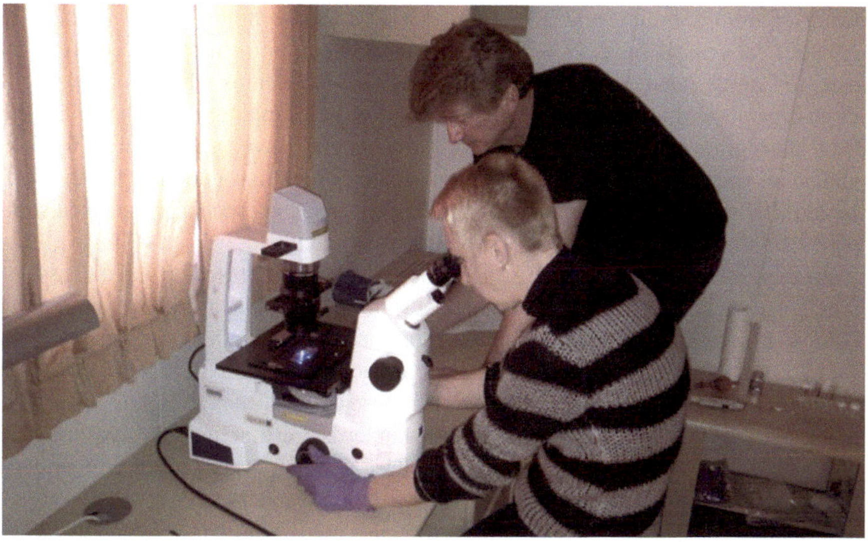

Fig. 24 Onboard use of an epifluorescence microscope

Discussion

Ballast water sampling may be conducted for various purposes: (a) to assess the biology and chemistry of ballast water (scientific research); (b) to identify potentially harmful or other organisms carried in ballast water (risk assessment); and, (c) to assess compliance with BWM requirements (monitoring and enforcement) which is in the focus of this chapter. Ballast water sampling is complex due to differences in organisms' dimensions and behaviour, as well as because of differences in ship construction including the availability of sampling points. These issues and the purpose of the ballast water sampling study have influence on the sampling method selection.

The sampling point is clearly related to the sampling purpose, e.g., indicative or in detail, D-1 or D-2 standards compliance sampling. The in-line sampling point will need to be installed on vessels according to the G2 and G8 Guidelines. However, there are no detailed provisions for in-tank sampling points so that ballast water to be sampled from a tank needs to be sampled via existing access points, i.e., manholes, sounding or air pipes. The availability of these sampling access points has proven to be critical. Employment of sampling equipment, modified for on board use, and a flexible approach are essentially needed to allow sampling via the different access points.

New methods were developed and are currently further developing to ease ballast water sampling on board ships including especially designed equipment for in-tank sampling through sounding pipes. Sounding pipe sampling was achieved by the use of a variety of methods including an air-driven well pump, hand pump, a water-column sampler, and a bottom and sediment sampler. Tests on board of commercial vessels have shown that this sampling equipment can successfully be used to sample most organisms despite some size limitations which may occur for large organisms due to the opening dimension of the sampling tool. The suitability tests also confirmed that all three water samplers considered can be safely used on almost all ships, while not disturbing normal ship operations conducted in the port. In addition to their deployment via sounding pipes these especially designed samplers can also be employed via manholes or tank hatches.

In-tank sampling may be more appropriate for scientific research and risk assessment analysis with the aim to assess ballast water biota, while at discharge sampling is more appropriate for the compliance monitoring with BWM requirements (e.g., the D-2 standard). However, certain tanks are not discharged through pipework on board, but may use gravity for emptying. In those cases in-tank sampling is the only approach to prove compliance with BWM standards. Further, in-tank sampling may also be used to confirm risk assessment results, e.g., to proof the presence or absence of target organisms before the ballast water is being discharged.

Sounding pipe sampling for zooplankton does not result in a representative sample of species in the tank as comparisons of sounding pipe and manhole samples from the identical tank showed that net samples were more diverse. Sounding pipe samples contained 0–60 % of the organisms of a net sample which highlights the need to sample ballast tanks via opened manholes. Further, pumps used via open manholes delivered more diverse samples compared to plankton net samples, so that pumps may also be considered when sampling via manholes.

In summary, future ballast water studies should consider that sampling via sounding pipes is inferior when selecting appropriate sampling techniques. However, in many occasions manholes cannot be opened due to overlaying cargo or cargo operations in the area where the manhole is located so that it is unsafe for sampling work. In these instances sounding pipe sampling might be the only possible option to sample the ballast water at all.

If the sampling has to document non-compliance, i.e., violations of the ballast water discharge standard, much less onerous sampling requirements are posed to the port State to demonstrate that an explicit organism concentration value is exceeded. For example, should a sample from the tank contain 1,000 viable organisms greater than or equal to 50 μm in minimum dimension and the tank capacity is 100 cubic metres, the organism concentration would exceed the D-2 standard when the water is discharged.

Recommendations

Several different ballast water sampling methods and equipments have been used for different sampling purposes. Shipboard sampling is also conducted for BWMS performance testing for type approval. Hence, shipboard sampling methods for testing BWMS exist, and these have been approved by different national responsible authorities. However, scientific studies have shown that sampling results may be biased by different sampling approaches because of, e.g., the patchy distribution of organisms in tanks, mortality of organisms during sampling etc. The lack of a commonly agreed ballast water sampling methodology or approach may impact representative ballast water sampling so that certain vessels may be found in compliance in one port, but not in another.

Different methods and sampling equipment may be used for different sampling goals, e.g., D-1 or D-2 standards, indicative or detailed sampling. The selection of appropriate sampling methods and equipment also depends on the ballast water access points, i.e., in-tank via manholes, sounding pipes or air vents, or in-line installed sampling points, and also on the target groups of organisms as stated in the D-2 standard.

It is of prime importance to evaluate the appropriateness of a selected sampling approach for compliance control according to the BWM Convention. The ballast water sampling methods recommended here for compliance with the D-1 and D-2 standards are based on the author's sampling experience gained on more than 80 shipboard tests for type approval of 18 different BWMS. Very importantly, these sampling methods were also scientifically tested, improved and validated during three studies on representative BWS for compliance monitoring. These studies were conducted for the Federal Maritime and Hydrographic Agency, Hamburg, Germany in 2009 and 2012, and for the European Maritime Safety Agency, Lisbon, Portugal in 2010 (Gollasch and David 2009, 2010a, b, 2013).

For D-2 standard compliance tests it is suggested that samples should be taken during discharge, i.e., from the ballast water discharge line after the pump prior to the discharge overboard. This approach delivers the most representative and accurate results regarding the organism concentration in the ballast water discharge from a vessel as a side stream of the discharge is sampled. Samples should be taken in two or more 10 min sampling sequences with a sample volume of 300–500 l in each sequence. With this approach, organisms are less exposed to negative impacts during sampling. Further, such a sampling approach is logistically more appropriate than sampling over the entire discharge time, and our scientific studies have shown this sampling method to be representative of the whole discharge. In reality, this sampling method as per Guidelines G2 is not exactly the same as for the type approval testing of BWMS according to Guidelines G8, but has all components of it, e.g., the same sampling equipment may be used for the two or three recommended sequential samples. This method is also in-line with the agreement at IMO that sampling methods applied by PSC for compliance checks, i.e., G2 Guidelines sampling, should be no more stringent than the methods applied for BWMS type approval, i.e., G8 Guidelines sampling.

However, one key problem remains with the in-line sampling approaches and this is that compliance or non-compliance can only be proven during discharge while the ballast water is being pumped overboard. Consequently, the ballast water may already have been discharged before it is clear whether or not it is in compliance with the BWM Convention standards. Should high risk organisms be suspected in the ballast water intended to be discharged, in-tank compliance control sampling may be the more appropriate method compared to the in-line approach to possibly avoid the discharge of not properly treated ballast water into the environment. Further, in-tank sampling may be the only possible sampling approach if the ballast tanks to be discharged have only direct discharge to the sea. For these reasons in-tank sampling remains important to be dealt with.

The previously conducted ballast water sampling studies tested also different in-tank sampling methods using different sampling equipment. However the method efficiency would need to be studied further in more detail, should be scientifically validated, and cross compared before conclusions on the method application can be drawn.

At the last meeting of BLG 17 (February 2013), the BWS Guidance was finalised in a form of an IMO Circular. It was recognized that many of the sampling and test methods in the BWS Guidance were not yet adequately validated, and were not yet fully integrated in PSC procedures in order to validate their practicality for determining compliance with the BWM Convention. Given that these methods are rapidly improving, IMO members and observers were encouraged to further develop sampling and analysis protocols, including the range of options outlined in the BWS Guidance.

Further it was accepted by IMO that once the BWM Convention enters into force, a trial period of 2–3 years would be initiated where PSC can test the approaches in the BWS Guidance to ensure they are practical and fit for purpose. During the trial period it is anticipated that port States share the results of on board ballast water sampling and analysis. The trial results and findings from sampling efforts should also be communicated to IMO which will likely result in an update of the IMO BWS Guidance documents and/or the G2 Guidelines. In continuation of the trial period, further sampling events will be conducted to determine if changes are needed to standardize the sampling options available. Lastly, in the end of the trial period recommendations are to be provided to MEPC on standardized sampling and analysis protocols and on possible advances in scientific knowledge which may be considered to update the IMO BWS documents accordingly.

For sample processing two different approaches to analyse ballast water samples to proof compliance with BWM requirements were implemented, i.e., an indicative or a detailed analysis. It was recognized by the authors that organism detection technologies are available today to conduct an indicative and detailed inspection of ballast water samples. Some of the recommended methods do not deliver organism counts, but give a semi-quantitative measurement or a presence/absence documentation which is a suitable way to document indicatively compliance with BWM requirements. Most organism detection tools for an indicative analysis are portable and deliver a result promptly so that PSC officers can use them on board at the sampling point.

For a detailed sample analysis, the recommended methods are more cumbersome and require more time to a compliance control result. The sample processing methods for a detailed analysis are not portable and require a high experience level of a trained biologist so that the samples either need to be brought to a laboratory for subsequent analysis or a van may be equipped with these methods and driven to the port for a sample analysis on the pier.

The studies the authors were involved in have shown that there are BWS methods and sample analysis methods available which were extensively used on board of commercial vessels to test BWMS to proof compliance especially with the D-2 standard. These methods were scientifically validated by additional tests and studies on board and in land-based experiments. Most of the recommended methods have also shown to be relatively simple, i.e., no special background education is needed for their application, they are cost effective, i.e., there is no need for very expensive equipment, there are no high running costs, and they are generally applicable on all vessel types and in all geographic regions. The authors believe that the sampling and sample analysis recommendations suggested in this chapter may result in a workable, equitable and pragmatic solution to support the entry into force of the BWM Convention.

Acknowledgements WWF International is thanked for having recently provided funds to generate a ballast water sampling report which was presented to IMO. This report to a large extend was considered when drafting this book chapter. Part of this publication has been produced with the financial assistance of the IPA Adriatic Cross-Border Cooperation Programme – strategic project Ballast Water Management System for Adriatic Sea Protection (BALMAS). The contents of this publication are the sole responsibility of authors and can under no circumstances be regarded as reflecting the position of the IPA Adriatic Cross-Border Cooperation Programme Authorities.

References

Antonov JI, Seidov D, Boyer TP, Locarnini RA, Mishonov AV, Garcia HE, Baranova OK, Zweng MM, Johnson FR (2010) World Ocean Atlas 2009, Volume 2: Salinity. In: Levitus S (ed) NOAA Atlas NESDIS 69

David M (2013) Ballast water sampling for compliance monitoring – ratification of the Ballast Water Management Convention. Final report of research study for WWF International. Project number 10000675 – PO1368

David M, Perkovic M (2004) Ballast water sampling as a critical component of biological invasions risk management. Mar Poll Bull 49:313–318

David M, Gollasch S, Cabrini M, Perkovic M, Bošnjak D, Virgilio D (2007) Results from the First Ballast Water Sampling Study in the Mediterranean Sea – the Port of Koper study. Mar Poll Bull 54:53–65

Gollasch S (1996) Untersuchungen des Arteintrages durch den internationalen Schiffsverkehr unter besonderer Berücksichtigung nichtheimischer Arten, Ph.D. thesis, University of Hamburg, Verlag Dr. Kovac

Gollasch S, David M (2009) Results of an on board ballast water sampling study and initial considerations how to take representative samples for compliance control with the D-2 Standard of the Ballast Water Management Convention. Report of research study of the Bundesamt für Seeschifffahrt und Hydrographie (BSH), Hamburg

Gollasch S, David M (2010a) Recommendations how to take a representative ballast water sample, emerging ballast water management systems. IMO-WMU Research and Development Forum, Malmö

Gollasch S, David M (2010b) Testing sample representativeness of a ballast water discharge and developing methods for indicative analysis. European Maritime Safety Agency (EMSA), Lisbon

Gollasch S, David M (2013) Recommendations for Representative Ballast Water Sampling. Final report of research study of the Bundesamt für Seeschifffahrt und Hydrographie (BSH), Hamburg, Germany. Order Number 4500025702

Gollasch S, Kacan S (2014) German national report. In: ICES (ed) 2014 Report of the ICES/IOC/IMO Working Group on Ballast and Other Ship Vectors (WGBOSV), ICES CM 2014/ACOM:31, Copenhagen, pp 54–73

Gollasch S, Lenz J, Dammer M, Andres HG (2000) Survival of tropical ballast water organisms during a cruise from the Indian Ocean to the North Sea. J Plankton Res 22(5):923–937

Gollasch S, Rosenthal H, Botnen H, Crncevic M, Gilbert M, Hamer J, Hülsmann N, Mauro C, McCann L, Minchin D, Öztürk B, Robertson M, Sutton C, Villac MC (2003) Species richness and invasion vectors: sampling techniques and biases. Biol Invasion 5:365–377

Gollasch S, Stehouwer PP, David M (2012) Technical outline and requirements for organism detection systems for establishing compliance enforcement with ballast water management requirements. NIOZ, Texel, Final report. Prepared for Interreg IVB North Sea Ballast Water Opportunity project

Hallegraeff GM, Bloch CJ (1991) Transport of toxic dinoflagellate cysts via ship's ballast water. Mar Poll Bull 22:27–30

Hay C, Handley S, Dogdshun T, Taylor M, Gibbs W (1997) Cawthron's ballast water research programme. Final report 1996–97, Cawthron Institute, Report No. 417

Hunt CD, Tanis D, Bruce E, Taylor M (2007) Optical signatures of seawater and potential use for verification of mid-ocean ballast water exchange. Mar Ecol Prog Ser 331:35–47

IMO (2008) Guidelines for ballast water sampling (G2), IMO Resolution MEPC.173(58). International Maritime Organization, London

IMO (2009) Report of the Ballast Water Working Group, BLG 13/WP.5. International Maritime Organization, London

IMO (2010a) Results of an Onboard Ballast Water Sampling Study and initial considerations on how to take representative samples for compliance control with the D-2 Standard of the Ballast Water Management Convention, BLG14/INF.6. International Maritime Organization, London

IMO (2010b) Additional guidance for port States and ships during enforcement of the standards and requirements of the Ballast Water Management Convention, BLG15/5/1. International Maritime Organization, London

IMO (2010c) Additional guidance on indicative analysis, BLG15/5/4. International Maritime Organization, London

IMO (2010d) Additional guidance on sampling methodologies and sampling protocols, BLG15/5/5. International Maritime Organization, London

IMO (2010e) Additional guidance for the analysis of a ballast water sample, BLG15/5/6. International Maritime Organization, London

IMO (2013) International Convention for the Control and Management of Ships' Ballast Water and Sediments, 2004, Guidance on ballast water sampling and analysis for trial use in accordance with the BWM Convention and Guidelines (G2), BWM.2/Circ.42. International Maritime Organization, London

King DM, Tamburri MN (2010) Verifying compliance with ballast water discharge regulations. Ocean Dev Int Law 41:152–165

Lemieux EJ, Wier TP, Steinberg MK, Robbins SH, Riley SC, Schrack EC, Hyland WB, Grant JF, Moser CS, Drake LA (2010) Design and preliminary use of a commercial filter skid to capture organisms ≥50 μm in minimum dimension (nominally zooplankton) for evaluating ships' ballast water management systems at land-based test facilities. NRL Letter Report 6130/1029, Washington, DC

Lenz J, Andres H-G, Gollasch S, Dammer M (2000) Einschleppung fremder Organismen in Nord- und Ostsee: Untersuchungen zum ökologischen Gefahrenpotential durch den Schiffsverkehr. Umweltbundesamt, Berlin

Macdonald E, Davidson R (1997) Ballast water project. Fisheries Research Services Report Number 3/97, FRS Marine Laboratory, Aberdeen

McCarthy SA, Khambaty FM (1994) International dissemination of epidemic *Vibrio cholerae* by cargo ship ballast and other non-potable waters. Appl Environ Microbiol 60:2597–2601

Medcof JC (1975) Living marine animals in a ships' ballast water. Proc Nat Shellfish Ass 65:54–55

Murphy KR, Ritz D, Hewitt CL (2002) Heterogeneous zooplankton distribution in a ship's ballast tanks. J Plankton Res 24(7):729–734
Murphy KR, Boehme J, Coble P, Cullen J et al (2004) Verification of mid-ocean ballast water exchange using naturally occurring coastal tracers. Mar Poll Bull 48:711–730
Murphy KR, Ruiz GM, Dunsmuir WTM, Waite TD (2006) Optimized parameters for fluorescence-based verification of ballast water exchange by ships. Environ Sci Technol 40:2357–2362
Murphy KR, Field MP, Waite TD, Ruiz GM (2008) Trace elements in ships' ballast water as tracers of mid-ocean exchange. Sci Total Env 393:11–26
Oemcke D, van Leeuwen J (1998) Chemical and physical characteristics of ballast water: implications for treatment processes and sampling methods. CRC Reef Research, Technical Report No. 23
Ruiz GM, Rawlings TK, Dobbs FC, Drake LA, Mullady T, Huq A, Colwell RR (2000) Global spread of microorganisms by ships. Ballast water discharged from vessels harbours a cocktail of potential pathogens. Nature 408:49–50
Schillack L (2013) Effective new technologies for the assessment of compliance with the Ballast Water Management Convention. Presentation by Germany, BLG17. International Maritime Organization, London
Sutton CA, Murphy K, Martin RB, Hewitt CL (1998) A Review and evaluation of ballast water sampling protocols. Centre for Research on Introduced Marine Pests (CRIMP), Technical report number 18, Hobart
Wier T, Moser C, Drake L, Grant J, First M, Robbins-Wamsley S, Riley S (2013) Third prototype shipboard filter skid (p3SFS): results of shipboard installation and testing. Abstract of the 18th International Conference on Aquatic Invasive Species. Niagara Falls, ON

Ballast Water Management Decision Support System

Matej David and Stephan Gollasch

Abstract A Decision Support System (DSS) is a supporting tool enhancing a decision-making process. Decision-makers are frequently faced with the problem to take decisions on very complex issues, which requires large data inputs, and a timely decision process. DSSs provide decision makers with a tool to reduce uncertainties, and to simplify and speed-up the decision process as well as to avoid subjectivism induced by the decision-maker and to guarantee transparency of a decision process. The DSS approach has been introduced in the ballast water management (BWM) field and the need primarily arose with the introduction of the selective BWM approach. More precisely, it was recognised that a supporting tool is needed to aid transparency and consistency when deciding on BWM requirements to achieve better environmental protection and lessen burden on vessels. The DSS process starts with communication and data input, continues with risk assessment, BWM decisions, vessel's action(s), and ends with monitoring and review processes. Throughout the entire decision process information needs to be exchanged with outer (e.g., vessel, other ports) and inner sources (e.g., vessel's particulars, compliance history), and therefore needs to be supported by adequate communication processes and data management. When required BWM measures were not conducted properly the BWM DSS endpoints range from situations where unmanaged ballast water can be discharged to cases where vessels may be turned away. The chapter provides a detailed step-by-step DSS model which may be used by administrations and other authorities involved in the decision making processes.

Keywords Decision support system • Ballast water management • Port State control • Port State authority • Risk assessment • Ballast water management convention • Harmful aquatic organisms and pathogens

M. David (✉)
Dr. Matej David Consult, Korte, Izola, Slovenia
e-mail: matej.david@siol.net

S. Gollasch
Gollasch Consulting (GoConsult), Hamburg, Germany
e-mail: sgollasch@aol.com

M. David, S. Gollasch (eds.), *Global Maritime Transport and Ballast Water Management*, Invading Nature - Springer Series in Invasion Ecology 8,
DOI 10.1007/978-94-017-9367-4_9

What Is a Decision Support System

A Decision Support System (DSS) is a supporting tool enhancing a decision-making process (Bhatt and Zaveri 2002). DSSs use a combination of models, analytical techniques, and information retrieval to help developing and evaluating appropriate decision alternatives (Adelman 1992; Sprague and Carlson 1982, Sojda 2007). Today DSSs are widely used to support decision-making processes in business, social sciences, medicine, politics, games, information technologies, transport (Marquez and Blanchar 2006), and they are major components in environmental management and science (Denzer 2005).

Decision-makers are frequently faced with the problem to take decisions on very complex issues, which requires large data inputs, and a timely decision process. DSSs provide decision makers with a tool to reduce uncertainties (Graham and Jones 1988), and to simplify and speed-up the decision process.

Decision Process and Decision Support System

One of the critical factors in the decision making process is subjectivism induced by the decision-maker (Paradice 2006). Decisions are influenced by subjectivity mostly because different decision-makers have varying levels and different background, knowledge, skills, moods, etc. The use of a DSS from this point of view is important, because, by principle, it eliminates subjectivity impacts of different decision-makers in the same process, which leads to more consistent results – i.e., decisions. It also ensures consistency of decisions taken by the same decision-maker. However, the DSS is exposed to subjectivity during the preparation/construction process. The results of a decision-making process may further be influenced (sometimes this is almost anticipated) by the authorities that order a DSS, i.e., they would like to achieve a certain result of their interest.

Another critical point is the transparency of a decision process. DSS should be constructed in such a way that decision models as well as decision steps are transparent, thereby enabling a review of the decision process at any time in the future. This is especially critical when a DSS is used in a regulatory framework.

Any errors possibly resulting from a decision process should also be known. Errors could occur in view of exactness and accuracy. Exactness means that a step of the process, or the process itself, in certain instances (e.g., lack of data, reliability of data, precision of data, subjective impacts), could produce a biased (false) result. Accuracy means that the result of a step in the decision process, or the process itself, may have a certain discrepancy or deviation as a consequence of certain instances (e.g., lack of data, reliability of data, exactness of data, subjective impacts). Hence, the DSS should produce exact answers with an acceptable accuracy.

Decision Support System Generic Structure

DSSs may have different elements and structures depending on their field of application and complexity (Denzer 2005). However, their very generic framework may be similar in different fields of application and generally contains decisions and data management. Decisions comprise of management decision steps and decision models (see Fig. 1) which represent the core elements of the DSS. The data management as component of a DSS comprises databases for data retrieval and data storage.

The integration of basic DSS elements is important for the preparation of a computer support architecture (Denzer 2005). As an example, the focus/application of a DSS may use different methodologies supporting the decision making process as, e.g., multicriteria decision making (e.g., Vincke 1993), fuzzy logic (e.g., Ru and Eloff 1996; Ekel 2002; David and Malej 2002), neural networks, etc. Once a computer model for a DSS process is prepared, this may also be used, with some adaptations, for another similar application and hence facilitate the development of a new DSS.

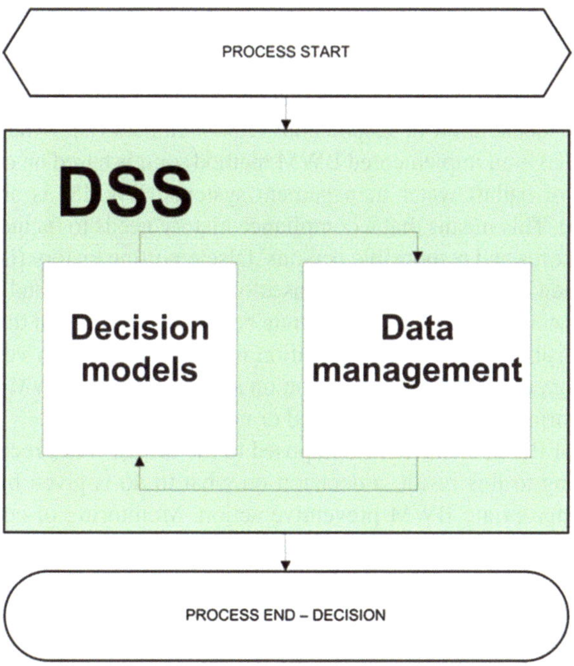

Fig. 1 Basic structure of a decision support system (DSS) showing how decision models and data management are related

Decision Support System in Ballast Water Management

The DSS approach has been introduced in the ballast water management (BWM) field and its need primarily arose with the introduction of the selective BWM approach. More precisely, it is a supporting tool needed to aid transparency and consistency when deciding on most efficient BWM requirements, and to lessen the burden on vessels (David 2007).

There are two different basic selective approaches in BWM, i.e., the "risk assessment (RA) approach" and "compliance history approach".

The RA approach is when the decision on BWM requirements relies purely on results from a scientifically based RA. For instance, a vessel which sails to a port where it needs to discharge ballast water, may be exempted from BWM requirements if the ballast water does not pose a risk or is of an acceptable level of risk to a recipient port. However, if the ballast water is found to be of (very) high risk, different additional measures may be introduced as a protective BWM measure. The RA approach could be either based on environmental matching, be species specific or use biogeographical aspects (see chapter "Risk Assessment in Ballast Water Management").

The compliance history approach relies on the documentation of vessels compliance or non-compliance with requested BWM practices, which is very much the regular practice of Port State Control (PSC) inspections. Vessels may not be in compliance with BWM requirements for different reason (e.g., technical failure, bad weather). However, the critical issue is that compliance monitoring in the first place is based on the declaration of responsible crew members (i.e., when ballast water exchange (BWE) is an implemented BWM method) or it is based on certificates (i.e., when the use of ballast water management systems (BWMS) is an implemented BWM method). This means that a compliance history needs to include vessels non-compliance records and responsible persons' false reporting history (i.e., trustworthiness) (Chad Hewitt pers. comm.). In cases of non-compliance and relative to the reason (e.g., history of technical failure may be treated less critical than false reporting of a responsible person), more attention may be paid to such vessels to ensure compliance, e.g., conduct PSC inspection on such vessels, or BWM measures may be even more stringent because of limited or no trustworthiness.

The result of RA is the level of risk posed to the ballast water receiving environment. According to this result, a decision on what to do is given by the DSS and followed by appropriate BWM preventive action. Monitoring of compliance with the implemented BWM regime (i.e., requested actions) is essential. Further, monitoring of compliance, as well as the DSS effectiveness, also needs to be conducted. If necessary, corrective actions are to be taken (see Fig. 2).

While the RA result is a simple answer in terms of the level of risk, in the following steps a more complex process is generated when a decision on "what to do" has to be taken considering the RA result, vessel trustworthiness, adequate and feasible BWM options, etc. DSS is the core part or, in other words, is the brain of the whole process.

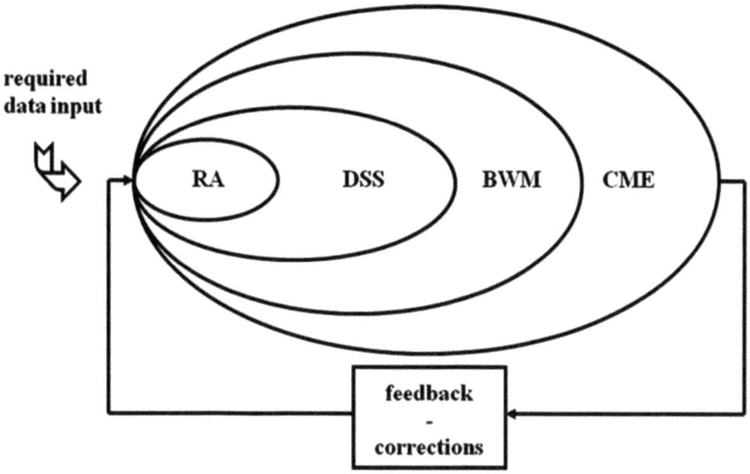

Fig. 2 The BWM process under the selective approach supported by the DSS (Enhanced after David 2007). *BWM* Ballast Water Management, *CME* Compliance Monitoring and Enforcement, *RA* Risk Assessment

Ballast Water Management Decision Support System Model

Model High Level Elements and Sequences

The DSS process starts with communication and data input, continues with RA, BWM decisions, vessel's action(s), and ends with a monitoring and review processes. Throughout the entire decision process information needs to be exchanged with outer (e.g., vessel, other ports) and inner sources (e.g., vessel's particulars, compliance history), and therefore needs to be supported by adequate communication processes and data management (see Fig. 3).

One of the critical issues is the position/situation/location of the vessels in relation to its ability to comply with requested BWM measures. In this regard we created four situations a vessel may be facing:

Situation (1), the vessel has left the last port of call and is able to conduct BWM on its intended route, and:

- has time and is in conditions to conduct the requested BWM measure(s);
- conducts BWM measures according to the requirements and enters the port with the permission to discharge ballast water.

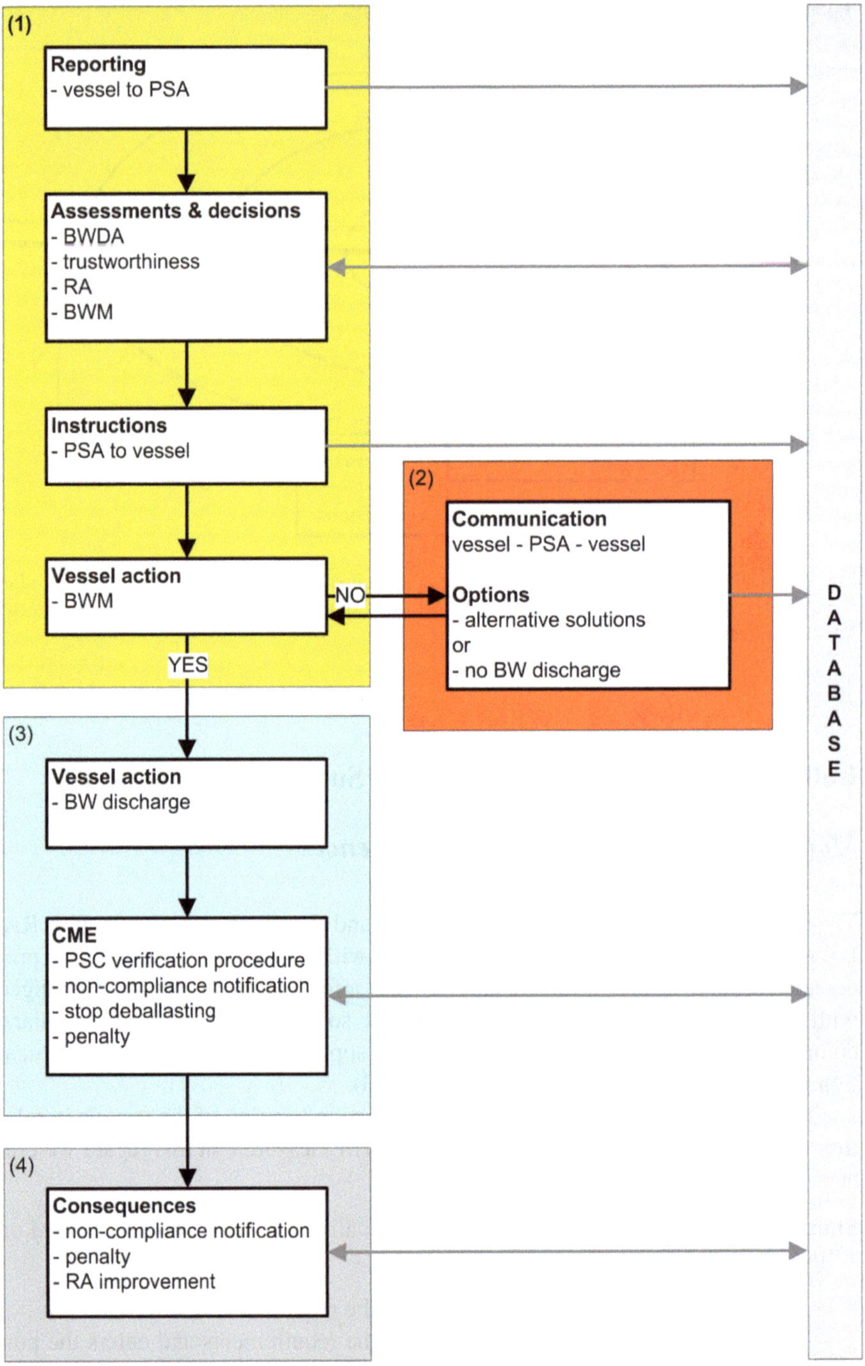

Fig. 3 DSS model high level elements (Enhanced after David 2007). *BW* Ballast Water, *BWDA* Ballast Water Discharge Assessment, *PSA* Port State Authority, *PSC* Port State Control. *Yellow box* is Situation (1) – vessel is on the way to port of call, BWM enabled; *orange box* is Situation (2) – vessel is on the way to port of call or even entered the port, no BWM enabled and the port entry permit is not yet issued; *light blue box* is Situation (3) – vessel is in the port, the port entry permit is issued; and *grey box* is Situation (4) – vessel has left port of call

Situation (2), the vessel has left the last port of call but is not able to conduct BWM on its intended route, and may already be in the port of arrival but the port entry permit[1] is not yet issued, and:

- did not use the BWMS;
- did not conduct BWM, but complies with the requirements (when the D-1 standard is required) because on its intended route the vessel does not exit the 50 nautical miles from nearest land and 200 m water depth limits to enable a BWE, nor it crosses a ballast water exchange area (BWEA);
- did not conduct BWM for other reasons;
- is deviated from its intended route to the BWEA and/or slowed down to conduct BWE and complies with the requirements;
- would need to be sent outside the 50 nautical miles and 200 m limits, or to a BWEA at a substantial change of her intended route, or use an alternative solution[2] to conduct BWM and comply with the requirements;
- depending on the RA result, may be allowed to discharge unmanaged ballast water,
- may be penalized, or
- may not be allowed to discharge ballast water without BWM.

Situation (3), the vessel is in the port of arrival and has received the port entry permit, and:

- may be targeted for different levels of compliance control;
- if the vessel is found non-compliant with BWM requirements, depending on the RA result, she may be allowed to discharge unmanaged ballast water, may be penalized, or may not be allowed to discharge unmanaged ballast water.

Situation (4), the vessel has received clearance[3] and left the port, and:

- the vessel's ballast water was sampled and was identified as non-compliant with BWM requirements, this is communicated to the vessel, the vessel's administration, the recognized organization responsible for the issue of certificates, and the next port of call; or
- the vessel's ballast water was sampled and was identified as compliant with BWM requirements, no action is required.

The high level elements with the four different vessel's situations are presented in Fig. 3.

The DSS integrates seven basic elements:

- data collection and management process;
- communication processes;
- ballast water RA process;

[1] Permit to start operations in a port, including anchorage, after having complied with port State requirements and submitted all required documents for port entry.

[2] e.g., alternative ballast water discharge area, port reception facilities.

[3] Permit to leave the port after having complied with port State requirements and submitted all required documents for leaving the port.

- BWM decision and action process;
- BWM action process;
- compliance monitoring process; and
- RA review process.

Each of these elements has its own function and the DSS structure provides for their effective integration, and supports their independent functioning as well as their mutual interrelations.

Data Collection and Management Process

The data collection process is critical simply because decisions are based on these; i.e., wrong data input would result in a wrong answer. This mostly relies on port States, as the BMW Convention does not provide requirements for reporting from vessels to ports regarding ballast water operations, but only requires an on board BWM log book. In this context two main aspects need to be considered: data availability; and data reliability.

Regarding data availability, correct data need to be available at the right time. This means all data needed for the whole process to enable taking all decisions are essential; e.g., biological data on ballast water source ports, environmental data from source and receiving ports, vessels data on previous reporting. It is important that the data are available timely to allow the vessel to conduct the requested BWM practice; i.e., time to conduct RA, take a decision on BWM requirements, communicate with vessel, conduct BWM or take appropriate action.

Data reliability has quality and quantity aspects. The quality of data in the first place means that the DSS input data are based on reliable sources. In terms of quantity, there should be enough comprehensive data to be statistically robust.

Most of the data received externally as well that from the decision process need to be managed properly, be safely stored and accessible, which may be best arranged in a DSS database. This database provides the DSS with the information needed, e.g., vessels particulars, historical data on vessels compliance, as well as it is serving the needs of outer sources; e.g., sharing information with other stakeholders, reviewing the DSS process and as back-up for a later review process of a single decision.

Communication Processes

The communication process consists of communications among the port State authority (PSA) and:

- the vessel;
- the vessel's administration;
- the vessel's recognized organization responsible for the issue of certificates

- the vessel's next port of call; and
- IMO.

The essentially needed communication regarding BWM is established between PSA in the ballast water recipient port and the vessel that intends to discharge ballast water as follows:

- the vessel intending to discharge ballast water submits requested information as ballast water reporting form (BWRF) to the PSA;
- PSA communicates to the vessel the decision on BWM requirements;
- other communication, e.g., in case the BWRF was not satisfactory completed or the vessel was not able to conduct the required BWM.

In case a vessel would be found non-compliant with the BWM Convention, PSA that established this, needs to communicate it to the related vessel, the vessel's Administration, the vessel's next port of call and the recognized organization responsible for the issue of certificates. Should additional BWM measures be introduced in a known epidemic or emergency situation, PSA needs to communicate this to all vessels in the area(s) under their jurisdiction where vessels should not uptake ballast water, and the ballast water uptake avoidance area(s) need to be communicated also to IMO.

The preferred communication pathway may be via electronic means, fully or partially automated, e.g., via internet application, email, fax, telex, vessels agent, telephone. Non-automated means of submitting information, i.e., on paper forms, are considered as impractical since the information would not be exchanged and implemented into DSS in a timely manner.

Ballast Water Risk Assessment

The RA forms a core part of DSS triggering different decisions regarding:

- BWM practice needed;
- compliance monitoring needs; and
- the level of inspection.

In the RA based DSS, the decisions on BWM practices mostly[4] rely on the results of the RA, e.g., high/extreme risk – the vessel must conduct BWM, medium risk – should conduct BWM, low risk – may conduct BWM. The RA results are further critical for taking decisions regarding compliance monitoring; i.e., targeting vessels for inspection, as well as taking decisions on the level of inspection, i.e., paper checks, indicative BWS, detailed BWS.

---

[4] Decision on the need for BWM practice may rely also on trustworthiness, *i.e.*, compliance history of a vessel, master or responsible officer.

Ballast Water Management Decision and Action

Management decisions in this context are required BWM practices which are selected on the basis of the RA result, vessels trustworthiness, and if the BWM measures have already been undertaken the acceptability of these. Based on the RA result and vessel's trustworthiness, the vessel may also be exempted from undertaking BWM, or may be exposed to additional measures according to the level of risk assessed. Such additional measures include to conduct BWE, deviate from its intended route or slow down to conduct full BWE, treat ballast water with active substances before discharge, discharge ballast water to a reception facility, or do not discharge unmanaged ballast water.

Compliance Monitoring

Compliance monitoring is a process needed to back-up the port State requirements. It is focused on the adequate and effective implementation of the requested BWM practices. This process may be triggered by suspected false ballast water reporting (e.g., ballast water discharge assessment (BWDA) result, vessels and/or crews trustworthiness), by suspected non-compliance, by RA (e.g., when high/extreme risk ballast water is to be discharged), or by random vessel selection as part of the regular inspections process. A vessel selected for compliance monitoring will be inspected, and if non-compliant the ballast water operation may have to be stopped, and the vessel may be penalized.

Risk Assessment Review Process

A review process needs to be implemented which is critical for further improvements of the BWM DSS process and results. The review process includes a re-assessment of the RA procedure based on ballast water sampling results.

Generic Ballast Water Management Decision Support System Model

The BWM DSS process starts with the vessel submitting the required data to enter the port, and through the RA and BWM ends with the monitoring process and, if necessary, result in corrective actions of the process. Throughout the entire process is a dynamic flow of information exchange supported by adequate communication processes and data management. Considering that there are a lot of different situations and issues (e.g., non-adequate or false reporting, non-ability to comply with required BWM practice, technical issues) that may arise during each vessel call to a port, the BWM DSS model was prepared to cover possibly all predictable events, as well to respond rapidly. The generic model is presented in the Fig. 4, followed by

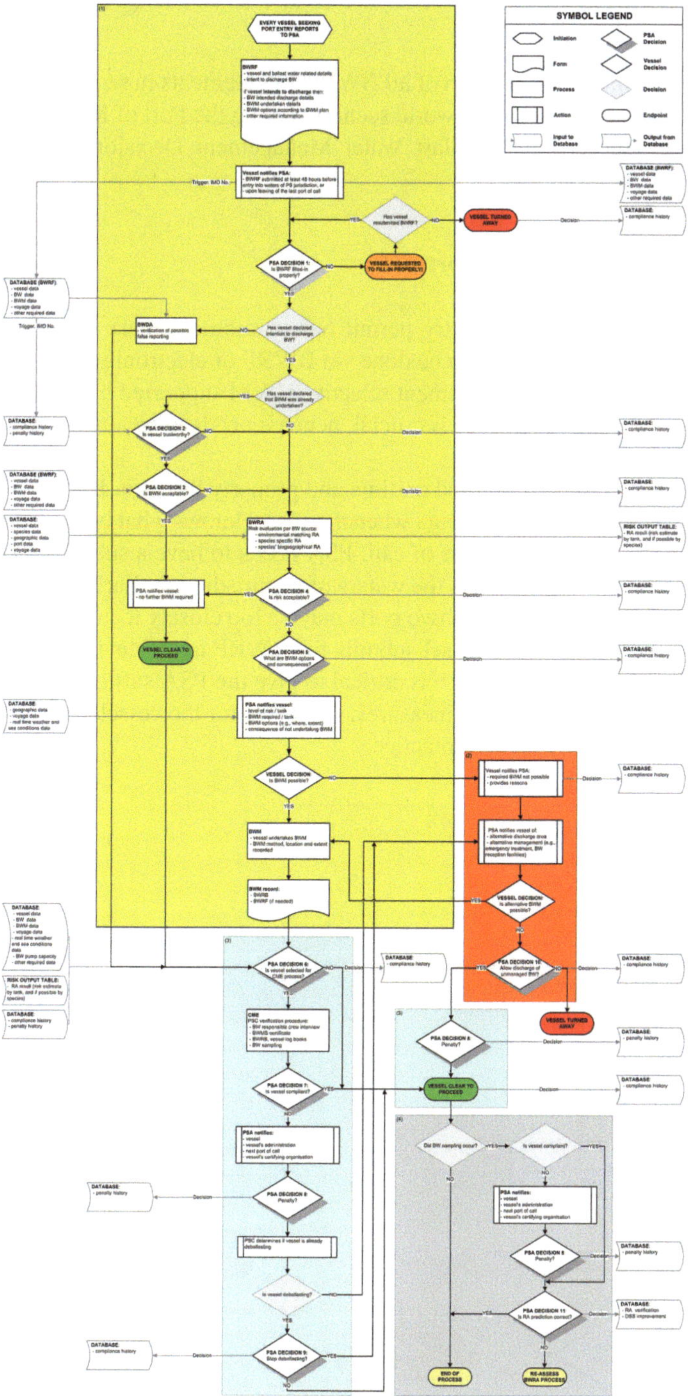

Fig. 4 BWM DSS generic model (Enhanced after David 2007). The *yellow box* is Situation (1) – vessel is on the way to port of call, BWM enabled); *orange box* is Situation (2) – vessel is on the way to port of call or even entered the port, no BWM enabled and the port entry permit is not yet issued; *light blue box* is Situation (3) – vessel is in the port, the port entry permit is issued; and *grey box* is Situation (4) – vessel has left the port (*BWRA* Ballast Water Risk Assessment, *BWRB* Ballast Water Record Book, *BWRF* ballast water reporting form) (This figure can be downloaded from http://extras.springer.com/)

the presentation and description of all BWM DSS elements in sequence. The BWM DSS was also applied to a real world scenario taking the Port of Koper, Slovenia, as an example (see chapter "Ballast Water Management Decision Support System Model Application").

Vessel Intended to Enter a Port

Each vessel seeking a port entry permit has to submit ballast water information requested by the PSA. This can be done via BWRF or electronic means, depending on PSA requirements. To implement selective BWM supported by BWM DSS, ballast water reporting in advance is crucial, hence it needs to be a mandatory requirement for port entry (Fig. 5).

BWRF needs to be submitted on time and properly filled-in. BWRF needs to be submitted as soon as possible; e.g., when the vessel knows what ballast water operation is expected in the next port of call. PSA needs to have a submission deadline, e.g., 48 h before a vessel enters the waters of its jurisdiction. Early submission may not always be possible because two ports may be too closely located. In such a case it is recommended that the vessel submits the BWRF upon leaving the last port of call. Early submission of BWRF is critical to give the PSA sufficient time to take a decision on appropriate BWM measures, as well as for the vessel to be in a position to conduct the required BWM practice.

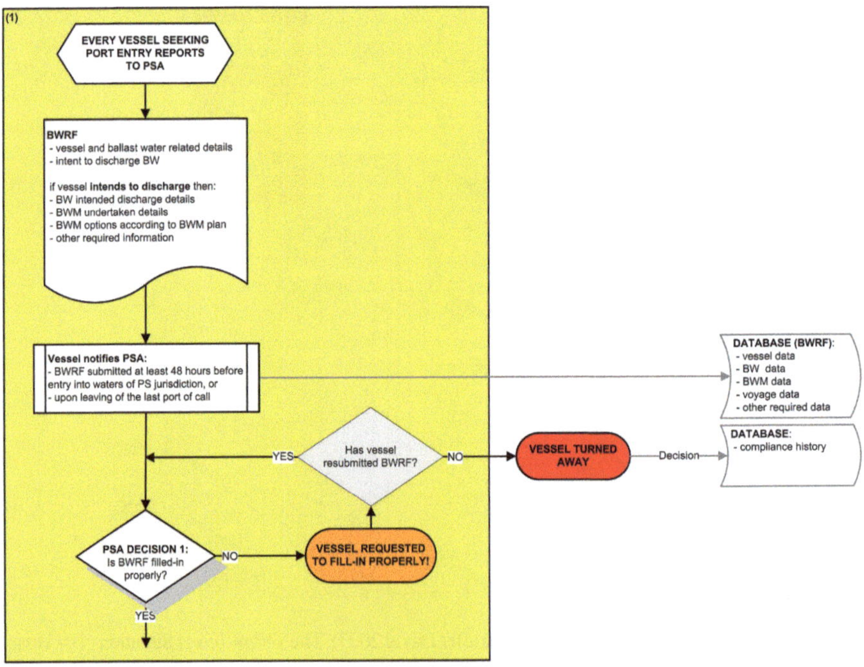

Fig. 5 BWRF submission process (*PS* Port State) (Enhanced after David 2007) (This figure can be downloaded from http://extras.springer.com/)

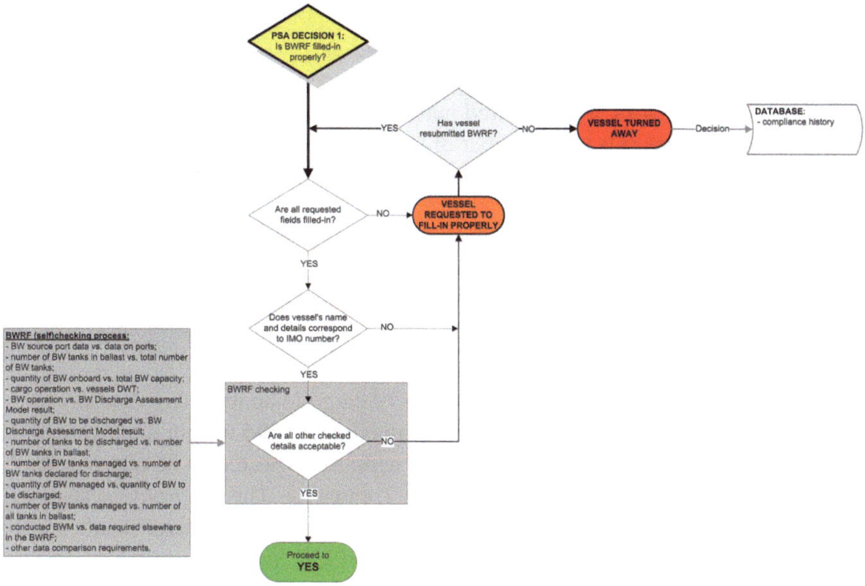

Fig. 6 Decision (1) on correct BWRF submission (*DWT* Dead Weight Tonnage) (This figure can be downloaded from http://extras.springer.com/)

Decision 1: Was BWRF Filled-In Properly?

BWRF has to be filled-in properly to start the DSS procedure. It is absolutely critical for the PSA, i.e., DSS, to have all requested data available to be able to take a proper BWM decision. Omissions, mistakes, as well as false-reporting can be anticipated. Therefore, the data provided need to be checked quantitatively as well as qualitatively. The vessel cannot obtain a port entry permit if it does not submit all required data (same practice as for other required reporting). Such cases are also registered in the "compliance history" database (see Fig. 6).

In the first two steps the BWRF is checked to ensure that all requested fields are filled-in and that the vessel's basic data correspond with the IMO number. If this is not satisfactory, the BWRF should not be accepted and the vessel becomes automatically turned away. In case an electronic submission system is used, this can be checked automatically and the BWRF does not need be submitted on paper.

In the third step, the submitted data is further checked qualitatively (see Fig. 6, grey box on the left).

The checking process includes:

The ballast water source port data (e.g., UN LOCODE port code, name, geographical position) needs to be confirmed to ensure true data for the source of ballast water intended to be discharged. This is important for the assessment of different vessel voyage related data, however this is absolutely critical for the RA data needs, and includes biological and environmental data. The UN LOCODE port code is the suggested triggering reference. An electronic system may be used to check this automatically which may also be done for elementary port data. In case there is no

UN LOCODE data provided for the relevant source port, the vessel needs to provide its name (in English) and geographical position.

The number of ballast tanks in ballast is to be compared with the vessel's total number of ballast tanks. The declared number of filled ballast tanks in no case can be larger than the vessels total number of ballast tanks. This can be compared automatically by the electronic system, which does not allow a higher tank number to be entered in the BWRF.

The quantity of ballast water on board is to be compared with the vessel's total ballast water capacity. The declared ballast water quantity on board in no case can be higher than the vessels total ballast capacity. The electronic system can compare these numbers automatically and does not allow a higher amount of ballast water to be entered in the BWRF.

The cargo operation (i.e., quantity in tons of cargo to be loaded or discharged) in no situation can be greater than the vessel's maximum dead weight tonnage (DWT) capacity.[5] The maximum cargo capacity is actually expected to be approximately 10 % lower than the vessels maximum DWT capacity. The electronic system can compare these numbers automatically and does not allow a greater number to be entered in the BWRF. This information is also critical for the assessment (verification) of expected (reported) ballast water operations in the related port, which is automatically done by the BWDA model.

A ballast water operation is to be expressed in terms of the expected ballast water quantity to be discharged or loaded in the related port. The declared operation, as well as the declared quantity of ballast water intended to be discharged, are to be compared with the BWDA model result. A mismatch in operation (i.e., no discharge declared but the model shows a discharge) as well in quantity (i.e., the model assessed discharge quantity of ballast water is substantially bigger than declared) triggers suspicion that there is a mistake in reporting, or even false reporting. However, it is not suggested that this would automatically prevent BWRF submission and the vessel to be turned away, but this information is to be used later as the trigger in the compliance monitoring process.

If the vessel has declared that it will discharge ballast water in the port, the number of ballast tanks to be discharged is to be compared with the number of tanks in ballast. The declared number of ballast tanks to be discharged in no case can be higher than the number of tanks in ballast. Again, the electronic system can compare these numbers automatically and does not allow a higher number to be entered in the BWRF.

If the vessel has declared to have already managed the ballast water intended for discharge, the number of ballast water tanks managed is to be compared with the number of ballast water tanks to be discharged. The declared number of ballast water tanks managed can be greater or lesser than the number of ballast water tanks declared for the discharge. In practice it is not expected that a vessel would conduct

[5] i.e., vessel's carrying capacity, which includes cargo and all weights (e.g., fuel, ballast water, stores), crew and passengers that may be loaded onboard a vessel up to her permissible limits, which is regulated by IMO international conventions, mainly the Load Lines Convention.

BWE for tanks if these are not intended to be discharged, hence a higher number of those BWE managed tanks would most likely be a mistake. However, as a consequence of using BWMS (i.e., treatment of ballast water to meet the D-2 standard) which treats ballast water on uptake, this would be a regular result. In case the declared number of ballast water tanks managed is lower than the number of ballast water tanks declared for the discharge, it is necessary to confirm whether this is a mistake or there are tanks with ballast water that need to be considered in the next steps by the RA process. The numbers need to be compared as follows:

- if the number of managed tanks is greater than the number of tanks declared for the discharge, allow submission of BWRF with no further questions;
- if the number of managed tanks is lower than the number of tanks declared for the discharge, the vessel needs to correct this to have the same numbers, or declare the tanks that have not been managed, but are to be discharged (i.e.; Number of tanks to be discharged = Number of managed tanks to be discharged + Number of unmanaged tanks to be discharged). Should these BWRF entries not match, then the BWRF should not be allowed to be sent or not be accepted by the PSA.

Whichever BWM method has been declared, it should be confirmed that tanks declared for the discharge are those which were managed. The electronic system can compare this automatically and act as appropriate.

The quantity of ballast water managed is to be compared with the quantity of ballast water to be discharged. This is an analogue process, a comparison of the number of ballast water tanks managed vs. the number of ballast water tanks declared for discharge as described above. Hence, the same procedure is to be applied using "quantity of ballast water" instead of "number of tanks".

The number of ballast water tanks managed is to be compared with the number of all tanks in ballast. The declared number of ballast water tanks managed in no case can be greater than the number of all tanks in ballast. The electronic system can compare these numbers automatically and does not allow a greater number of tanks with managed ballast water compared to all tanks in ballast be entered in the BWRF to be submitted.

The conducted BWM is to be compared with the data required elsewhere in the BWRF. If a vessel has declared that it has already conducted BWM also stating the BWM method used, there is a need also to report the number of managed tanks with the quantity of ballast water managed, and if BWE was used as BWM method, it needs to be reported where this was conducted. The BWM method declared and further information requirements need to be related quantitatively, i.e., all fields related need to be filled-in, and when possible also qualitatively.

Since this is a generic DSS model, it is expected that when it is applied, regional and national specific requirements may result in a need to add different 'other' data comparison requirements.

If a vessel reports satisfactory, then it enters the next phase of the DSS process, in which she is being selected to enter the RA process.

Selection of a Vessel for the RA Process

The selection of a vessel that will need to enter the RA process is done on the basis of data submitted via BWRF. Basically, all vessels which have declared an intention to discharge unmanaged ballast water in the port are selected for the RA process.

If a vessel has declared that it has ballast water on board which will not be discharged, such a vessel will not enter the RA process but will be notified that no BWM requirements apply to her and she is cleared to proceed. However, theoretically every vessel carrying ballast water and coming into a port has the potential to discharge ballast water, and in view of possible false reporting, such a vessel is checked with the BWDA model (see chapter "Vessels and Ballast Water"). The foreseen ballast water operation is assessed on the basis of expected cargo operations and vessel's particulars. If the BWDA model result disagrees with the declaration, the vessel will be targeted for the verification process.

If a vessel declared that she has already managed the ballast water intended for discharge, then she will be, in the next two steps, checked for her trustworthiness and the acceptability of the BWM method used. If she is found not trustworthy or the BWM used was not acceptable, than she will enter the RA process. If a vessel was not selected for RA process she is clear to proceed (see Fig. 7).

Decision 2: Is Vessel Trustworthy?

The main reason for introducing trustworthiness is the human factor. It is known that false reporting occurs and that it is very difficult to survey it. There are also many other reasons, some of the outstanding are low quality of vessel systems maintenance, low crew skill level, sometimes also ignorance. These, however, are also critical for proper and safe functioning of vessel systems.

Trustworthiness is focussed on the history of the false reporting of responsible crew members, as well as on the vessel compliance history. False BWM reporting related to a person may be kept in the records lifelong or time dependent, i.e., valid for a certain period of time, e.g., 10 years. The vessel BWM compliance history and general compliance is time dependent (see Fig. 8).

Decision 3: Is Ballast Water Management Acceptable?

If a vessel declares that it has already conducted BWM, this needs to be compared with the port State BWM requirements. The decision relies on the information provided in the BWRF.

All ballast water tanks that are intended for discharge need to be managed and the BWM method used is generally accepted if it fulfils the requirements of the BWM Convention and/or those of the port State. It is also important that the vessel follows procedures and requirements of the BWMS manufacturer and classification society (see Fig. 9).

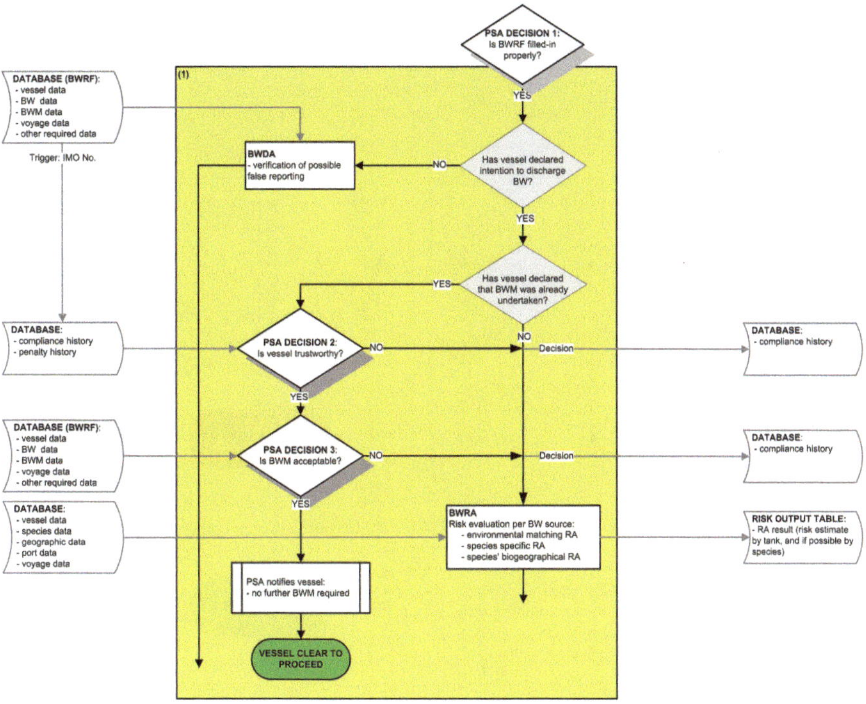

Fig. 7 Selection of a vessel for the RA process (Enhanced after David 2007). The *open arrow* going down from the BWDA box is directed to connect to the *Decision 6*: *Is vessel selected for CME process*? (see section "Decision 6: Vessel Selected for Compliance Monitoring and Enforcement Process?"). The *open arrow* going down from the BWRA box is directed down to connect to the *Decision 4*: *Is risk acceptable*? (see section Decision 4: Is Risk Acceptable?) (This figure can be downloaded from http://extras.springer.com/)

Ballast Water Risk Assessment Process

In this phase of the DSS process RA is undertaken to provide for adequate BWM based on the acceptability of the risk level assessed.[6] If the level of risk is acceptable, then the vessel will be cleared to proceed without conducting BWM. However she may still be selected for the verification process. If the level of risk is not acceptable, the vessel will need to undergo a BWM procedure (see Fig. 10).

Decision 4: Is Risk Acceptable?

The risk assessment process is in detail described in chapter "Risk Assessment in Ballast Water Management" and covers RA background, principles, RA end points, RA methods, RA errors and the application of RA under the BWM Convention.

[6] this is dependent on the port State environmental legislation, and the perception, values and ethics of the assessors.

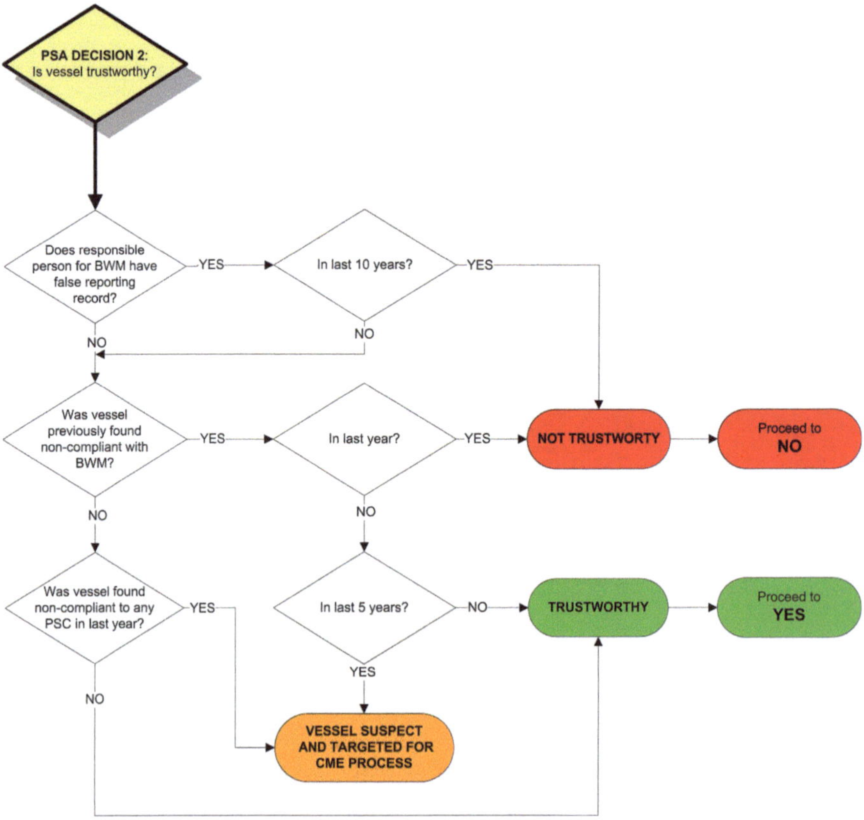

Fig. 8 Decision (2) on vessel trustworthiness (Enhanced after David 2007) (This figure can be downloaded from http://extras.springer.com/)

For the purpose of this BWM DSS, the BWRA model presented in chapter "Risk Assessment in Ballast Water Management" is to be applied to assess the level of risk for selective BWM measures. When the level of risk assessed is extreme, high or intermediate, it is deemed not to be acceptable (see Fig. 11).

Ballast Water Management Process

BWM requirements apply to a vessel when the risk posed by the ballast water intended for discharge is deemed unacceptable. This includes the selection of a feasible (for the vessel) and acceptable (for PSA) BWM method according to the level of risk posed, which is followed by consequences if the required BWM measure is not applied.

There may be different instances when a vessel may not be able to conduct BWM (e.g., route too close to the shore, bad weather and sea conditions, some issue with the BWMS). In those instances, the PSA needs to take a decision whether to allow the vessel to discharge unmanaged ballast water, or use (if available) some alternative

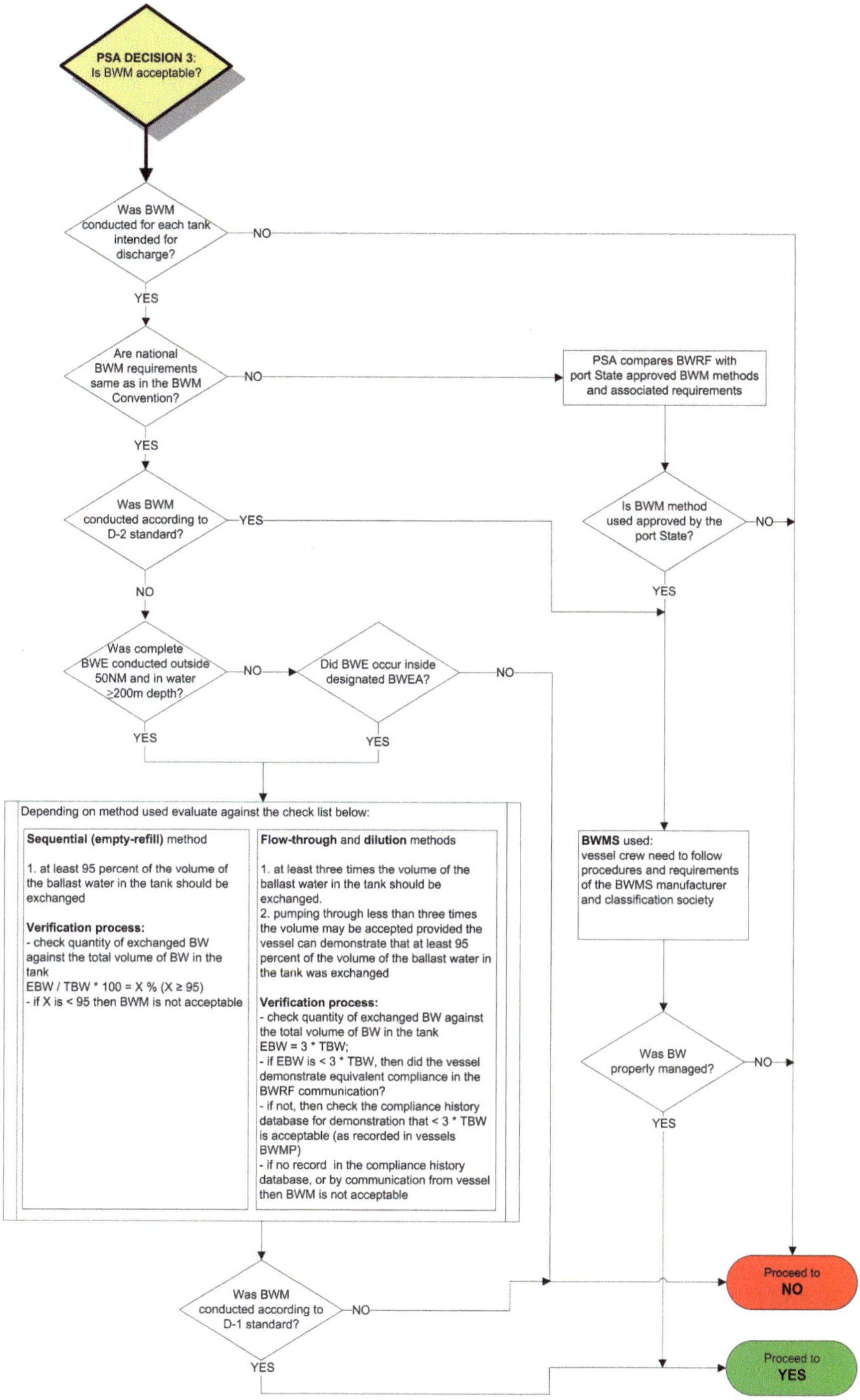

Fig. 9 Decision (3) on acceptability of BWM conducted (Enhanced after David 2007). *BWE* Ballast Water Exchange, *BWEA* Ballast Water Exchange Area, *BWMP* Ballast Water Management Plan, *BWMS* Ballast Water Management System, *D-2 standard* D-2 standard of the BWM Convention, *Reg. B4 and D-1 standard* Regulation B4 and D-1 standard of the BWM Convention, *EBW* Exchanged Ballast Water, *TBW* Total Ballast Water (This figure can be downloaded from http://extras.springer.com/)

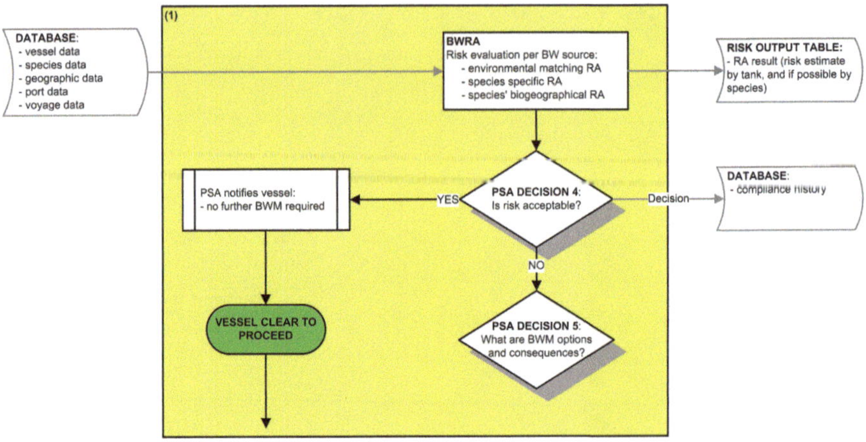

Fig. 10 The BWRA process (Enhanced after David 2007). The *open arrow* going down from the *green box* is directed to connect to the *Decision 6*: *Is vessel selected for CME process*? (see section "Decision 6: Vessel Selected for Compliance Monitoring and Enforcement Process?") (This figure can be downloaded from http://extras.springer.com/)

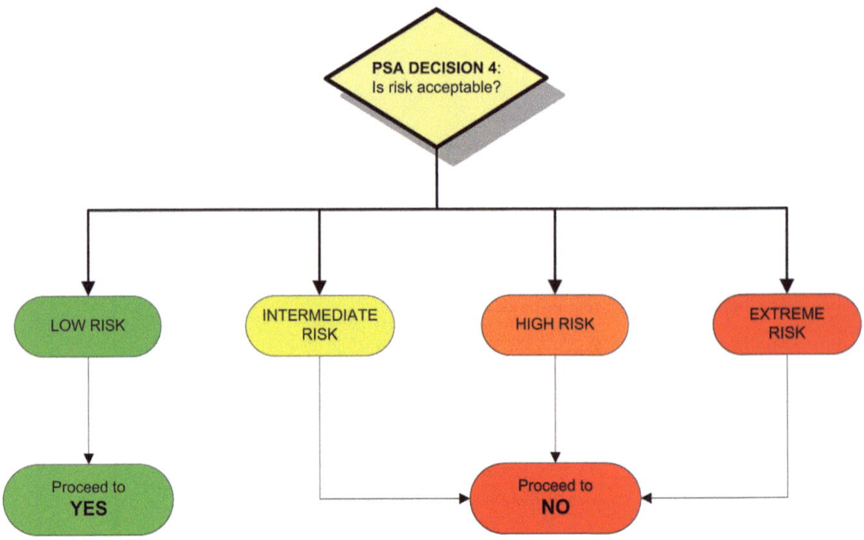

Fig. 11 Decision (4) on whether or not the risk posed by the ballast water intended for discharge is acceptable (This figure can be downloaded from http://extras.springer.com/)

option, retain the ballast water onboard, or in most critical situations to turn the vessel away. All these decisions are dependent on the risk level posed by the ballast water intended for discharge, by the vessel BWM options and the availability of alternative BWM options (see Fig. 12).

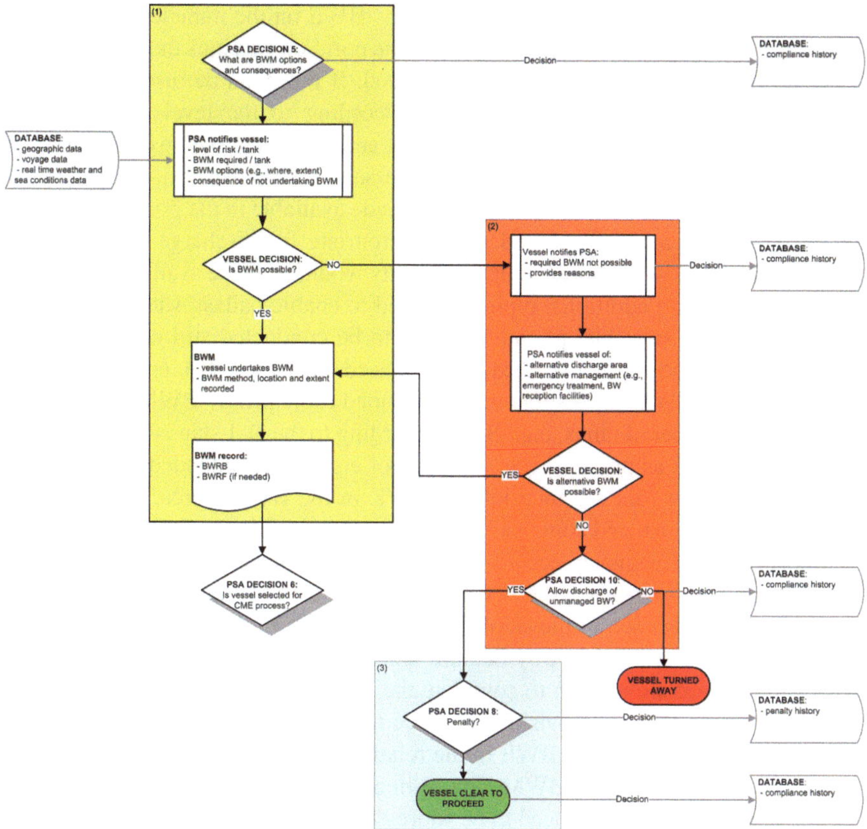

Fig. 12 BWM process (Enhanced after David 2007) (This figure can be downloaded from http://extras.springer.com/)

Decision 5: What Are Ballast Water Management Options and Consequences?

Retaining ballast water on board the vessel is considered as the first BWM option. This is only a feasible option for some vessel types with smaller ballast water capacity and especially in cases when vessels are only partially loading. If necessary the vessel might manage it by pumping ballast water from one tank to another without any discharge to the port. If this cannot be done, that ballast water would need to be managed.

If a vessel is capable of managing ballast water according to the BWM Convention D-2 standard, then it does so and is clear to proceed. If a vessel does not have BWMS installed, then BWE would need to be conducted as minimum BWM measure.

If the vessel is capable of properly conducting BWE on the intended route[7] for all the ballast water intended for discharge, and the port State accepts the BWE method, then the vessel does so and is clear to proceed. If not, then the use of BWEA or alternative options are to be considered according to the level of risk posed. Certainly, BWEA needs first to be designated, and/or the alternative options need to be studied, be available, and be feasible for vessels. Alternative options include ballast water reception facilities which may be made available in the port or an alternative discharge area may be found more appropriate for discharge of unmanaged ballast water. If a ballast water reception facility would be made available, the vessel would need to have additional piping installed to enable ballast water discharge in such a facility. If possible, partial BWM is to be conducted still on the intended route, and may be then continued and finalised in the BWEA or as alternative method. Partial BWM means that on the intended route proper BWM is conducted on a limited number of tanks, e.g., BWE according to the D-1 standard is conducted for as many tanks as possible, e.g., four out of eight tanks intended for discharge, and the remaining four are then left for BWE in the BWEA area, for alternative management options, and some ballast if necessary may also be retained on board.

If a BWEA is designated according to the BWM Convention provisions, vessels may use it if they sail through it on their intended route or if they choose to deviate, though this is a decision of the ship's Master. Hence, it can be anticipated that vessels will unlikely use BWEA by default; and even less so can it be expected that vessels deviate or slow down to complete the BWE within a BWEA. Therefore, a port State needs to have provisions in place to advise the vessel what to do. The requirements to regulate the BWE in the relation to the BWEA are not deemed as additional measures by the BWM Convention. However, most alternative options will be deemed as such and need to be addressed according to the provisions of the BWM Convention for additional measures (see chapter "Policy and Legal Framework and the Current Status of Ballast Water Management Requirements").

One of the very important aspects for appropriate BWM is that a vessel does not exchange the ballast water on board with water that is of a greater risk, e.g., areas with toxic algae blooms, which may occur in the BWEA. Even if the water in the BWEA is of the same risk level, BWE should not be conducted since the "older water" in the tanks is expected to be of lesser risk than the "new" exchanged water and can therefore lead to increased risk by adding, e.g., new nutrients or new organisms to the ballast tank. In consequence, as by the IMO *Guidelines on designation of areas for ballast water exchange* (G14) a BWEA should be monitored for HAOP. In case of HAOP presence in the BWEA, the vessels need to be instructed as appropriate to avoid BWE in this area (IMO 2006).

The following BWM options and consequences have been included if BWEA and/or alternative BWM options are available:

---

[7] The vessel may also consider a slight deviation and change "the shortest" route to be able to conduct BWE according to the BWM Convention limits, i.e., >50 NM distance of shore and >200 m of depth.

If the ballast water to be discharged was assessed as posing an intermediate risk, then:

- if a vessel crosses BWEA on its intended route, then she is requested to conduct BWE, but only if the water in the BWEA poses a low risk; however
- if a vessel is not able to fully complete BWE in a BWEA, here a deviation or slowing down is not meant to be requested.

If it was not appropriate or the vessel was not able to conduct or fully complete BWE in the BWEA, she will be:

- allowed to discharge unmanaged ballast water in the port; and
- targeted for a verification process.

If the ballast water to be discharged was assessed as posing a high risk, then:

- if a vessel crosses BWEA on its intended route, then she is requested to conduct BWE, but only if the water in the BWEA poses low or intermediate risk;
- if a vessel does not cross BWEA on its intended route, then she is requested to deviate a reasonable distance[8] to use the BWEA;
- if a vessel is not able to complete full BWE while crossing the BWEA, then she is requested to slow down or take other measures to fully complete BWE; and
- if a vessel was not able to conduct BWE or fully complete BWE in the BWEA, then she is requested to conduct alternative BWM.

If it was not appropriate or the vessel was not able to conduct or fully complete BWE in the BWEA, and the vessel has no further option to conduct alternative BWM, she will be:

- allowed to discharge unmanaged ballast water in the port; and
- targeted for a verification process.

If the ballast water to be discharged was assessed as posing an extreme risk, then:

- if a vessel crosses BWEA on its intended route, then she is requested to conduct BWE, if the water in the BWEA poses low, intermediate or high risk;
- if a vessel does not cross BWEA on its intended route, then she is requested to deviate in a reasonable distance to meet the BWEA;
- if a vessel is not able to complete full BWE while crossing BWEA, then she is requested to slow down or take other measures to fully complete BWE; and
- if a vessel was not able to conduct BWE or fully complete BWE in the BWEA, then she is requested to conduct alternative BWM.

If it was not appropriate or the vessel was not able to conduct or fully complete BWE in the BWEA, and has no further option to conduct alternative BWM, she will be turned away, as at this stage it is assumed that operations in ports cannot be completed without discharging unmanaged ballast water.

The BWM options and consequences are shown in Fig. 13.

[8] Reasonable distance is to be decided based upon regional specifics and deviation related costs.

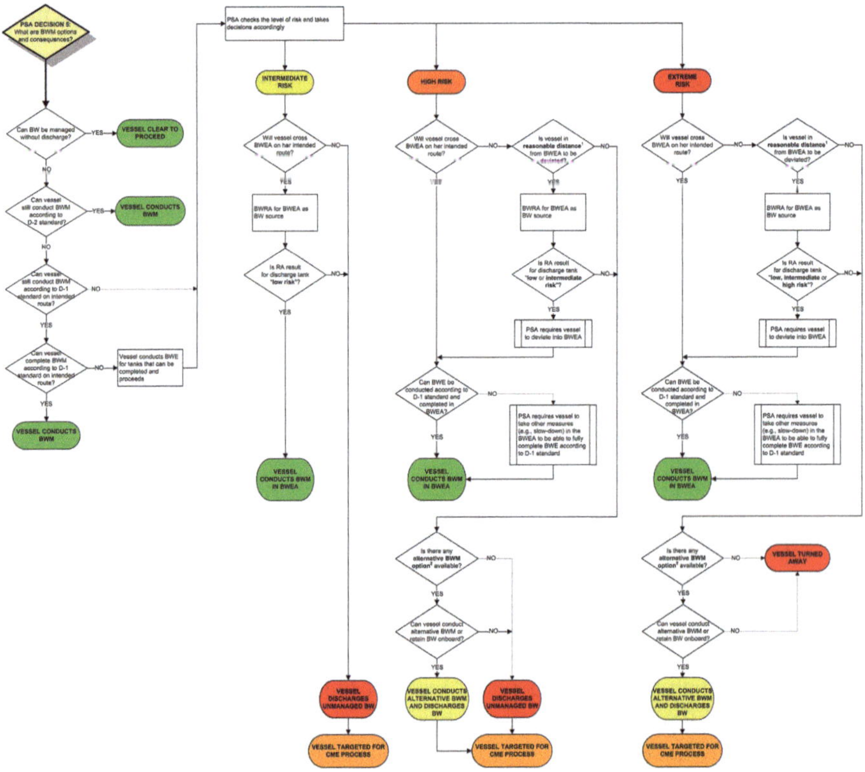

Fig. 13 Decision (5) on BWM options and consequences (Enhanced after David 2007). (*1*) reasonable distance – to be decided based upon regional specifics and deviation related costs; (*2*) alternative BWM option – alternative ballast water discharge area or alternative management, e.g., emergency treatment, BW reception facilities (This figure can be downloaded from http://extras.springer.com/)

After the BWM action has ended and the vessel has undertaken (or not) the required BWM practice, she may be selected for the verification process to verify compliance with the requirements, i.e., compliance monitoring and enforcement (CME) process.

Compliance Monitoring and Enforcement

The CME process is in the first place intended to back-up the management process, and to support the full implementation of the BWM Convention and any further port State BWM requirements. Every vessel allowed to enter the port may be selected for compliance monitoring. If a vessel is found not compliant, but has already entered the port and started deballasting, it may be stopped from deballasting and may be requested to take alternative BWM measures. The compliance monitoring process is followed by the penalty process for non-compliant vessels. A penalty

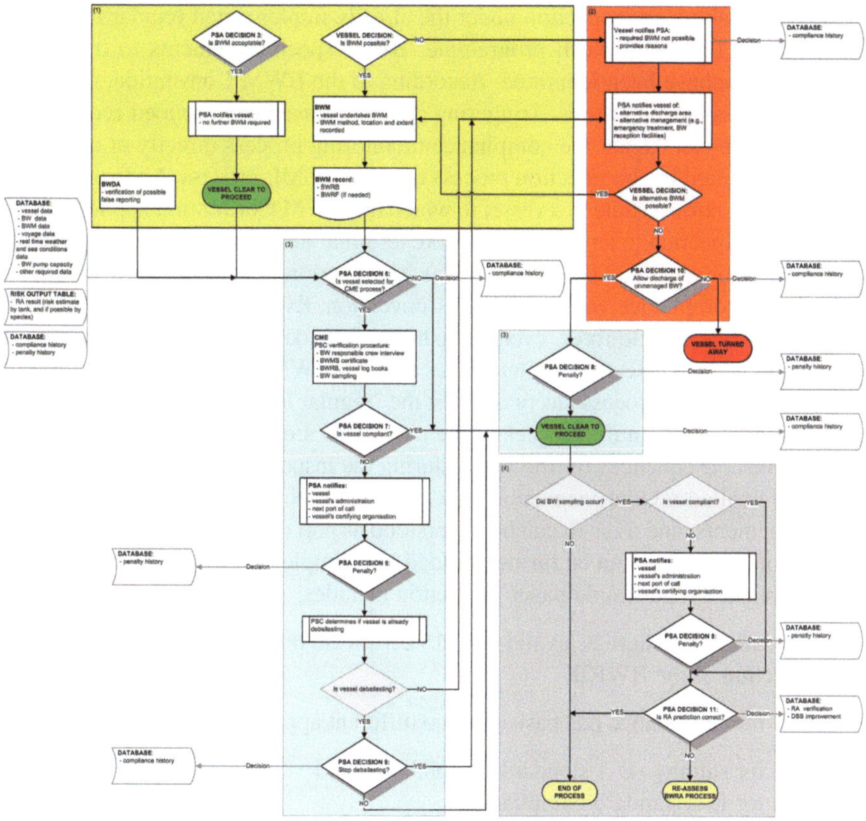

Fig. 14 Compliance monitoring and enforcement process (Enhanced after David 2007) (This figure can be downloaded from http://extras.springer.com/)

may be applied to a non-compliant vessel when she is still in the port or even when she has already left the port, depending on when the non-compliance is identified/confirmed. A vessel may be found non-compliant when in the port, e.g., when not carrying a valid BWMS certificate, or the non-compliance may be confirmed when the vessel has already left the port, e.g., when BWS for compliance monitoring was undertaken however analyses took longer then her stay in the port (see Fig. 14).

Decision 6: Vessel Selected for Compliance Monitoring and Enforcement Process?

The CME process is conducted by PSC and starts with the vessel selection. If PSC has a separate BWM CME programme, a random selection with a minimum number of vessels targeted, may be conducted. However, if there is no BWM specific programme adopted, then PSC may select a vessel for the BWM CME process while undertaking an inspection under the already implemented regular inspection programme. Further to such programme, BWM specific elements to trigger the CME process have been identified. According to the BWM Convention, the verification process has two levels. Triggering elements have been grouped accordingly. Each of these can trigger the compliance monitoring process directly or randomly. A vessel targeted by the selection process enters the CME process. According to the BWM Convention Article 9, a vessel to which the BWM Convention applies may be subject to inspection in any port or offshore terminal of the port State that is party to the BWM Convention. The purpose of such inspection is determining whether the vessel is in compliance with the BWM Convention. Even if the BWM Convention has not yet entered into force, every state has to provide for an effective verification process to support effective implementation of the BWM measures.

The verification process has two levels, the "regular inspection" and the "detailed inspection". The main differences of the two levels are the triggering elements, as well as the consequences for the vessel during the inspection process.

The so called regular inspection does not need special justification for the triggering elements, and as such can be understood as part of the basic and regular PSC inspection process. It can be further divided into simple paper inspection and BWS for compliance. The simple paper inspection includes:

- verification that there is a valid BWMS certificate on board the vessel;[9] and
- inspection of the BWRB.

BWS for compliance has basically two different approaches:

- BWS for salinity (D-1 standard compliance); and
- BWS for D-2 standard compliance.

The BWS for salinity is generally intended to be used for a verification of the BWE process, and specifically for the verification of the RA process when a decision was taken based on environmental matching salinity. The BWS for compliance with the D-2 standard requires analyses of viable aquatic organisms present in the ballast water.

The BWS for compliance should be conducted according to the Guidelines for ballast water sampling (G2) (IMO 2008) and its related guidance documents. If BWS is conducted as a part of the regular inspection, the vessel shall not be unduly delayed for the time required to analyse the ballast water samples. For more details about BWS see chapter "Ballast Water Sampling and Sample Analysis for Compliance Control".

A PSC may also decide to carry out a detailed inspection when a ship does:

- not carry a valid BWMS certificate; or there are
- clear grounds for believing that:

[9] If valid, it shall be accepted.

- the vessel or its equipment does not correspond substantially with the particulars of the certificate; or
- the master or the crew are not familiar with essential shipboard procedures relating to BWM, or these have not been implemented.

The detailed inspection includes, as appropriate:

- the inspection of all needed documents and log books;
- the inspection of the vessel (e.g., BWMS);
- indicative BWS.

When a PSC decides to carry out the detailed inspection, the vessel shall not discharge ballast water until it is confirmed that it can do so without risk of harm to the environment, human health, property or resources (see Fig. 15).

Decision 7: Is Vessel Compliant?

PSC has conducted an inspection to check if the vessel has complied with the BWM requirements. PSC checks if the vessel is carrying a valid BWMS certificate, if the conditions of the vessel and the BWMS correspond with the BWMS certificate, interview the BWM responsible crew members if they are familiar with the BWM procedures and if these were implemented. Even if all these checks were satisfactory for PSC, they may decide to proceed with conducting BWS to ascertain that BWM measures implemented are acceptable and efficient (Fig. 16).

Decision 8: Penalty?

National legislation would need to provide for the prevention of unwanted impacts caused by discharges of HAOP via ballast water. Legislation would also need to cover unlawful acts of vessels flying their flag (i.e., Flag state), as well as those occurring in their jurisdictional waters (i.e., Port State). The penalty process in this DSS is focussed only on port State requirements.

If a violation has been detected, the PSC should see whether national legislation has provided for such an act and proceed accordingly. If a vessel is penalised, this needs to be recorded in the penalty history database. The sanctions provided should be of adequate severity to discourage further violations (see Fig. 17).

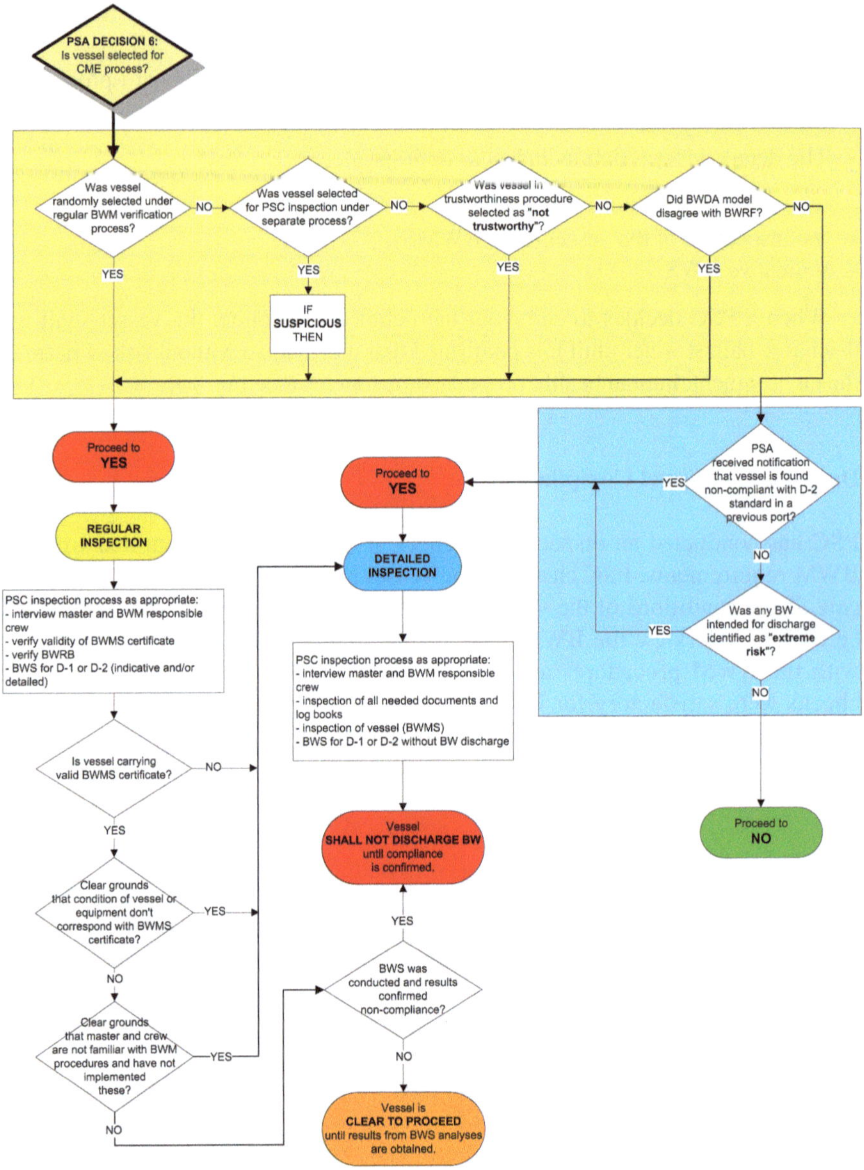

Fig. 15 PSA decision (6) on vessel selection for CME process, including the two different levels of inspection according to the BWM Convention, i.e., so called regular and detailed inspection (Enhanced after David 2007). The *light yellow box* includes elements that trigger the simple inspection; the *light blue box* includes elements that directly trigger the detailed inspection (This figure can be downloaded from http://extras.springer.com/)

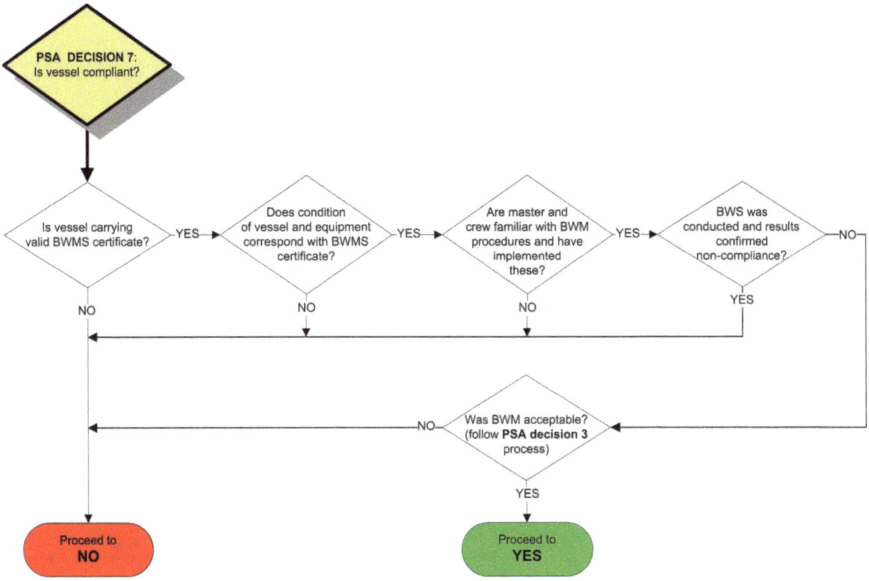

Fig. 16 PSA decision (7) on vessels compliance with the BWM requirements (This figure can be downloaded from http://extras.springer.com/)

Decision 9: Stop Deballasting?

If a vessel is found non-compliant with BWM requirements, PSC may decide to prevent deballasting. The decision regarding the prevention of a vessel from deballasting is basically related to the risk posed by the ballast water intended for discharge.

In case a non-compliant vessel has already started deballasting and the risk posed is unacceptable, such a vessel will be stopped from deballasting (see Fig. 18).

When a vessel was required to stop deballasting, the PSA authority notifies that vessel regarding possible alternative BWM options available. If feasible, the vessel conducts alternative BWM.

Decision 10: Allow Discharge of Unmanaged Ballast Water?

This is a position where none of the "regular" or alternative BWM options was implemented. A vessel in this situation would be one that:

- has declared to have on board unmanaged ballast water intended for discharge;
- did everything in her capability to comply with the requirements;
- was not able to conduct requested regular BWM practice; as well as
- was not able to conduct alternative BWM practices.

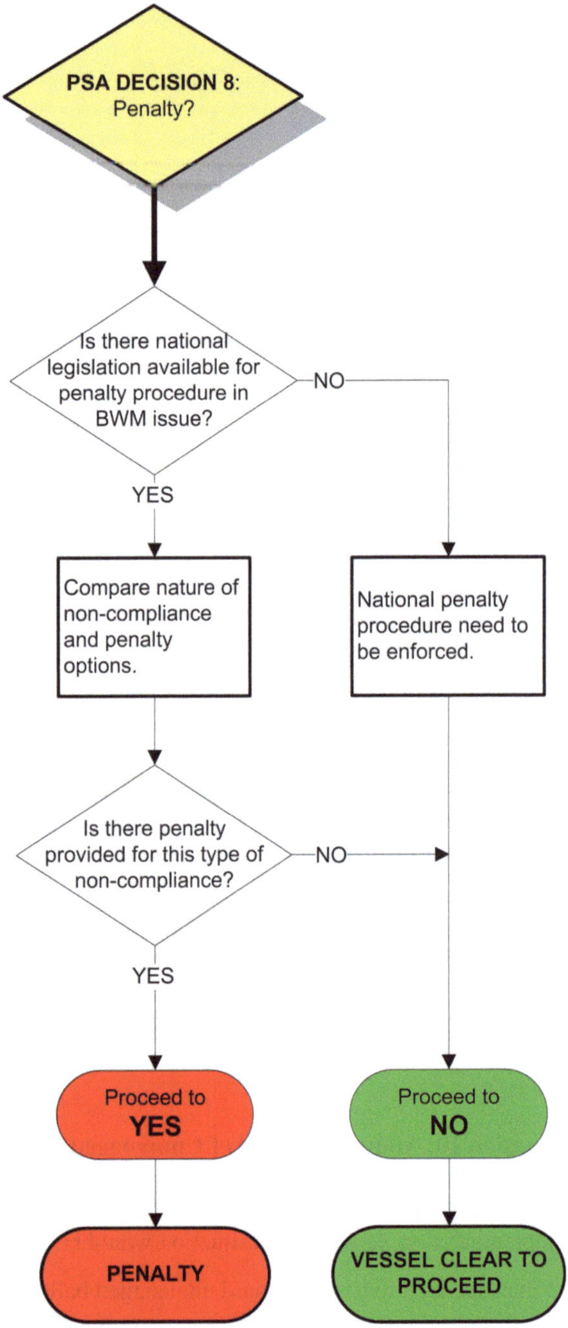

Fig. 17 Decision (8) on issuing a penalty to the non-compliant vessel (Enhanced after David 2007) (This figure can be downloaded from http://extras.springer.com/)

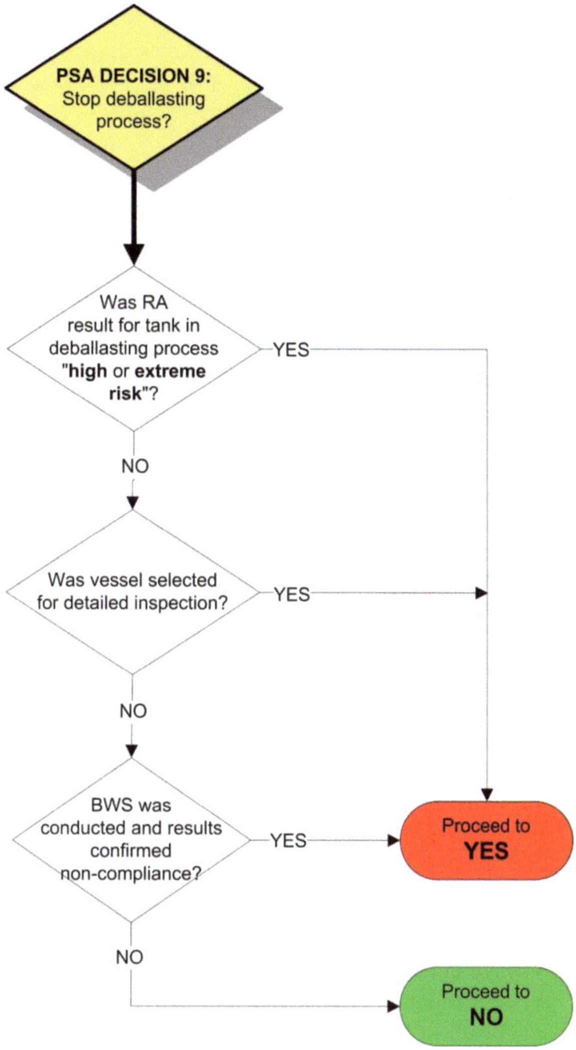

Fig. 18 Decision (9) on stopping a vessel to deballast (Enhanced after David 2007) (This figure can be downloaded from http://extras.springer.com/)

The PSA needs to take a decision whether or not to allow such a vessel to discharge unmanaged ballast water in the port or to turn it away. Such a decision should certainly be taken considering the risk posed by the ballast water intended for discharge. However, for the general practice and effectiveness of BWM measures it is also important that the vessel did everything in her capability to comply with the requirements. In this situation the PSA should check:

- BWM requirements according to the legislation;
- vessels' BWM options according to the BWM plan;
- intended route;

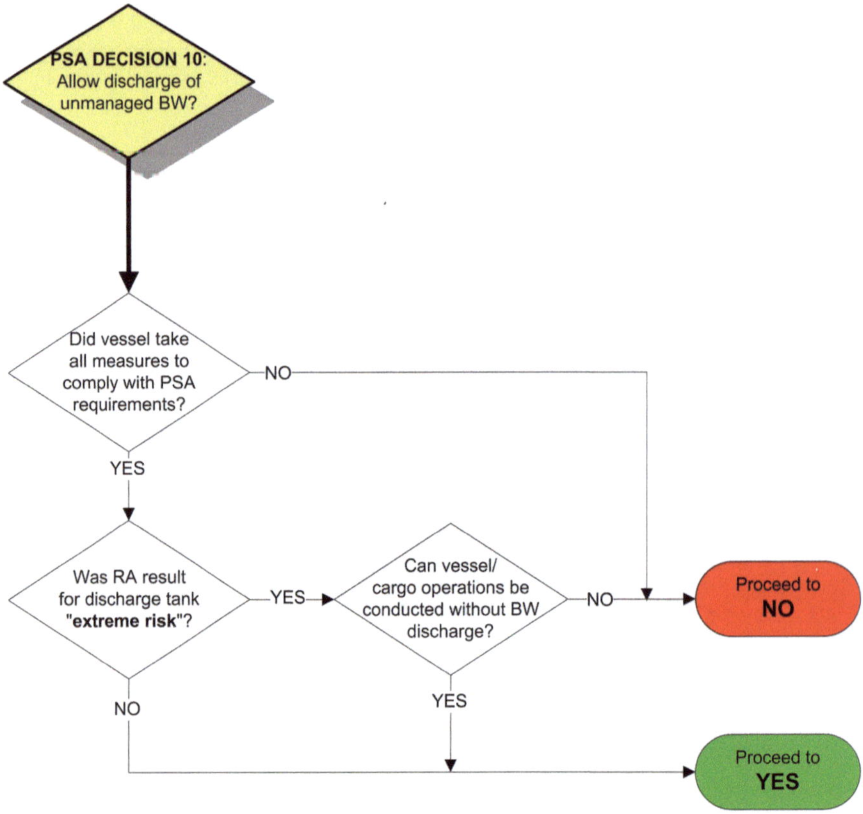

Fig. 19 Decision (10) on allowing or not a vessel to discharge unmanaged ballast water (Enhanced after David 2007) (This figure can be downloaded from http://extras.springer.com/)

- voyage duration and other conditions;
- vessels capability to conduct alternative BWM options; and
- the result of RA.

In case a vessel took all measures to comply with the requirements, including alternative BWM options, then the level of risk posed by the ballast water intended for discharge needs to be verified. If the ballast water was assessed as of extreme risk, than the vessel should not be allowed to discharge ballast water, however in cases when the risk level assessed was intermediate or high, the vessel may still be allowed to discharge ballast (see Fig. 19).

Certainly, this should be understood only as minimum criteria. It is up to each PSA to decide whether or not to apply a more stringent approach and possibly not allow discharge of unmanaged ballast water that was assessed as high or even intermediate risk which would be desirable especially from an environmental perspective.

BWRA Review Process

BWRA is a relatively new field of work and will certainly need to be improved over time. The basis for improvement should be found when more knowledge and information becomes available by experience. Especially the results of BWS for compliance may be a very valuable source to be used for the review process of BWRA, and findings may support BWRA improvements (see Fig. 20).

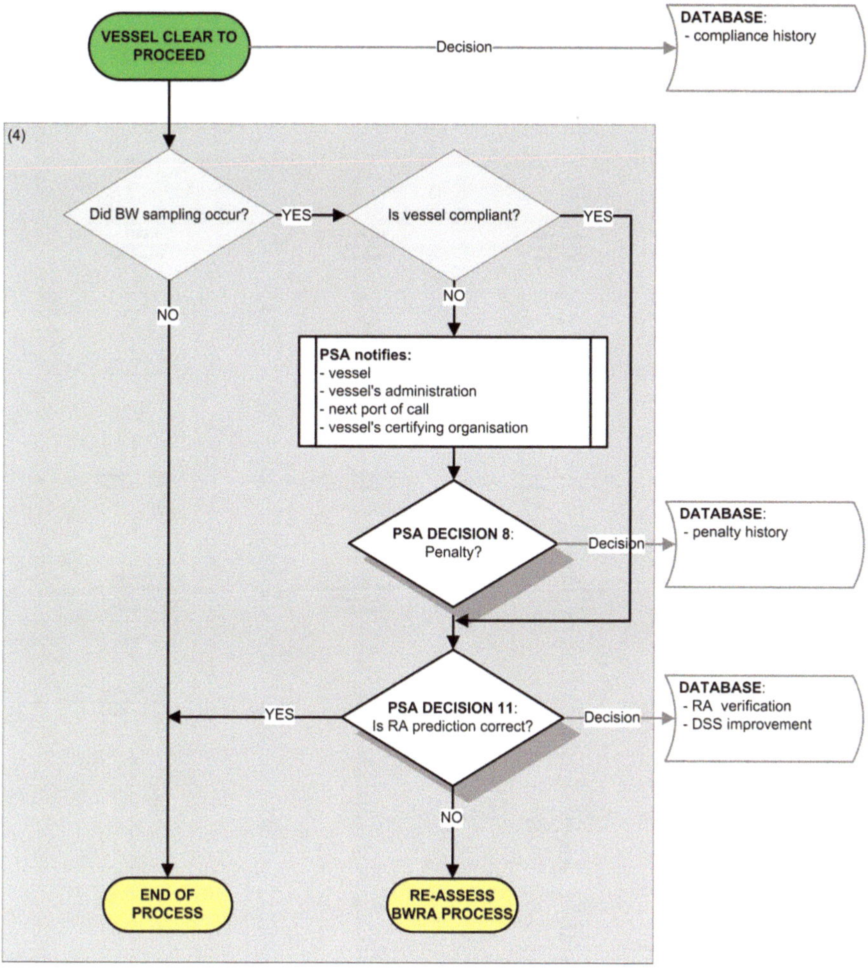

Fig. 20 The BWRA review process (Enhanced after David 2007) (This figure can be downloaded from http://extras.springer.com/)

Decision 11: Risk Assessment Prediction Correct?

The process is based on the comparison of the BWRA result with the BWS result. BWS may be conducted just for salinity, may encompass biological analysis focussed on the presence of viable organisms as per the D-2 standard, or may also include identification of HAOP. If only a salinity test was undertaken, then the results may be used only for the review of the BWRA that was based on environmental matching, while also an identification of HAOP is needed for a complete review of BWRA (see Fig. 21).

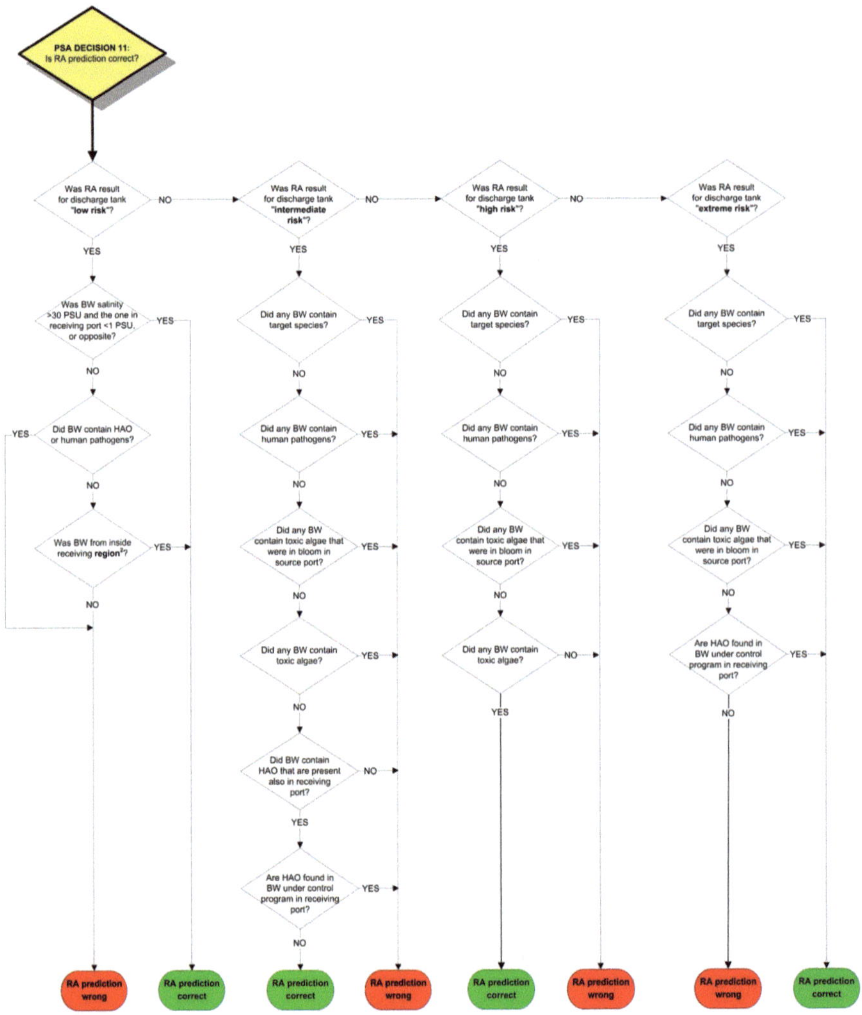

Fig. 21 Management decision (11) on correctness of the RA result (Enhanced after David 2007). *PSU* Practical Salinity Unit, *HAO* Harmful Aquatic Organisms (This figure can be downloaded from http://extras.springer.com/)

End-Points of the DSS

The selective approach in the process of BWM based on this DSS may result in one of the following situations:

- vessel is turned away because she has not submitted required data;
- vessel does not need to discharge ballast water;
- vessel may conduct BWM in advance;
- vessel is exempted from BWM requirements based on BWRA;
- vessel requested to conduct BWM may be able to comply or not;
- vessel requested to conduct BWM may do it properly or not;
- vessel may be selected for CME;
- vessel may be allowed to discharge unmanaged ballast water;
- vessel was able to comply with BWM requirements, but did not conduct BWM at all so she is turned away;
- vessel was able to comply with BWM requirements, but did not conduct BWM properly so she is turned away;
- sampling from CME reveals that BWM standards are not met so that the deballasting has to be stopped; or
- vessel found not in compliance may be penalized.

In addition to the decisions relating to BWM, a reassessment of the RA procedure is provided in the DSS process, which is important for further improvement of RA results.

Acknowledgements Part of these results were obtained in the framework of the research project Decision Model and Control of Ballast Water Management in the Slovenian Sea (L2-6291), which was financially supported by the Slovenian Research Agency and the Port of Koper (Luka Koper d.d.) and as Matej David's PhD work, supervised by Prof. Livij Jakomin, Slovenia and Prof. Chad Hewitt, Australia. The research leading to part of these results has received funding from the European Community's Seventh Framework Programme (FP7/2007-2013) under Grant Agreement No. [266445] for the project Vectors of Change in Oceans and Seas Marine Life, Impact on Economic Sectors (VECTORS).

References

Adelman L (1992) Evaluating decision support and expert systems. J Multi Crit Dec Anal 2(2):63–124

Bhatt GD, Zaveri J (2002) The enabling role of decision support systems in organizational learning. Dec Sup Sys 32:297–309

David M (2007) A decision support system model for ballast water management of vessels: doctoral dissertation. University of Ljubljana, Portorož

David M, Malej A (2002) Bio-pollution risk assessment with fuzzy logic. In: Jencek P, Štrubelj M (eds) ICTS 2002 proceedings. University of Ljubljana, Faculty of Maritime Studies and Transport, Portorož, pp 116–123

Denzer R (2005) Generic integration of environmental decision support systems – state-of-the-art. Environ Model Softw 20:1217–1223

Ekel PY (2002) Fuzzy sets and models of decision making. Comp Math Appl 44:863–875
Graham I, Jones PL (1988) Expert systems: knowledge, uncertainty, and decision. Chapman and Hall, New York
IMO (2006) Guidelines on designation of areas for ballast water exchange (G14). International Maritime Organization, Marine Environment Protection Committee, Resolution MEPC.151(55), 13 Oct 2006. London
IMO (2008) Guidelines for ballast water sampling (G2). International Maritime Organization, Marine Environment Protection Committee, Resolution MEPC.173(58), 10 Oct 2008. London
Marquez AC, Blanchar C (2006) A decision support system for evaluating operations investments in high-technology business. Dec Supp Sys 41:472–487
Paradice D (2006) Expanding the boundaries of DSS. Dec Sup Sys 43:1549–1552
Ru WG, Eloff JHP (1996) Risk analysis modelling with the use of fuzzy logic. Elsevier Comp Secur 15(3):239–248
Sojda RS (2007) Empirical evaluation of decision support systems: needs, definitions, potential methods, and an example pertaining to waterfowl management. Environ Model Software 22(2): 269–277
Sprague RH, Carlson ED (1982) Building effective decision support systems. Prentice-Hall, Englewood Cliffs
Vincke P (1993) Multicriteria decision-aid. Wiley, New York

Ballast Water Management Decision Support System Model Application

Matej David and Stephan Gollasch

Abstract In this chapter, the generic ballast water management (BWM) decision support systems (DSS) model presented in the chapter "Ballast Water Management Decision Support System" is validated by using one year real ballast water discharge data of the Port of Koper, Slovenia. All possible BWM options as outlined in the BWM Convention have been studied, and data on vessel voyages were collected or assessed, including vessel movements, main routes, navigational constraints and ballast water patterns, i.e., amount of ballast water to be managed per vessel and type, ballast water exchange (BWE) capacity rates per vessel type and source ports relevant for risk assessment (RA). The ballast water discharge data were analysed to assess (a) the number of vessels that would be able to conduct BWE on their intended routes according to the BWM Convention, and (b) the quantity of ballast water which would be discharged (managed versus unmanaged). It is most likely that only vessels from outside the Adriatic are enabled to conduct BWE before they call at the Port of Koper. A ballast water exchange area in the Adriatic would open more options to conduct BWE. The RA results from source ports were related to each vessel to assign the level of risk to each vessel discharging ballast water. A critical situation arises when ballast water is assessed as to pose an extreme risk as the BWM DSS would conclude that these vessels would not be allowed to discharge unmanaged ballast water.

Keywords Port of Koper • Ballast water management • Decision support system application • Risk assessment result • Ballast water discharge quantity

---

M. David (✉)
Dr. Matej David Consult, Korte, Izola, Slovenia
e-mail: matej.david@siol.net

S. Gollasch
Gollasch Consulting (GoConsult), Hamburg, Germany
e-mail: sgollasch@aol.com

M. David, S. Gollasch (eds.), *Global Maritime Transport and Ballast Water Management*, Invading Nature - Springer Series in Invasion Ecology 8,
DOI 10.1007/978-94-017-9367-4_10

Introduction

In the previous chapter ("Ballast Water Management Decision Support System"), a generic BWM DSS model was presented which was developed in a generic nature that it can be used for any port worldwide, if necessary with adaptations to address local specifics. In this chapter, the BWM DSS model is validated by using real ballast water discharge data of the Port of Koper, Slovenia, for the entire year 2005 (David 2007).

The implementation of BWM measures was considered in the view of possible options in the framework of the BWM Convention. All possible BWM options as outlined in the BWM Convention have been studied, and data on vessel voyages and operations were collected or assessed, including vessel movements, main routes, navigational constraints, and ballast water patterns, i.e., amount of ballast water to be managed per vessel and type, BWE capacity rates per vessel type, source ports relevant for RA[1] (David 2007). The BWM DSS model also addresses an information system management and the work of the relevant port State authority (PSA).

Studied Area Local Specifics

The Adriatic Sea is part of the Mediterranean Sea, situated between the Balkan and Apennine Peninsulas. The southern opening extends from the Strait of Otranto to the Cape of Santa Maria di Leuca (Italy) to the north coast of the island of Corfu (Greece) and where the mouth of the river Butrinit (Albania) is located (IMO 2003). The length of the Adriatic Sea, from the southern end (river Butrinit) to the Porto di Lido (Venice, Italy) in the north is 475 NM and its width, from the Port of Omišalj (Croatia) to the Port of Vasto (Italy) is 117 NM; its surface is 138,595 km^2 (IMO 2003), as shown in Fig. 1.

The Mediterranean Sea has numerous ports open for international shipping, but not so many of them are of intercontinental importance. These "hub-ports" are connected with a variety of hub-ports globally, while local (short sea) shipping connects them with secondary Mediterranean ports. The very intense traffic within the Mediterranean also includes transiting vessels which do not call for any Mediterranean port (see Fig. 2).

The Port of Koper is located in the very north of the Adriatic Sea and it is the only major Slovenian merchant port open to international shipping. Studies of cargo flows/shipping patterns have shown that Koper is very well connected with Mediterranean Sea hub-ports and also directly with different parts of the world[2] (Perkovič et al. 2003; David et al. 2007a) (see Fig. 3).

[1] The data on traffic patterns and BW discharges were collected with BWRF as part of the national research projects *Harmful Introductions and Ballast Water Management in the Slovenian Sea* and *Decision Model and Control of Ballast Water Management in the Slovenian Sea*, and taken for further analysis in this chapter.

[2] Sea transport connections, http://www.luka-kp.si/eng/vsebina.asp?IDpm=118#sea, last accessed January 2014.

Fig. 1 Location and extension of the Adriatic Sea, showing the southern opening (*dotted line*) and two examples of distances as *blue* and *red lines*

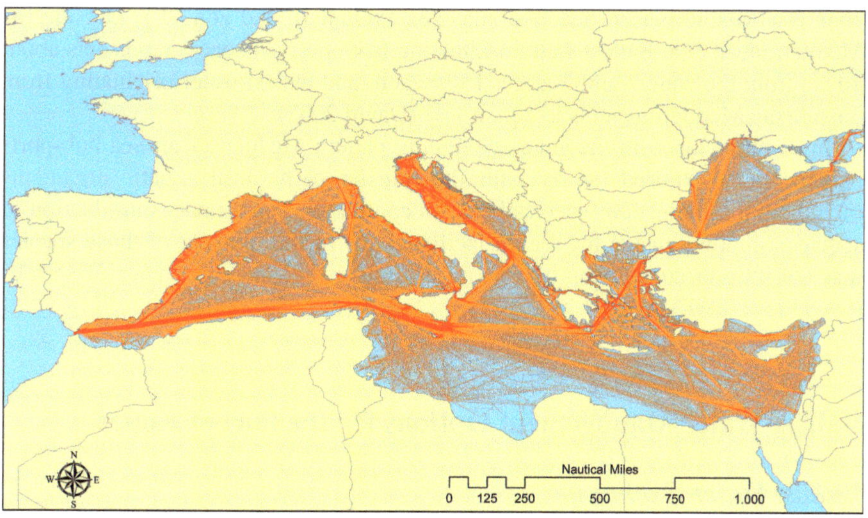

Fig. 2 Maritime traffic in the Mediterranean region also showing the transiting shipping routes e.g. from the Black Sea and Suez Canal to northern Europe (Source: Gašper Zupančič and Leon Gosar, Institute for Water of Republic of Slovenia)

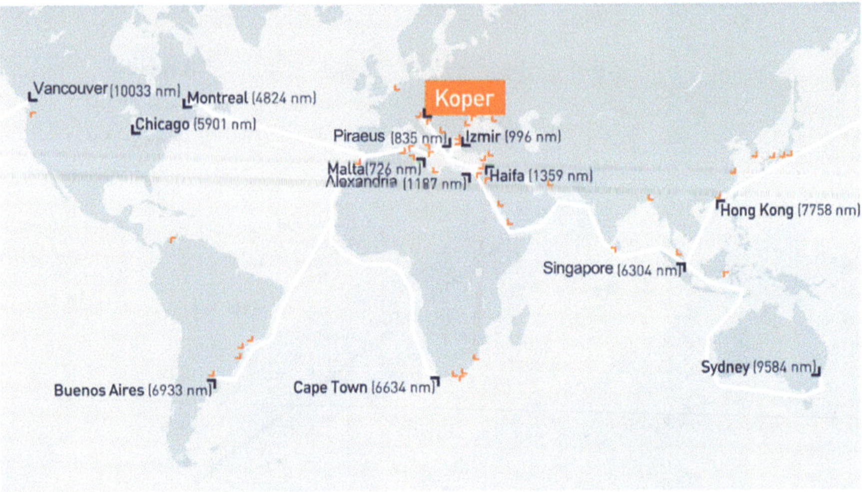

Fig. 3 Main direct Sea transport connections of the Port of Koper with other continents (Source: Luka Koper/Hal Interactive, http://www.luka-kp.si/eng/interactive-map, last accessed January 2014)

Ballast water discharges in the Port of Koper were studied using ballast water reporting forms and the ballast water discharge assessment model presented in the chapter "Vessels and Ballast Water". Results have shown that discharged ballast water in the Port of Koper originates almost exclusively from inside the Mediterranean region. For the purpose of BWM considerations the ballast water uptake ports or areas were divided in four zones: zone 1 = North Adriatic; zone 2 = South Adriatic; zone 3 = Mediterranean Sea (Adriatic Sea excluded); and zone 4 = Outside the Mediterranean Sea. Ballast water originating from ports in zone 1 represents about 70 %, from the zone 2 and 3 about 15 % each, and ballast water originating from outside the Mediterranean is less than 1 % (David 2007).

In combination with the general shipping pattern Mediterranean Sea hub-ports become most exposed (at risk) for the transfer of harmful aquatic organisms including non-indigenous species between continents, i.e., primary introduction of species. Shipping inside the region facilitates the further transfer of those species that are introduced into the hub-ports resulting in secondary transfer of species (David et al. 2007a, b).

Ballast Water Management Options for the Port of Koper

Ballast Water Exchange

Ballast water exchange (BWE) has been used since the 1980s to reduce the risks of coastal organisms being transferred and discharged. Though BWE is considered to be of a limited efficacy, the BWM Convention includes BWE as a BWM option

(D-1 standard). Furthermore, BWE is still the only widely applicable BWM option which should be used before certified BWMS are installed and in operation on board vessels (D-2 standard).

Ballast Water Exchange as a Blanket Ballast Water Management Approach for the Port of Koper

BWE has limitations in its effectiveness and especially in its applicability in semi-enclosed or enclosed areas. Ships in such areas usually sail within 50 NM from nearest land, as well as in too shallow waters (<200 m depth), therefore according to the BWM Convention they should not conduct BWE. As a result, unmanaged ballast water may be discharged in ports (see Fig. 4).

The ballast water discharge data for the Port of Koper from 2005 were analysed to assess (a) the number of vessels that would be able to conduct BWE on their intended routes according to the BWM Convention, and (b) the quantity of ballast water which would be discharged (managed versus unmanaged). As a result, in 2005 a total 448 vessels discharged 544,133 m^3 of ballast water in the Port of Koper.

It is most likely that only vessels from outside the Adriatic Sea are enabled to conduct BWE, therefore only vessels that discharged ballast from zone 3 or 4 were considered. Firstly, all source ports from zones 3 and 4 were identified and connected with intended shipping routes to the Port of Koper. This was done to identify which of the intended routes would enable BWE according to the IMO requirements (see Fig. 5).

The minimum distances needed to complete BWE were also taken into account considering the quantity of ballast water to be exchanged and the vessels BWE

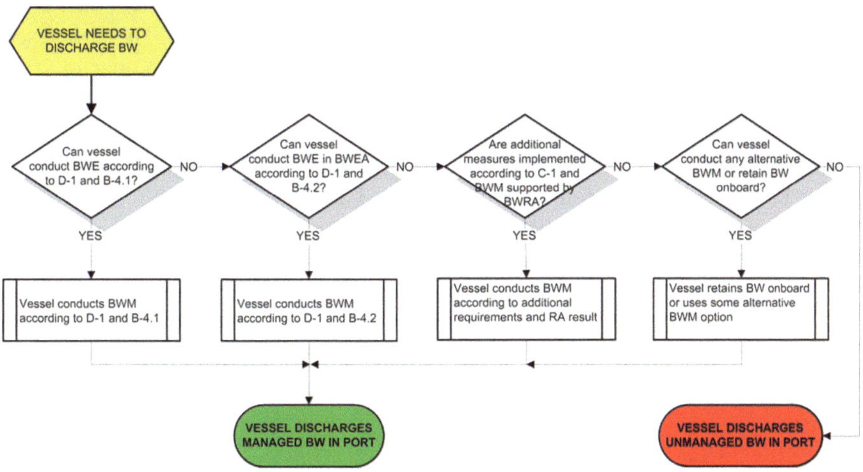

Fig. 4 Flowchart showing BWE options according to the BWM Convention. The references D-1, B-4.1, B-4.2 and C-1 refer to Regulations of the BWM Convention. *BW* ballast water, *BWEA* ballast water exchange area, *BWRA* ballast water risk assessment

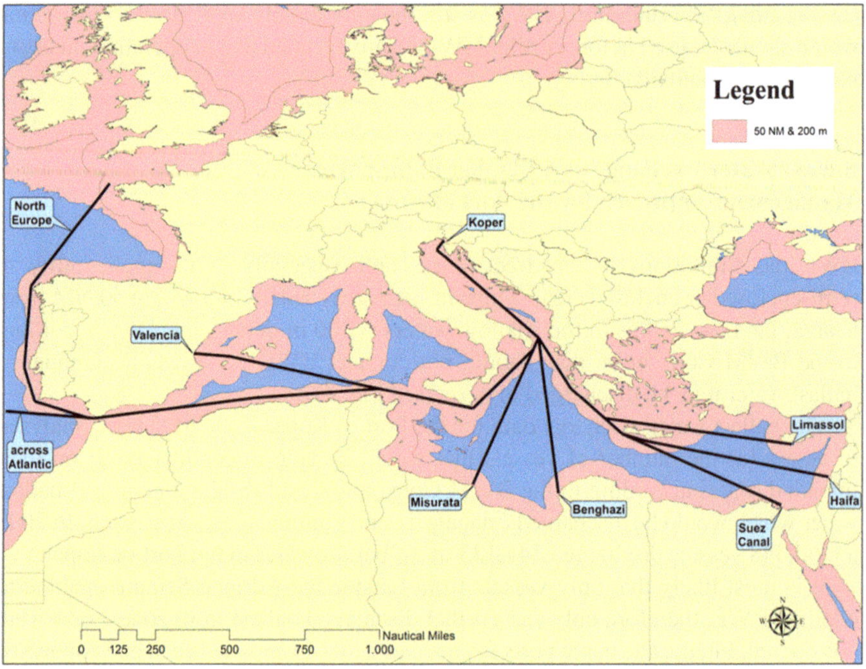

Fig. 5 Identified ballast water source ports and intended vessel routes to the Port of Koper where BWE could be conducted. The area less than 50 NM from nearest land and less than 200 m water depth is shown in *pink* (Enhanced after David 2007)

pump rate capacity (PRC) (see section "Time and Distance Needed to Complete Ballast Water Exchange"). Only 12 vessels were identified as able to conduct BWE out of the 48 vessels with ballast water source ports in zones 3 and 4. These represent 2.68 % of all vessels that discharged ballast water in the Port of Koper in 2005. Considering the quantity of ballast water discharged, only 10,866 m^3 of ballast water could have been managed out of the 49,385 m^3 originating from zones 3 and 4. This represents only 2 % of the total amount of ballast water discharged, and 22.00 % of the ballast water from source ports in zones 3 and 4 (see Table 1).

In consequence, the application of a blanket approach for the Port of Koper would result in more than 97 % of all vessels in the "do nothing" option of 'compliance' with the BWM Convention, i.e., 98 % of the discharged ballast water would be unmanaged.

Other BWM Options

A comprehensive review of BWMS is presented in chapter "Ballast Water Management Systems for Vessels". More than 100 such systems were identified which make use of different treatment technologies mostly in combination. BWMS

Table 1 Number of vessels and quantities of ballast water discharged in the Port of Koper in 2005 which could potentially be managed with BWE under the BWM Convention blanket approach

Vessels		BW discharged (m^3)	
No. vessels BWE	12	SUM BWE	10,866
No. vessels no BWE	36	SUM no BWE	38,519
No. vessels Zone 3, 4	48	BW disch. Zone 3, 4	49,385
Total No. 2005	448	Total BW disch. 2005	544,133
% BWE/Zone 3, 4	25.00	% BWE/Zone 3, 4	22.00
% BWE/Total 2005	2.68	% BWE/Total 2005	2.00

After David (2007)

are in different developmental stages, but more than 30 were already type approved by responsible authorities. This makes certified systems available for sales to the shipping industry, however the uncertainty remains if the production capacities will be able to accommodate the needs on the shipping side. Furthermore, shipyard installation capacities may become a bottleneck to meet the demand. This is a fast developing field as the interest is triggered by a worldwide market of close to 70,000 vessels that will need to be equipped with such systems. However, in the absence of the BWM Convention being into force, BWMS are rarely installed on vessels so that this chapter focuses on BWE as currently the widely applicable BWM option.

Designation of a Ballast Water Exchange Area in the Adriatic

Being aware of the difficulties in the BWM Convention implementation in the Adriatic Sea the Ballast Water Management Sub-Commission (BWMSC) for the Adriatic has considered to designate a BWEA in the Adriatic Sea according to the BWM Convention (Regulation B-4.2). In the absence of IMO Guidelines how to designate a BWEA at that time, it was proposed that the designation should be based on common criteria/aspects including:

- navigational, e.g., shipping patterns among ports, ships routeing;
- ballast water discharge, e.g., quantity, source, frequency;
- hydrological, e.g., currents, water depths;
- biological, e.g., presence of invasive alien species (IAS) and harmful aquatic organisms and pathogens (HAOP), plankton densities;
- anthropogenic, e.g., pollution;
- important resources and protected areas; and
- legal aspects. e.g., conflict with national or international law.

There were different BWEA options considered by the BWMSC, however there was no final agreement on its adoption. The considered options for BWEA in the Adriatic Sea are shown in Fig. 6.

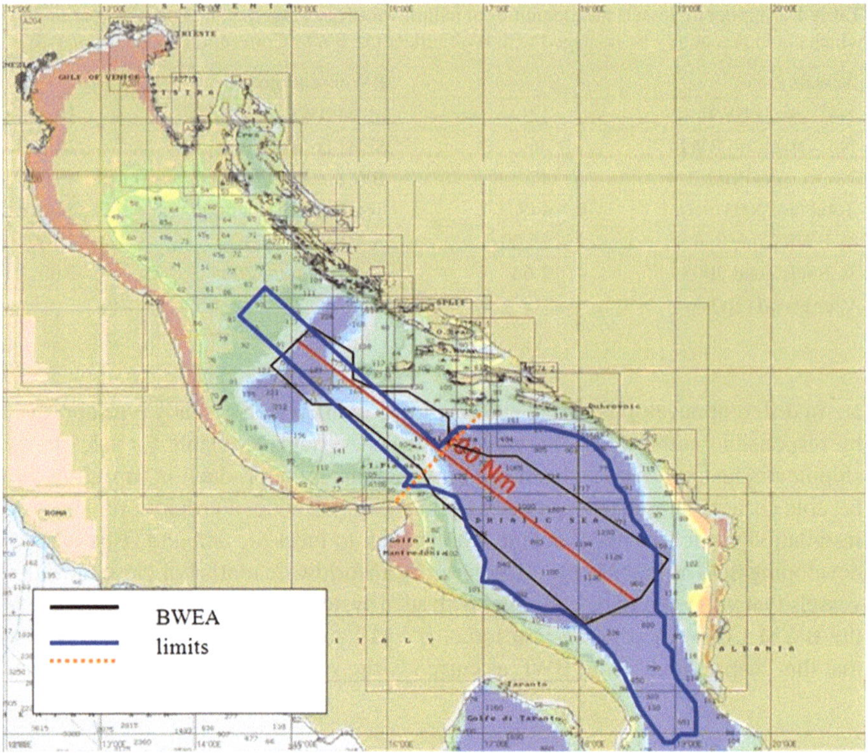

Fig. 6 Different considered options for the BWEA in the Adriatic Sea. The *blue*, *black* and *orange dotted lines* show different options for the limits of BWEA considered for the Adriatic Sea. The *red line* is a scale bar of 200 NM (After David 2007)

Ballast Water Management Requirements Related to the Ballast Water Exchange Area

The BWM Convention does not require ships to use a designated BWEA, because it states that a "ship may conduct ballast water exchange". Therefore, a port State has to inform/require the vessels what to do and the BWM options should be specified considering advantages and weaknesses in light of effective environmental protection and costs induced by such measures.

For the purpose of the Koper study, the BWM options were selected based on the BWM DSS model (see chapter "Ballast Water Management Decision Support System"), which is supported by the BWRA. This means that the BWM requirements are related to the result of the BWRA, i.e., when a higher level of risk from ballast water to be discharged is identified, more stringent requirements are considered, e.g., a vessel may need to deviate and/or slow down to fully complete BWE of all ballast water intended for discharge, and vice versa when a low risk is identified.

The vessel deviation and/or slowing down requirements cause a delay and this is addressed by the BWM Convention. Article 12 – *Undue delay to Ships* does not actually refer directly to Regulation B-4.2 (i.e., BWEA), but to Articles 7.2, 8, 9 or 10 and through these to additional measures, surveys, certification and inspections. The "additional measures" could be implicitly related to a BWEA, because it is expected that a BWEA will not only be designated, because even when designated, ships may use it. With this, BWEA needs to be regulated with the BWEA specific BWM requirements (i.e., selective requirements). However, it remains unclear whether the deviation of a vessel requested by a port State to use a BWEA will be considered as an additional measure. Actually, Regulation B-4.3 prevents a vessel's deviation and delay, however this is clearly and solely related to Regulation B-4.1, i.e., not applying to BWEA, which is under Regulation B-4.2. This implies that vessels may be deviated and slowed down to conduct BWE in a designated BWEA. As a result there is no direct reference between BWEA and the BWM Convention's definition of "undue delay". The undue delay is partially addressed in Article 2.3, which indicates that when port States consider taking a decision to request the deviation of a vessel they should take into account the costs associated with this requirement. In cases when the costs are low relative to the risks posed a delay should not be understood to be undue. Therefore, the undue delay needs to be reasonably assigned on a case-by-case basis relative to the balance of the impact of the measures requested and the risk posed.

The BWM DSS as outlined in detail in the chapter "Ballast Water Management Decision Support System" facilitates such a selective decision making process regarding BWM measures according to the level of risk posed, as well as to available and feasible BWM options. The application of the selective BWM approach supported by the BWM DSS is elaborated in detail hereafter, and the results are compared with the results when the blanket approach would be applied.

Application of the Ballast Water Management Decision Support System Model for the Port of Koper

Ballast Water Management Requirements Under the Decision Support System Model

The BWM measures are selected based on the risk posed by the ballast water intended for discharge. The main BWM measures are:

- ballast water reporting;
- request the vessel to conduct BWM;
- request the vessel to conduct BWE on its intended route;
- request the vessel to conduct BWE when crossing the BWEA on its intended route;
- request the vessel to deviate into the BWEA to conduct BWE;
- request the vessel to slow down to complete BWE in the BWEA;
- request the vessel not to discharge unmanaged ballast water.

A critical situation arises when ballast water is assessed as to pose an extreme risk (e.g., because of the presence of harmful algae blooms or indicator microbes in the source port) as the BWM DSS model decision would conclude that these vessels would not be allowed to discharge unmanaged ballast water. Special consideration was given to the vessels sailing to the Port of Koper from ports situated north of the Palagruža Island. It was recognised that a deviation of these vessels to conduct BWE in the BWEA would not be reasonable, therefore this would be considered as undue delay limit.

When planning feasible BWM options for the DSS in relation to the BWEA, it is critical to know:

- the BWEA location and dimensions;
- vessel traffic patterns, i.e., main intended routes, vessels speed, navigational constraints;
- ballast water patterns, i.e., ballast water quantity per vessel, ballast water PRC per vessel, and
- data on source ports, i.e., position, data needs for RA.

In general weather conditions in the Adriatic BWEA were assessed as not critical to prevent BWE in the majority of time in a year. There is also no monitoring programme for HAOP in the simulated BWEA, and there was also no previous record of HAOP in the simulated BWEA. Based on this DSS assumes for the Adriatic Sea BWEA that there are always good weather conditions and no HAOP are present in the BWEA. This provides a solid data framework for the application of the BWM DSS in case of the Port of Koper.

The Main "Intended" and "Deviated" Routes in the Adriatic in Relation to the Ballast Water Exchange Area

In the northern Adriatic Sea, the vessel traffic is regulated by the traffic separation schemes adopted by IMO (2008). In the central part, there is an additional separation scheme next to Palagruža Island which is also used by vessels.

The movements of vessels in the Adriatic are observed by coastal states primarily using radar and Automatic Identification Systems (AIS). To assess the intended routes of vessels, AIS data were used. It was recognized that the main routes for vessels sailing from Mediterranean ports to the northern Adriatic, e.g., Port of Koper, from the Otranto Straight northwards tend to go close to the Italian coast and then through the separation scheme next to Palagruža Island, and continue to the separation schemes in the northern Adriatic.

To assess the necessary deviations from vessels intended routes to conduct BWE in the BWEA, a map was prepared showing the BWEA and the intended routes of vessels sailing to the Port of Koper from source ports located outside the Adriatic Sea, as well as from those within the Adriatic. It became clear that vessels sailing from ports situated northwest of the BWEA would need to deviate substantially to

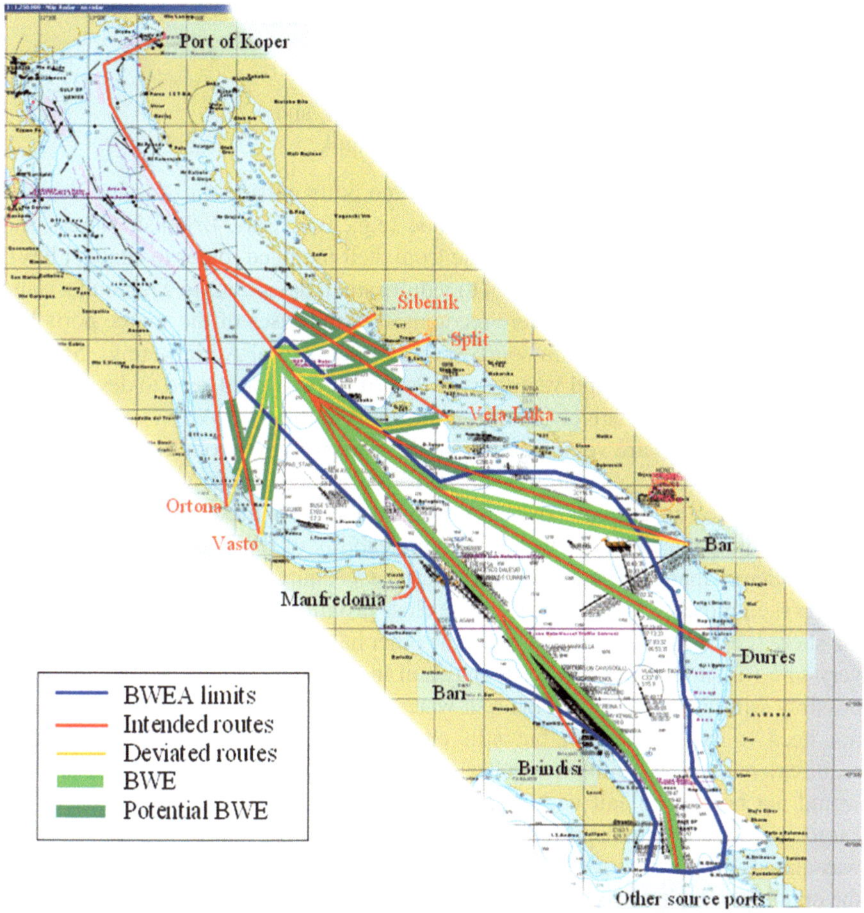

Fig. 7 The main "intended" and "deviated" vessel routes from ballast water source ports through the BWEA to the Port of Koper; the potential extension for BWE on routes was considered for intra-Adriatic traffic (David 2007)

be able to conduct BWE in the BWEA, hence, this was considered as not feasible as it would cause an undue delay to the vessel (see Fig. 7).

The approximate lengths of the **intended** routes through the BWEA are:

- 290 nautical miles for vessels sailing from outside the Adriatic (zones 3 and 4);
- 230 nautical miles for vessels sailing from Durres (zone 2);
- 200 nautical miles for vessels sailing from Brindisi (zone 2);
- 150 nautical miles (in two sections) for vessels sailing from Bar (zone 2); and
- 100 nautical miles for vessels sailing from Bari and Manfredonia (zone 2).

The approximate lengths of the **deviated** routes through the BWEA are:

- 160 nautical miles for vessels sailing from Bari;
- 190 nautical miles for vessels sailing from Bar;

- 60 nautical miles for vessels sailing from Vela Luka;
- 30 nautical miles for vessels sailing from Split;
- 30 nautical miles for vessels sailing from Vasto;
- 20 nautical miles for vessels sailing from Ortona; and
- 15 nautical miles for vessels sailing from Šibenik.

Figure 7 shows that vessels from all source ports outside the Adriatic Sea and most source ports in the southern Adriatic (e.g., Bar, Bari, Brindisi, Duress and Manfredonia) would not need to deviate from their intended routes to conduct BWE in the designated BWEA. The option to require small deviations to gain longer route distances through the BWEA were recognised in the cases of the source ports Bar and Bari. Vessels sailing from source ports in the central Adriatic (e.g., Šibenik, Split, Vela Luka, Ortona and Vasto) to the Port of Koper would need to deviate more substantially to gain relatively short route-lengths in the BWEA.

The most critical are vessels sailing to the Port of Koper from the ports situated north of the BWEA, i.e., northern Adriatic. When their ballast water intended for discharge poses an extreme risk according to the DSS these are not allowed to discharge unmanaged ballast water. The applicability of BWE as the BWM method is considered inappropriate on these routes especially because of very shallow waters (i.e., mostly <24 m depth) and the close proximity of their intended routes to the shore (i.e., <30 nautical miles). On the other hand such very enclosed sea area is very safe for sailing, hence vessels may sail in minimum (light) ballast condition, possibly without loading extreme risk water or to load only a quantity that would not need to be discharged, but still considering the vessels' minimum safety requirements, e.g., stability, propeller immersion. The vessels considered in 2005 would not be able to sail without ballast. However, instead of using vessels, the cargo could also be transported by barges without doing ballast operation, which was already in practice for the same cargo at these routes years ago. Technically[3] the option to substitute vessels with barges exists. Unfortunately, this is not a feasible BWM option for the selective approach because the DSS decision is taken after the BWRF is submitted, i.e., after ballast water is already loaded on the relevant vessel. Therefore, all vessels discharging ballast water from the source ports which are situated north of the BWEA are excluded from the assessment.

Time and Distance Needed to Complete Ballast Water Exchange

Certain large vessels need up to 2 days in navigation to complete full BWE. At a speed of 15 knots this would mean that more than 700 NM voyage length may be needed to complete BWE. However, the length of the required route to complete

[3] This approach has also other implications (e.g., availability, financial) that would need to be studied first.

BWE depends very much on the BWE method used, as well as on the quantity of ballast water to be exchanged.

To calculate the possibility of vessels to conduct BWE in the BWEA, the length of the route through the area, the quantity of ballast water to be exchanged, the vessel's speed and BWE PRC are factors to be considered. Different approaches have been used in previous studies to prepare assessments of the time needed to conduct BWE (e.g., AQIS 1993; Royal Haskoning 2001; Dragsund et al. 2005). However the results are not helpful for the Port of Koper situation because the vessels considered are mostly of different profile than those discharging ballast in or when approaching this port.

One of the options to assess BWE PRC is to calculate it as the minimum ballast water pump capacity of the vessels' ballast water system (i.e., theoretical capacity) according to the shipbuilding rules (e.g., ABS 2006), or by the ballast water pump capacity reported with BWRF (i.e., the one based on vessels documentation). It was recognised that the theoretical BWE capacity was always lower than the practical capacity. In reality, BWE PRC is expected to be lower then the ballast water pump capacity based on ships documentation, because:

- when using the sequential method, more than 95 % of the water needs to be exchanged to comply with the D-1 standard and this may require partially closing of valves to avoid suction of air on pumps, or even stripping may need to be conducted;
- when using the pump-through method the vents through which water is pumped out may slow down the pumping rate, and
- additional time is needed to switch between tanks when more tanks need to be exchanged, i.e., usually tanks would be exchanged in pairs in diagonal to lower the stresses on vessels hull.

To illustrate this, a vessel with a ballast water pump capacity of 500 cubic metres per hour (two pumps each with 250 m^3/h) and 5,000 m^3 of ballast water on board to be discharged would need approximately 21–22 h to complete the BWE when utilizing the sequential BWE method. With an average speed of 15 knots, the vessel will sail for 330 NM in 22 h. The same vessel, when using the pump-through method, would need approximately 50 % time in addition than needed for the sequential method (i.e., 33 h), and consequently a larger BWEA distance to complete the exchange (i.e., 495 NM). Some mainly very large dry bulk carriers and tankers may need to conduct BWE according to their BWMP using a combination of the sequential and pump-through methods, i.e., for some tanks the sequential method is selected, for others the pump-through method is used.

Because of this BWE PRC was based on the minimum calculated ballast water pump capacity, i.e., the theoretical capacity. For the sequential method ballast water needs to be pumped out of the tank and thereafter that tank needs to be filled again, what would approximate to two times the water volume to be intended for discharge. For the pump-through method three times the volume of the ballast water to be discharged needs to be pumped through.

To assess the time and distance vessels would need to complete a full BWE the following equations were used:

- for the sequential method: Time= Amount of ballast water intended for discharge * 2/BWE PRC; Distance = Speed * Time
- for the pump-through method: Time=Amount of ballast water intended for discharge * 3/BWE PRC; Distance = Speed * Time

In 2005, 448 ballast water discharges occurred. Among these, only 63 ballast water discharges were identified as originating from source ports located in the BWE (BWM) relevant area, i.e., located out of northern Adriatic Sea. These vessels are studied as BWM relevant. The vessels' speeds were approximated for different vessel types based on logs from AIS. Vessels with ballast water source ports in the northern Adriatic were not assessed because BWE was excluded as a BWM option (see Table 2).

The available length of intended and deviated routes in the BWEA per vessel was compared with (divided by) the calculated route distance needed for this vessel to complete BWE with the sequential and pump-through methods, and expressed as factors. The factor result is >1 when the available length of route to conduct BWE was larger than the calculated route distance needed for this vessel to complete BWE for each of the BWE methods (see Table 3).

Vessel intended routes from the central Adriatic source ports do not cross the BWEA, therefore the potential extensions (see Fig. 7) of the length of the routes were taken into account as an alternative option. The applicability has been assessed to obtain a delay prognosis for vessels sailing from central Adriatic source ports. The option to deviate and slow down was considered in relation to the BWE PRC and the quantity of ballast water that would need to be exchanged per each vessel. It was recognised that all such vessels would need to deviate and slow down to conduct BWE in the BWEA. The deviation does not seem to be so critical (i.e., 5–15 NM longer routes). However, due to the short distance of the route in the BWEA (i.e., 15–60 NM) vessels would need to slow down during that part of their voyage on average to approximately 1/3 of the normal speed (i.e., range from 6 to 74 %) to complete the BWE with the sequential method, or even to 1/6 in case the flow through method will be used (i.e., range from 4 to 50 %) (see Table 4).

The relatively intensive slow-down needed to meet the BWE requirements and the high variation among vessels suggests that this is inappropriate to be used for all vessels. However, it is a feasible option for those vessels which need to exchange ballast water which was assessed as extreme and high risk under the selective approach.

Risk Assessment

The BWRA model presented in the chapter "Risk Assessment in Ballast Water Management" has been applied within the BWM DSS. As there was no target species list available for Port of Koper waters, this approach is not applied. Therefore,

Table 2 BWE PRC calculated for each vessel based on theoretical pump capacities, and time and route distances needed to complete BWE calculated for the sequential (SEQ) and pump-through (PT) methods per vessel, excluding vessels which originated from northern Adriatic ports

Vessel	Vessel type	LOA (m)	B (m)	Ts (m)	BW pipe diam. (mm)	BWE PRC (m^3/h)	Vessel speed (knots)	BW for disch. (m^3)	SEQ time (h)	SEQ dist. (NM)	PT time (h)	PT dist. (NM)
1	BC	185.21	22.80	10.10	177.18	355.35	14	3,426	19.28	269.96	28.92	404.93
2	BC	81.00	11.33	5.31	96.41	105.22	14	348	6.61	92.61	9.92	138.91
3	GC	74.60	12.80	7.86	100.48	114.29	15	720	12.60	188.99	18.90	283.48
4	GC	87.03	13.75	5.65	105.34	125.61	15	400	6.37	95.54	9.55	143.31
5	GC	45.55	9.40	4.10	73.37	60.94	15	470	15.43	231.38	23.14	347.06
6	GC	114.00	13.00	3.67	111.34	140.34	15	751	10.70	160.54	16.05	240.82
7	GC	107.35	13.00	3.85	109.11	134.77	15	700	10.39	155.82	15.58	233.74
8	GC	88.30	12.50	5.17	102.21	118.26	15	352	5.95	89.30	8.93	133.95
9	GC	113.80	13.00	5.50	114.61	148.69	15	200	2.69	40.35	4.04	60.53
10	GC	91.37	13.26	5.11	105.29	125.50	15	701	11.17	167.57	16.76	251.35
11	BC	114.00	13.00	3.81	111.60	140.99	14	1,200	17.02	238.31	25.53	357.46
12	GC	67.00	11.65	4.27	89.12	89.92	15	470	10.45	156.81	15.68	235.22
13	GC	78.65	12.25	5.04	97.10	106.72	15	361	6.77	101.48	10.15	152.22
14	BC	84.48	13.80	1.68	98.04	108.80	14	489	8.99	125.84	13.48	188.76
15	GC	81.57	14.00	5.50	103.11	120.35	15	828	13.76	206.40	20.64	309.61
16	GC	72.45	11.20	4.53	91.04	93.83	15	506	10.79	161.78	16.18	242.68
17	GC	69.77	12.80	4.06	92.68	97.23	15	213	4.38	65.72	6.57	98.58
18	GC	82.17	11.92	5.71	98.97	110.87	15	618	11.15	167.22	16.72	250.83
19	GC	72.45	11.20	4.53	91.04	93.83	15	500	10.66	159.87	15.99	239.80
20	RR	139.50	23.60	6.84	153.96	268.33	17	476	3.55	60.31	5.32	90.47
21	OT	165.80	27.40	10.39	180.23	367.70	16	225	1.22	19.58	1.84	29.37
22	CS	181.44	27.80	10.59	188.64	402.82	18	477	2.37	42.63	3.55	63.94

(continued)

Table 2 (continued)

Vessel	Vessel type	LOA (m)	B (m)	Ts (m)	BW pipe diam. (mm)	BWE PRC (m^3/h)	Vessel speed (knots)	BW for disch. (m^3)	SEQ time (h)	SEQ dist. (NM)	PT time (h)	PT dist. (NM)
23	CS	100.63	15.20	5.63	114.85	149.32	18	383	5.13	92.34	7.69	138.51
24	GC	100.58	16.21	6.38	118.35	158.56	15	250	3.15	47.30	4.73	70.95
25	GC	100.58	16.21	6.38	118.35	158.56	15	130	1.64	24.60	2.46	36.89
26	CS	260.05	32.25	12.62	236.58	633.61	18	3,318	10.47	188.52	15.71	282.78
27	GC	169.03	25.40	9.96	176.43	352.35	15	5,010	28.44	426.57	42.66	639.85
28	RR	113.00	16.00	5.11	121.35	166.71	17	350	4.20	71.38	6.30	107.07
29	BC	63.18	15.30	4.94	95.52	103.28	14	600	11.62	162.66	17.43	243.99
30	GC	72.45	11.20	4.53	91.04	93.83	15	500	10.66	159.87	15.99	239.80
31	BC	151.94	24.00	9.69	164.98	308.10	14	5,364	34.82	487.48	52.23	731.22
32	BC	151.94	24.00	9.69	164.98	308.10	14	5,261	34.15	478.12	51.23	717.18
33	CS	166.14	28.54	11.61	184.74	386.36	18	752	3.89	70.07	5.84	105.10
34	CS	166.14	28.54	11.61	184.74	386.36	18	680	3.52	63.36	5.28	95.04
35	GC	158.91	23.05	10.10	166.55	314.01	15	470	2.99	44.90	4.49	67.35
36	GC	158.91	23.05	10.10	166.55	314.01	15	376	2.39	35.92	3.59	53.88
37	RR	113.00	16.00	5.11	121.35	166.71	17	200	2.40	40.79	3.60	61.19
38	GC	158.59	24.80	11.53	172.69	337.58	15	1,576	9.34	140.06	14.01	210.09
39	GC	130.80	17.70	8.10	138.10	215.89	15	48	0.44	6.67	0.67	10.00
40	GC	79.77	12.83	5.33	99.38	111.81	15	520	9.30	139.52	13.95	209.29
41	GC	69.10	10.00	5.03	87.50	86.67	15	121	2.79	41.88	4.19	62.82

42	GC	139.81	16.40	4.67	132.47	198.65	15	1,740	17.52	262.78	26.28	394.16
43	GC	130.70	17.69	8.10	138.03	215.68	15	830	7.70	115.45	11.54	173.17
44	GC	110.00	16.30	5.02	120.65	164.79	15	1,146	13.91	208.64	20.86	312.95
45	GC	100.50	20.42	8.18	129.93	191.09	15	900	9.42	141.29	14.13	211.94
46	GC	82.17	11.92	5.71	98.97	110.87	15	716	12.92	193.74	19.37	290.61
47	GC	74.60	12.80	7.86	100.48	114.29	15	329	5.76	86.36	8.64	129.53
48	GC	169.33	26.60	10.92	180.84	370.21	15	662	3.58	53.65	5.36	80.47
49	GC	182.81	31.15	12.04	199.12	448.84	15	462	2.06	30.88	3.09	46.32
50	GC	165.86	23.70	10.02	171.06	331.25	15	3,996	24.13	361.90	36.19	542.85
51	GC	99.82	18.02	6.39	122.03	168.58	15	106	1.26	18.86	1.89	28.30
52	CS	208.30	29.80	11.40	206.61	483.22	18	1,543	6.39	114.95	9.58	172.43
53	CS	100.63	15.20	5.63	114.85	149.32	18	1,050	14.06	253.15	21.10	379.72
54	CS	192.50	32.26	12.80	207.33	486.61	18	430	1.77	31.81	2.65	47.72
55	BC	224.95	32.28	13.91	223.93	567.64	14	400	1.41	19.73	2.11	29.60
56	CS	208.30	29.80	11.40	206.61	483.22	18	434	1.80	32.33	2.69	48.50
57	CS	187.40	28.40	11.12	193.56	424.12	18	1,118	5.27	94.90	7.91	142.35
58	CS	87.60	13.10	5.07	103.18	120.51	18	298	4.95	89.02	7.42	133.53
59	CS	208.30	29.80	11.40	206.61	483.22	18	725	3.00	54.01	4.50	81.02
60	CS	213.00	32.20	11.52	214.63	521.47	18	2,272	8.71	156.85	13.07	235.27
61	CS	213.00	32.20	11.52	214.63	521.47	18	2,839	10.89	195.99	16.33	293.99
62	CS	213.00	32.20	11.52	214.63	521.47	18	1,436	5.51	99.14	8.26	148.70
63	CS	260.60	32.25	12.52	236.63	633.88	18	982	3.10	55.77	4.65	83.66

Enhanced after David (2007)

LOA length over all, *B* breadth, *Ts* summer draft, *BC* bulk carrier, *GC* general cargo vessel, *RR* Ro-Ro vessel, *CS* container vessel, *OT* oil tanker

Table 3 Feasibility of BWE sequential (*SEQ*) and pump-through (*PT*) methods from selected ports through the BWEA

Vessel	Source Port	Area	Vessel speed (knots)	BW for disch. (m^3)	BWEA distance (NM)	SEQ factor	PT factor	BWEA deviated distance (NM)	Deviated SEQ factor	Deviated PT factor
10	BAR	SA	15	701	150	0.90	0.60	190	1.13	0.76
11	BARI	SA	14	1,200	100	0.42	0.28	160	0.67	0.45
12	BARI	SA	15	470	100	0.64	0.43	160	1.02	0.68
13	DURRES	SA	15	361	230	2.27	1.51	/	/	/
14	MANFREDONIA	SA	14	489	100	0.79	0.53	/	/	/
15	BRINDISI	SA	15	828	200	0.97	0.65	/	/	/
16	ALEXANDRIA	MED	15	506	290	1.79	1.20	/	/	/
17	ALEXANDRIA	MED	15	213	290	4.41	2.94	/	/	/
18	ANNABA	MED	15	618	290	1.73	1.16	/	/	/
19	BENGHAZI	MED	15	500	290	1.81	1.21	/	/	/
20	BENGHAZI	MED	17	476	290	4.81	3.21	/	/	/
21	CONSTANZA	BS	16	225	290	14.81	9.87	/	/	/
22	IZMIR	MED	18	477	290	6.80	4.54	/	/	/
23	LATTAKIA	MED	18	383	290	3.14	2.09	/	/	/
24	MALTA	MED	15	250	290	6.13	4.09	/	/	/
25	MALTA	MED	15	130	290	11.79	7.86	/	/	/
26	MEDITERRANEAN	MED	18	3,318	290	1.54	1.03	/	/	/
27	MEDITERRANEAN	MED	15	5,010	290	0.68	0.45	/	/	/
28	MEDITERRANEAN	MED	17	350	290	4.06	2.71	/	/	/
29	MISURATA	MED	14	600	290	1.78	1.19	/	/	/
30	MISURATA	MED	15	500	290	1.81	1.21	/	/	/
31	PIOMBINO	MED	14	5,364	290	0.59	0.40	/	/	/
32	PIOMBINO	MED	14	5,261	290	0.61	0.40	/	/	/
33	PIRAEUS	MED	18	752	290	4.14	2.76	/	/	/
34	PIRAEUS	MED	18	680	290	4.58	3.05	/	/	/
35	PIRAEUS	MED	15	470	290	6.46	4.31	/	/	/
36	PIRAEUS	MED	15	376	290	8.07	5.38	/	/	/
37	PIRAEUS	MED	17	200	290	7.11	4.74	/	/	/
38	MEDITERRANEAN	MED	15	1,576	290	2.07	1.38	/	/	/
39	SAN NICOLAS	MED	15	48	290	43.48	28.99	/	/	/
40	MEDITERRANEAN	MED	15	520	290	2.08	1.39	/	/	/
41	SKIKDA	MED	15	121	290	6.92	4.62	/	/	/
42	TARANTO	MED	15	1,740	290	1.10	0.74	/	/	/
43	TARANTO	MED	15	830	290	2.51	1.67	/	/	/
44	TORRE ANUNZIATA	MED	15	1,146	290	1.39	0.93	/	/	/
45	TUZLA	MED	15	900	290	2.05	1.37	/	/	/
46	VALENCIA	MED	15	716	290	1.50	1.00	/	/	/
47	VALLETTA	MED	15	329	290	3.36	2.24	/	/	/
48	ATLANTIC OCEAN	O	15	662	290	5.41	3.60	/	/	/
49	ATLANTIC OCEAN	O	15	462	290	9.39	6.26	/	/	/
50	BANDAR ABBAS	O	15	3,996	290	0.80	0.53	/	/	/
51	EREGLI	BS	15	106	290	15.37	10.25	/	/	/
52	INDIAN & PACIFIC	O	18	1,543	290	2.52	1.68	/	/	/
53	LATTAKIA	MED	18	1,050	290	1.15	0.76	/	/	/
54	OCEAN MIX	MED	18	430	290	9.12	6.08	/	/	/
55	PACIFIC	MED	14	400	290	14.70	9.80	/	/	/
56	PACIFIC	MED	18	434	290	8.97	5.98	/	/	/
57	PIRAEUS, LIMASSOL	MED	18	1,118	290	3.06	2.04	/	/	/
58	MARSAXLOK	MED	18	298	290	3.26	2.17	/	/	/
59	TARANTO	MED	18	725	290	5.37	3.58	/	/	/
60	TARANTO	MED	18	2,272	290	1.85	1.23	/	/	/
61	TARANTO	MED	18	2,839	290	1.48	0.99	/	/	/
62	INDIAN OCEAN	MED	18	1,436	290	2.93	1.95	/	/	/
63	VANCOUVER, HONG KONG, PUSAN, HAIFA, ISTANBUL	O	18	982	290	5.20	3.47	/	/	/

Enhanced after David (2007)

SA southern Adriatic, *MED* Mediterranean Sea, *BS* Black Sea, *O* other. *Red marked cells* indicate that the vessel would not be able to complete the BWE when travelling at her normal speed

Table 4 Feasibility of BWE sequential (SEQ) and pump-through (FT) methods for vessels bound for the Port of Koper arriving from selected ports in the central Adriatic (CA) via the BWEA and the potential extended length

Vessel	Source Port	Area	Vessel speed (knots)	BW for disch. (m^3)	BWEA distance (NM)	SEQ factor	PT factor	Potential + BWEA distance (NM)	Potential + BWEA SEQ factor	Potential + BWEA PT factor
1	ŠIBENIK	CA	14	3,426	15	0.06	0.04	45	0.17	0.11
2	SPLIT	CA	14	348	30	0.32	0.22	60	0.65	0.43
3	SPLIT	CA	15	720	30	0.16	0.11	60	0.32	0.21
4	SPLIT	CA	15	400	30	0.31	0.21	60	0.63	0.42
5	VELA LUKA	CA	15	470	60	0.26	0.17	90	0.39	0.26
6	ORTONA	CA	15	751	20	0.12	0.08	50	0.31	0.21
7	ORTONA	CA	15	700	20	0.13	0.09	50	0.32	0.21
8	ORTONA	CA	15	352	20	0.22	0.15	50	0.56	0.37
9	VASTO	CA	15	200	30	0.74	0.50	60	1.49	0.99

Where the field is marked *red* the vessel would not be able to complete the BWE when travelling at her normal speed

data on salinity as key environmental matching parameter and the presence of HAOP in the ballast water source ports become relevant. The BWRA in addition considers data reliability. In this regard, port baseline surveys and monitoring for HAOP are considered as critical factors for data reliability.

Data Availability and Related Assumptions

For the environmental matching RA, salinity data for the Port of Koper and for all source ports were collected. It was recognised that the Port of Koper is a marine port with a salinity always being above 30 psu. None of the source ports had salinity levels below 1 psu (i.e., freshwater port), therefore, none of the ports was found enough environmentally incompatible to trigger a low risk.

When searching for data on port baseline surveys and/or monitoring programmes for HAOP in source ports it was recognized that no source port has these data available.[4] In the light of the precautionary approach and considering the results from the BWS study in the Port of Koper (David et al. 2007b), even if ballast water originates from the same biogeographic region as the Port of Koper, the presumption is taken that the ballast water may contain non-indigenous species. With this, by default a high risk RA result is triggered for all considered source ports.

As the next step possibilities to trigger intermediate risk (i.e., if the same HAO are present in the source and receiving ports) or extreme risk (i.e., if some of the HAO present in the source port are under a control programme in the receiving port or Slovenian sea) have been considered. The option of identifying a source port as

[4] i.e., no baseline surveys or monitoring programs for HAO were conducted in source ports.

an intermediate risk is excluded for the same reason that the ports were identified as high risk (previous paragraph), hence this would become an option if a source port(s) would have a port baseline survey and/or monitoring programme for HAOP. The option of extreme risk in this regards is also excluded since there is no control programme for HAOP introduced in the Slovenian sea.

Further, the presence of potentially toxic algae in the blooming state (i.e., harmful algae bloom – HAB) or/and human pathogens (i.e., indicator microbes)[5] in the source port remain the triggering points for extreme risk. HABs as well as the presence of human pathogens are not permanent phenomena but vary throughout the year (e.g., Žohar-Čretnik and Gubina 2002; WGHAB 2006; GEOHAB 2012, Dean Bošnjak, pers. comm.). HABs are monitored in different parts of the world under different programmes.[6] Indicator microbe monitoring is mostly related to the monitoring of the water quality in bathing and aquaculture areas. None of these programmes is focussed on the dispersal of these organisms with vessels and most programmes lack sampling stations in port environments (Žohar-Čretnik and Gubina 2002; Dean Bošnjak, pers. comm.).

The occurrence data on HABs and indicator microbes was searched and available data were located on websites of the International Oceanographic Committee (IOC) regional HAB networks. Regarding indicator microbes, in addition to the literature, World Health Organisation (WHO) data[7] and national data on web pages were examined; however, there was no relevant data found for the source ports for 2005.

According to the BWM Convention, information on the presence of, e.g., potentially toxic algae in bloom state or/and human pathogens (i.e., indicator microbes) in the source port water would need to be communicated to vessels so that they can avoid ballast water uptake in these areas. In real conditions, if ballast water would anyway be loaded in such conditions, this would need to be reported to the port authorities of the ballast water recipient port.

Due to the lack of reliable data for the time and source of ballast water loading, historical data on HABs in the source port areas were considered. The presence of HABs in a port was simulated based on distribution maps[8] prepared by Zingone et al. (2004) in the framework of the International Society for the Study of Harmful Algae (ISSHA) project HAB-MAP.[9] This was studied separately for ballast water source ports out-

[5] i.e., indicator microbes, as a human health standard, shall include:

1. Toxicogenic *Vibrio cholerae* (O1 and O139) with less than 1 colony forming unit (cfu) per 100 ml or less than 1 cfu per 1 g (wet weight) zooplankton samples;
2. *Escherichia coli* less than 250 cfu per 100 ml;
3. Intestinal Enterococci less than 100 cfu per 100 ml.

Results reported as MPN per 100 ml.

[6] http://ioc.unesco.org/hab/activit.htm

[7] http://www.who.int/csr/don/en, last accessed January 2014.

[8] i.e., limited to Mediterranean; ports outside the Mediterranean are were considered.

[9] HANA, IOC Network on Harmful Algae in North Africa.

side the Mediterranean (i.e., Bandar Abbas,[10] Hong Kong,[11] Pusan[12] and Vancouver)[13] (PICES 2002, 2006). A source port was considered as in HAB state at the time of ballast water loading when a HAB occurrence was known to occur at the time of ballasting or in case HABs occurred several times in the area of that port.

Risk Assessment Results

The relevant data and assumptions as described above were used to obtain RA results for all ballast water source ports (see Table 5). An extreme risk was assessed for many Mediterranean ports and also for ports located overseas.

Final Results

Vessels discharging ballast water in the Port of Koper which originates from source ports in the northern Adriatic represent a very critical issue because BWE is not an appropriate BWM method for this shallow water and enclosed area, and vessels do not have BWMS installed onboard as the BWM Convention is not yet into force. From the quantitative perspective, i.e., water volumes, these vessels represent the majority of ballast water discharges in the Port of Koper, i.e., usually more than 80 % of all ballast water discharges, in 2005 even 85 % of all ballast water discharges in the Port of Koper (Perkovič et al. 2003). Also from the qualitative perspective (i.e., HAO HAB), ballast water originating from these ports was found to be critical (David et al. 2007b). It is concluded that these critical situations may only be managed with the installation of BWMS on vessels or by substituting ships with barges. Therefore, this specific ballast water discharge profile in the Port of Koper has to be considered according to the RA results as shown above, i.e., differences among the approaches studied should be considered in comparison to the ballast water discharges from those source ports outside the northern Adriatic.

The RA results from source ports were related to each vessel arriving in Koper to assign the level of risk to each vessel discharging ballast water. Vessels that could manage ballast water before discharge under each studied approach were identified (see Table 6).

---

[10] Found no data that would confirm HABs.

[11] i.e., Considered as HAB based on Yan et al. 2002, A national report on harmful algal blooms in China. (http://www.pices.int/publications/scientific_reports/Report23/HAB_China.pdf, last accessed January 2014).

[12] i.e., considered as HAB based on Lee et al. 2002. Harmful Algal Blooms (Red Tides): Management and Mitigation in the Republic of Korea. (http://www.pices.int/publications/scientific_reports/Report23/HAB_Korea.pdf, last accessed January 2014).

[13] i.e., considered as HAB based on Taylor and Harrison 2002. Harmful Algal Blooms in Western Canadian Coastal Waters. (http://www.pices.int/publications/scientific_reports/Report23/HAB_Canada.pdf, last accessed January 2014).

Table 5 RA results of the ballast water source ports

Source Port	Location	ENV. M.	HAO	HAB	P	Tg.sp.	CP	RA result
ŠIBENIK(Croatia)	CA	HIGH	YES	YES	NO	N/A	N/A	extreme risk
SPLIT (Croatia)	CA	HIGH	YES	NO	NO	N/A	N/A	high risk
VELA LUKA (Croatia)	CA	HIGH	YES	NO	NO	N/A	N/A	high risk
ORTONA (Italy)	CA	HIGH	YES	NO	NO	N/A	N/A	high risk
VASTO (Italy)	CA	HIGH	YES	NO	NO	N/A	N/A	high risk
BAR (Montenegro)	SA	HIGH	YES	NO	NO	N/A	N/A	high risk
BARI (Italy)	SA	HIGH	YES	NO	NO	N/A	N/A	high risk
BRINDISI (Italy)	SA	HIGH	YES	NO	NO	N/A	N/A	high risk
DURRES (Albania)	SA	HIGH	YES	NO	NO	N/A	N/A	high risk
MANFREDONIA (Italy)	SA	HIGH	YES	NO	NO	N/A	N/A	high risk
ALEXANDRIA (Egypt)	MED	HIGH	YES	YES	NO	N/A	N/A	extreme risk
ANNABA (Algeria)	MED	HIGH	YES	NO	NO	N/A	N/A	high risk
BENGHAZI (Libya)	MED	HIGH	YES	NO	NO	N/A	N/A	high risk
HAIFA (Israel)	MED	HIGH	YES	NO	NO	N/A	N/A	high risk
ISTANBUL (Turkey)	MED	HIGH	YES	NO	NO	N/A	N/A	high risk
IZMIR (Turkey)	MED	HIGH	YES	YES	NO	N/A	N/A	extreme risk
LATTAKIA (Syria)	MED	HIGH	YES	NO	NO	N/A	N/A	high risk
LIMASSOL (Cyprus)	MED	HIGH	YES	NO	NO	N/A	N/A	high risk
MALTA (Malta)	MED	HIGH	YES	NO	NO	N/A	N/A	high risk
MARSAXLOK (Malta)	MED	HIGH	YES	NO	NO	N/A	N/A	high risk
MISURATA (Libya)	MED	HIGH	YES	NO	NO	N/A	N/A	high risk
PIOMBINO (Italy)	MED	HIGH	YES	NO	NO	N/A	N/A	high risk
PIRAEUS (Greece)	MED	HIGH	YES	YES	NO	N/A	N/A	extreme risk
SKIKDA(Algeria)	MED	HIGH	YES	NO	NO	N/A	N/A	high risk
TARANTO (Italy)	MED	HIGH	YES	NO	NO	N/A	N/A	high risk
TORRE ANNUNZIATA (Italy)	MED	HIGH	YES	YES	NO	N/A	N/A	extreme risk
VALENCIA (Spain)	MED	HIGH	YES	YES	NO	N/A	N/A	extreme risk
VALLETTA (Malta)	MED	HIGH	YES	NO	NO	N/A	N/A	high risk
CONSTANZA (Romania)	BS	HIGH	YES	NO	NO	N/A	N/A	high risk
EREGLI (Turkey)	BS	HIGH	YES	NO	NO	N/A	N/A	high risk
BANDAR ABBAS (Iran)	O	HIGH	YES	NO	NO	N/A	N/A	high risk
HONG KONG	O	HIGH	YES	YES	NO	N/A	N/A	extreme risk
PUSAN(Korea)	O	HIGH	YES	YES	NO	N/A	N/A	extreme risk
VANCOUVER (Canada)	O	HIGH	YES	YES	NO	N/A	N/A	extreme risk

Enhanced after David (2007)
ENV.M. Environmental Matching, *HAO* Harmful Aquatic Organisms, *HAB* Harmful Algae Bloom, *P* Pathogens and Indicator Microbes, *Tg.sp.* Target Species, *CP* species under a Control Programme

Application of the Blanket Approach

In case the BWM Convention would be applied under the blanket approach in total 12 vessels were identified which would have been enabled to conduct BWE out of 63 vessels that had a ballast water source port in the BWM relevant area, i.e., outside the northern Adriatic. This represents 2.68 % of all vessels that discharged ballast water in the Port of Koper in 2005, and 19.05 % of those with a source port in the BWM relevant area. Looking at the quantity of ballast water discharged, 10,866 m^3 of ballast water could have been managed out of 64,754 m^3 from the relevant area, representing only 2.00 % of all ballast water discharged in the Port of Koper in 2005, and 16.78 % of the ballast water from source ports in the BWM relevant area (see Table 7). None of the vessels would be delayed since

Table 6 RA results related to vessels and discharges of ballast water under the blanket, blanket with BWEA and DSS approaches

Vessel	Vessel type	Source Port	Area	BW for disch. (m^3)	BWRA risk level	Blanket	Blanket BWEA	Selective BWM DSS
1	BC	SIBENIK	CA	3,426	extreme	no BWM	no BWM	BWM
2	BC	SPLIT	CA	348	high	no BWM	no BWM	BWM
3	GC	SPLIT	CA	720	high	no BWM	no BWM	BWM
4	GC	SPLIT	CA	400	high	no BWM	no BWM	BWM
5	GC	VELA LUKA	CA	470	high	no BWM	no BWM	BWM
6	GC	ORTONA	CA	751	high	no BWM	no BWM	BWM
7	GC	ORTONA	CA	700	high	no BWM	no BWM	BWM
8	GC	ORTONA	CA	352	high	no BWM	no BWM	BWM
9	GC	VASTO	CA	200	high	no BWM	no BWM	BWM
10	GC	BAR	SA	701	high	no BWM	BWM	BWM
11	BC	BARI	SA	1,200	high	no BWM	BWM	BWM
12	GC	BARI	SA	470	high	no BWM	BWM	BWM
13	GC	DURRES	SA	361	high	no BWM	BWM	BWM
14	BC	MANFREDONIA	SA	489	high	no BWM	BWM	BWM
15	GC	BRINDISI	SA	828	high	no BWM	BWM	BWM
16	GC	ALEXANDRIA	MED	506	extreme	BWM	BWM	BWM
17	GC	ALEXANDRIA	MED	213	extreme	BWM	BWM	BWM
18	GC	ANNABA	MED	618	high	no BWM	BWM	BWM
19	GC	BENGHAZI	MED	500	high	BWM	BWM	BWM
20	RR	BENGHAZI	MED	476	high	BWM	BWM	BWM
21	OT	CONSTANZA	BS	225	high	no BWM	BWM	BWM
22	CS	IZMIR	MED	477	extreme	no BWM	BWM	BWM
23	CS	LATTAKIA	MED	383	high	no BWM	BWM	BWM
24	GC	MALTA	MED	250	high	no BWM	BWM	BWM
25	GC	MALTA	MED	130	high	no BWM	BWM	BWM
26	CS	MEDITERRANEAN	MED	3,318	high	no BWM	BWM	BWM
27	GC	MEDITERRANEAN	MED	5,010	high	no BWM	BWM	BWM
28	RR	MEDITERRANEAN	MED	350	high	no BWM	BWM	BWM
29	BC	MISURATA	MED	600	high	BWM	BWM	BWM
30	GC	MISURATA	MED	500	high	BWM	BWM	BWM
31	BC	PIOMBINO	MED	5,364	high	no BWM	BWM	BWM
32	BC	PIOMBINO	MED	5,261	high	no BWM	BWM	BWM
33	CS	PIRAEUS	MED	752	extreme	no BWM	BWM	BWM
34	CS	PIRAEUS	MED	680	extreme	no BWM	BWM	BWM
35	GC	PIRAEUS	MED	470	extreme	no BWM	BWM	BWM
36	GC	PIRAEUS	MED	376	extreme	no BWM	BWM	BWM
37	RR	PIRAEUS	MED	200	extreme	no BWM	BWM	BWM
38	GC	MEDITERRANEAN	MED	1,576	high	no BWM	BWM	BWM
39	GC	SAN NICOLAS	MED	48	high	no BWM	BWM	BWM
40	GC	MEDITERRANEAN	MED	520	high	no BWM	BWM	BWM

(continued)

Table 6 (continued)

Vessel	Vessel type	Source Port	Area	BW for disch. (m³)	BWRA risk level	Blanket	Blanket BWEA	Selective BWM DSS
41	GC	SKIKDA	MED	121	high	no BWM	BWM	BWM
42	GC	TARANTO	MED	1,740	high	no BWM	BWM	BWM
43	GC	TARANTO	MED	830	high	no BWM	BWM	BWM
44	GC	TORRE ANUNZIATA	MED	1,146	extreme	no BWM	BWM	BWM
45	GC	TUZLA	MED	900	high	no BWM	BWM	BWM
46	GC	VALENCIA	MED	716	extreme	BWM	BWM	BWM
47	GC	VALLETTA	MED	329	high	no BWM	BWM	BWM
48	GC	ATLANTIC OCEAN	O	662	high	no BWM	BWM	BWM
49	GC	ATLANTIC OCEAN	O	462	high	no BWM	BWM	BWM
50	GC	BANDAR ABBAS	O	3,996	high	BWM	BWM	BWM
51	GC	EREGLI	BS	106	high	no BWM	BWM	BWM
52	CS	INDIAN & PACIFIC OCEAN	O	1,543	high	BWM	BWM	BWM
53	CS	LATTAKIA	MED	1,050	high	no BWM	BWM	BWM
54	CS	OCEAN MIX	MED	430	high	no BWM	BWM	BWM
55	BC	PACIFIC	MED	400	high	BWM	BWM	BWM
56	CS	PACIFIC	MED	434	high	BWM	BWM	BWM
57	CS	PIRAEUS, LIMASSOL	MED	1,118	extreme	no BWM	BWM	BWM
58	CS	MARSAXLOK	MED	298	high	no BWM	BWM	BWM
59	CS	TARANTO	MED	725	high	no BWM	BWM	BWM
60	CS	TARANTO	MED	2,272	high	no BWM	BWM	BWM
61	CS	TARANTO	MED	2,839	high	no BWM	BWM	BWM
62	CS	INDIAN OCEAN	MED	1,436	high	no BWM	BWM	BWM
63	CS	VANCOUVER, HONG KONG, PUSAN, HAIFA, ISTANBUL	O	982	extreme	BWM	BWM	BWM

Enhanced after David (2007)
no BWM ballast water would have been discharged unmanaged, *BWM* ballast water for discharge would have been managed

Table 7 Potential for BWM under the BWM Convention blanket approach in number of vessels and quantities of ballast water discharged from the relevant area, i.e., outside the northern Adriatic

Vessels		BW discharged (m³)	
Nr. vessels BWM	12	BW BWM	10,866
Nr. vessels no BWM	51	BW no BWM	53,888
Nr. vessels from BWM relevant area	63	BW discharge from BWM relevant area	64,754
Total Nr. 2005	448	Total BW discharge 2005	544,133
% BWM / BWM relevant area	19.05	% BWM / BWM relevant area	16.78
% BWM / Total 2005	2.68	% BWM / Total 2005	2.00
Delay OK	N/A	RA high untreated	45,243
Undue delay	N/A	RA extreme untreated	8,645

Enhanced after David (2007)
no BWM ballast water would have been discharged unmanaged, *BWM* ballast water for discharge would have been managed

the blanket BWM approach does not foresee measures such as deviation and/or slowing-down of vessels.

RA was conducted for the purposes of the DSS model. When combining the RA results with the ballast water discharges under the blanket approach it was recognized that 45,243 m^3 of high risk, and 8,645 m^3 of ballast water assessed as extreme risk would be discharged unmanaged (see Table 7). Without conducting BWRA this would be the actual ballast water discharge situation, but the risk posed would remain unknown.

Application of the Blanket Approach with the Designated Ballast Water Discharge Area

The blanket approach assumes that a BWEA would be designated as discussed above, and all vessels would be required to use it to complete BWE. It was recognised that all vessels sailing from ballast water source ports in the central Adriatic (CA) would need to deviate from their intended routes and be delayed for more than 8 h on average (i.e., ranging from less than 1 h up to 18.21 h) to complete BWE in the BWEA. Therefore, this approach was considered as inappropriate especially considering the relatively short voyage distances and the need for intensive slow-down of certain vessels to fully complete BWE (i.e., vessel 1 would need to slow down approximately to 1/10 of her normal speed) (see Table 8).

The same calculations were applied for all remaining vessels sailing from source ports in the southern Adriatic and Mediterranean Sea area close to the Adriatic Sea. It was recognised that eight vessels would need to slow-down, with three of them

Table 8 Vessels sailing to the Port of Koper from the source ports in the central Adriatic (CA) and expected delay time in hours to complete BWE in the BWEA

Vessel	Source Port	Area	Vessel speed (knots)	BW for disch. (m^3)	BWEA distance (NM)	SEQ factor	PT factor	Potential + BWEA distance (NM)	Potential + BWEA SEQ factor	Potential + BWEA PT factor	Delay (h)
1	ŠIBENIK	CA	14	3,426	15	0.06	0.04	45	0.17	0.11	18.21
2	SPLIT	CA	14	348	30	0.32	0.22	60	0.65	0.43	4.47
3	SPLIT	CA	15	720	30	0.16	0.11	60	0.32	0.21	10.60
4	SPLIT	CA	15	400	30	0.31	0.21	60	0.63	0.42	4.37
5	VELA LUKA	CA	15	470	60	0.26	0.17	90	0.39	0.26	11.43
6	ORTONA	CA	15	751	20	0.12	0.08	50	0.31	0.21	9.37
7	ORTONA	CA	15	700	20	0.13	0.09	50	0.32	0.21	9.05
8	ORTONA	CA	15	352	20	0.22	0.15	50	0.56	0.37	4.62
9	VASTO	CA	15	200	30	0.74	0.50	60	1.49	0.99	0.69

Enhanced after David (2007)

Table 9 Vessels sailing to the Port of Koper from source ports in the southern Adriatic (SA) and Mediterranean Sea (MED) close to the Adriatic Sea, and expected delayed time in hours to complete BWE before and in the BWEA

Vessel	Source Port	Area	Vessel speed (knots)	BW for disch. (m^3)	BWEA distance (NM)	SEQ factor	PT factor	Potential + BWEA distance (NM)	Potential + BWEA SEQ factor	Potential + BWEA PT factor	Delay (h)
10	BAR	SA	15	701	150	0.90	0.60	190	1.13	0.76	1.17
11	BARI	SA	14	1,200	100	0.42	0.28	160	0.67	0.45	9.88
12	BARI	SA	15	470	100	0.64	0.43	160	1.02	0.68	3.79
14	MANFREDONIA	SA	14	489	100	0.79	0.53	/	/	/	1.85
15	BRINDISI	SA	15	828	200	0.97	0.65	/	/	/	0.43
27	MEDITERRANEAN	MED	15	5,010	290	0.68	0.45	/	/	/	9.10
31	PIOMBINO	MED	14	5,364	290	0.59	0.40	/	/	/	14.11
32	PIOMBINO	MED	14	5,261	290	0.61	0.40	/	/	/	13.44

Enhanced after David (2007)

Table 10 Potential for BWM under the BWM Convention blanket approach in number of vessels and quantities of ballast water discharged

Vessels		BW discharged (m^3)	
Nr. vessels BWM	54	BW BWM	57,387
Nr. vessels no BWM	9	BW no BWM	7,367
Nr. vessels from BWM relevant area	63	BW discharge from BWM relevant area	64,754
Total Nr. 2005	448	Total BW discharge 2005	544,133
% BWM / BWM relevant area	85.71	% BWM / BWM relevant area	88.62
% BWM / Total 2005	12.05	% BWM / Total 2005	10.55
Delay OK	4	RA high untreated	3.941
Undue delay	4	RA extreme untreated	3.426

Enhanced after David (2007)

having the option to deviate (i.e., vessels 10, 11 and 12). Their average delay compared to the intended routes would amount to 6.72 h (i.e., ranging from less than half an hour up to 14.11 h). These delays are assessed as less critical since they are related to longer voyages. Nevertheless, the four red-highlighted delays may be considered as undue (see last column in Table 9).

As a result, under this approach 54 vessels were identified as being able to conduct BWE out of 63 vessels that had a ballast water source port in the BWM relevant area i.e., outside the northern Adriatic. This represents 12.05 % of all vessels that discharged ballast water in the Port of Koper in 2005, and 88.62 % of those with a source port in the BWM relevant area. In total 57,387 m^3 of ballast water discharged could have been managed out of 64,754 m^3 from the BWM relevant area. This represents 10.55 % of all ballast water discharged in the Port of Koper in 2005, and 88.62 % of the ballast water from source ports in the BWM relevant area (see Table 10).

When combining the RA results with the blanket approach with available BWEA it was recognized that 3,941 m^3 of high risk, and 3,426 m^3 of ballast water assessed as extreme risk would be still discharged unmanaged (see Table 10). As in the previous approach, without conducting BWRA this would be the actual ballast water discharge situation, but the risk posed would remain unknown.

Application of the Proposed Decision Support System Model with the Designated Ballast Water Discharge Area

In general, as a result of the DSS model implementation, all 63 vessels studied were identified as able to conduct BWE under different levels of BWM requirements which would have been selected according to the level of risk assessed per each vessel ballast water discharge. These represent 14.06 % of all vessels that discharged ballast water in the Port of Koper in 2005, and 100 % of those with a source port in the BWM relevant area. All 64,754 m^3 of ballast water discharged originating from source ports in the BWM relevant area could have been managed. This represents 11.90 % of all ballast water discharged in the Port of Koper in 2005 (see Table 11).

Here below the application of the proposed DSS model is also to be seen in view of the results according to the end-points of the BWM DSS.

Vessels Turned Away Because They Have Not Submitted Required Data

Based on data of the Port of Koper PSA, 2,368 vessels entered Slovenia in 2005, and 989 of these did not submit BWRF. According to the BWM DSS, as this is a port entry requirement, these could be understood as to be "turned away". At the time of this study the submission of BWRF was not legally binding, but for an efficient implementation of a selective BWM approach supported by BWM DSS, reporting in advance on ballast water operations need to be a port entry requirement.

None of the vessels studied, however, could be simulated as a "turned away" case since this study was based on vessels that have submitted BWRF with relevant (critical) data on ballast water intended to be discharged in the Port of Koper.

Table 11 Result of the implementation of the BWM DSS model

Vessels		BW discharged (m^3)	
Nr. Vessels BWM	63	BW BWM	64,754
Nr. vessels no BWM	0	BW no BWM	0
Nr. vessels from BWM relevant area	63	BW discharge from BWM relevant area	64,754
Total Nr. 2005	448	Total BW discharge 2005	544,133
% BWM / BWM relevant area	100	% BWM / BWM relevant area	100
% BWM / Total 2005	14.06	% BWM / Total 2005	11.90
Delay OK	14	RA high untreated	0
Undue delay	0	RA extreme untreated	0
Undue delay RA OK	8		

Enhanced after David (2007)

When a mandatory reporting is introduced it is expected that turning a vessel away because of the lack of submitted BWRF would not occur, same as this is not the case with other documents required for port entry. However, turning away of a vessel still has to be considered as a last resource in case a vessel would not provide BWRF or not report as required.

Vessels Exempted from Ballast Water Management Requirements

According to the RA results none of the vessels was exempted from BWM requirements. Considering that most of the ballast water discharged in the Port of Koper originates from the same biogeographic region, exemptions of vessels could be found suitable. This option remains potentially as a component of the DSS implementation in the future, if reliable data on ballast water source ports required for RA becomes available.

Vessels Requested to Conduct Ballast Water Management May or May Not Be Able Do It

According to the RA results all vessels were required to conduct BWM. The BWE sequential and pump-through methods were taken into account as feasible BWM options. However, due to the inapplicability of available BWM methods in the northern Adriatic 385 vessels out of 448 were not able to conduct BWM. For these, other management measures were proposed. Furthermore, BWM DSS provides requirements also for vessels in this area when the BWM Convention enters into force.

The remaining 63 vessels were assessed as able to conduct BWE using one of the available options. In total 17 vessels would not be able to complete BWE on their intended route and at their intended speed; therefore, BWE was required as an additional measure. For 12 of these vessels a deviation and/or slowing-down was assessed as an option, and the other 5 vessels would only need to slow-down. Since additional measures were required based on RA in a case by case approach, none of the vessels concerned was unduly delayed.

Navigational safety aspects were also considered. It was recognised that for those vessels which have to slow-down, some of these would be sailing during BWE at less than half of their usual speed on their intended routes, therefore additional attention of vessels on the same routes would be required.

Vessels Requested to Conduct Ballast Water Management May or May Not Do It Properly

Whether or not a vessel conducted BWM properly could not have been assessed. Therefore, for the results of this study it was assumed that all vessels that were assessed as able to conduct BWM have done it properly.

Vessels Selected for Compliance Monitoring

Out of six elements available in the DSS model for selecting a vessel for compliance monitoring there are only two triggering elements feasible to be addressed for this study. These are: (1) The BWDA model disagrees with the declared ballast water operation; and (2) The ballast water intended for discharge was identified as posing an extreme risk.

1. The assessment included all vessels that have submitted BWRF in 2005; i.e., 1,379 vessels. As a consequence of suspecting that vessels wrongly declared the ballast water discharge operations, 49 vessels were selected for compliance monitoring.
2. The ballast water of 13 vessels was identified as posing an extreme risk, and these were also selected for compliance monitoring.

These results also need to be seen from the perspective of the different levels of monitoring (i.e., PSC inspection) triggered by each DSS model result. The suspected false-declaration based on the disagreement of the BWDA model triggers in the first step the "general" inspection; meanwhile the extreme risk ballast water triggers the "detailed" inspection. This is important information because according to the BWM Convention if a vessel is selected for a detailed inspection, it should not be allowed to discharge any ballast water until it is ascertained that this can be done without a risk of harm to the environment, human health, property or resources.

Vessels Allowed to Discharge Unmanaged Ballast Water

As a consequence of BWE being currently the only available BWM option, but which is not applicable to vessels sailing to Port of Koper from the ports located in the northern Adriatic, 385 vessels out of 448 were not able to conduct BWM and they were allowed to discharge unmanaged ballast water in the Port of Koper. However, this should not be considered as the result of DSS since these were excluded from DSS BWM options.

All the remaining 63 vessels were able to conduct BWM properly. Therefore, none of the vessels considered discharged unmanaged ballast water.

Vessels May Be Turned Away Because Not All Measures to Conduct Ballast Water Management were Taken Properly

None of the vessels studied was "turned away" since all vessels conducted BWM properly.

The study also showed that the designation of an adequate BWEA would offer the possibility for most of the vessels to comply with BWM requirements without the need to deviate from their intended routes or to slow-down. Consequently, few delays are expected.

Vessels Found Not in Compliance Could Be Penalized

This element could not have been assessed, because this option is not yet addressed in Slovenian legislation.

Acknowledgements Part of these results were obtained in the framework of the research project Decision Model and Control of Ballast Water Management in the Slovenian Sea (L2-6291), which was financially supported by the Slovenian Research Agency and the Port of Koper (Luka Koper d.d.) and as Matej David's PhD work, supervised by Prof. Livij Jakomin, Slovenia and Prof. Chad Hewitt, Australia. This contribution was developed to support the project Vectors of Change in Oceans and Seas Marine Life, Impact on Economic Sectors (VECTORS), which has received funding from the European Community's Seventh Framework Programme (FP7/2007–2013) under Grant Agreement No. 266445.

References

ABS (2006) Rules for building and classing steel vessels 2006. American Bureau of Shipping, Houston

AQIS (1993) Australian quarantine & inspection service. Ballast Water Management. Ballast Water Research Series, Report No. 4, AGPS, Canberra

David M (2007) A decision support system model for ballast water management of vessels: doctoral dissertation. University of Ljubljana, Portorož

David M, Gollasch S, Hewitt C, Jakomin L (2007a) Ballast water management for European seas – is there a need for a decision support system? In: IEEE Conference proceedings, Aberdeen

David M, Gollasch S, Cabrini M, Perkovič M, Bošnjak D, Virgilio D (2007b) Results from the first ballast water sampling study in the Mediterranean Sea – the Port of Koper study. Mar Pollut Bull 54:53–65

Dragsund E, Andersen AB, Gollasch S, ten Hallers-Tjabbes CC, Skogen K (2005) Ballast Water Scoping Study. Det Norske Veritas, Report 2005-0638, Hövik

GEOHAB (2012) Global ecology and oceanography of harmful algal blooms, GEOHAB Core Research Project: HABs in benthic systems. In: Berdalet E, Tester P, Zingone A (eds) IOC, UNESCO and SCOR, Paris and Newark

IMO (2003) Routeing of ships, ship reporting and related matters; establishment of new recommended traffic separation schemes and other new routeing measures in the Adriatic Sea; Submitted by Albania, Croatia, Italy, Slovenia and Serbia and Montenegro. International Maritime Organization (IMO), 49th session of the Sub-Committee on Safety of Navigation (NAV), NAV 49/3/7, International Maritime Organization, London

IMO (2008) Ships' routeing 2008 edition (consolidated). International Maritime Organization, London

Perkovič M, David M, Suban V (2003) Škodljivi vnosi in upravljanje balastnih vod v slovenskem morju : raziskovalni project (L-3208), Model za ugotavljanje balastnih kapacitet in statistični prikaz izpusta balasta v slovenskem morju za obdobje 1990–2002: projektno poročilo št. 1. Portorož: Fak. za pomorstvo in promet

PICES (2002) Harmful algal blooms in the PICES region of the North Pacific, Report 23. The North Pacific Marine Science Organization (PICES), Sidney, BC

PICES (2006) Annual Report 2005. The North Pacific Marine Science Organization (PICES), Sidney, BC

Royal Haskoning (2001) Global market analysis of Ballast Water Treatment Technology. Haskoning Netherland BV Environmental Management, Northeast-Midwest Institute, Netherlands

WGHAB (2006) Report of the ICES-IOC working group on harmful algal bloom dynamics (WGHABD), ICES CM 2006/OCC 04, Gdynia

Zingone A, Honsell G, Marino D, Montresor M, Socal G (2004) Fitoplancton. In: Innamorati M, Ferrari I, Marino D, Ribera D'Alcalà M (eds) Metodi dell'ecologia del plancton marino. Nova Thalassia 11:183–198

Žohar-Čretnik T, Gubina M (eds) (2002) Sanitarna mikrobiologija v javnem zdravstvu 2002: zbornik predavanj. Zavod za zdravstveno varstvo, Ljubljana

Overall Conclusions on the Ballast Water Issue and Its Management Options

Matej David and Stephan Gollasch

Abstract Ballast water management was demonstrated to be a complex issue, hence there are no simple solutions. The BWM Convention was adopted to support globally a uniform approach to prevent harmful aquatic organisms and pathogens to be further spread around the world by ballast water and sediment releases, considering the aspects of safe and efficient operations of shipping, while at the same time providing for the protection of natural environments, human health, property and resources. The conclusions and the current state of knowledge is summarized here and presented thematically sorted as per the book chapters. The overall final conclusions are presented at the end including an outlook highlighting future ballast water management related issues which need to be solved.

Keywords Ballast water • Harmful aquatic organisms and pathogens • Invasive species • Transfer • Ballast water management • Ballast water risk assessment • Ballast water management decision support system

Vessels and Ballast Water

When a vessel is not fully laden, i.e., a situation when she is not at her maximum allowed draft, additional weight is required to compensate for the increased buoyancy in order to provide for the vessel's seaworthiness. This implies that not only commercial vessels, but also other vessels use ballast water to provide for adequate seaworthiness. Even when a vessel is fully laden ballast water operations may be needed due to a non-equal distribution of weights on the vessel. Other dynamic factors may also require ballast water operations, such as weather and sea conditions

M. David (✉)
Dr. Matej David Consult, Korte, Izola, Slovenia
e-mail: matej.david@siol.net

S. Gollasch
Gollasch Consulting (GoConsult), Hamburg, Germany
e-mail: sgollasch@aol.com

M. David, S. Gollasch (eds.), *Global Maritime Transport and Ballast Water Management*, Invading Nature - Springer Series in Invasion Ecology 8,
DOI 10.1007/978-94-017-9367-4_11

on the route, an approach to shallow waters, and the consumption of fuel during the voyage. As a result, vessels fundamentally rely on ballast water for safe operations as a function of their design and construction.

Transfer of Organisms via Ballast Water

Many ballast water studies conducted in different parts of the world proved that ships substantially facilitate the transfer of aquatic organisms across natural barriers. Almost all species types have been found in ballast water samples ranging from unicellular algae, macroalgae, invertebrates to fish. It has also been confirmed that human pathogens are being transferred with ship's ballast water and at least every 9 weeks a new species is found along the coasts of ICES member countries, which includes secondary species introductions. Voyage length critically affects the survival rate of organisms in ballast water. However, the organisms can survive in ballast water for a relatively long time. Some algae, in particular dinoflagellates, can form cysts which sink to the ballast water sediment and may remain viable for several years. There are also known cases when organisms have reproduced and expanded their population inside a tank so that a single ballast water discharge from a ship can be potentially threatening.

One might think that ballast water was moved with ships since more than 100 years and all species which may become ballast water transported have reached the areas they can colonise, but this is not the case. Studies have shown that the number of new non-indigenous species records is increasing since the last 50 years. This can also be due to the focus of scientists on this subject starting at that time and because of intensified research especially over the last two decades. The increase of newly found non-indigenous species by ballast water since the last 50 years may also be related to ever increasing ship speed and sizes. With increased speed the unfavourable conditions an organism is exposed to inside a ballast tank during transit get shorter thereby increasing the en-route survival potential. With increasing vessel size ballast tanks also tend to get bigger, which may further support organism survival due to longer lasting favourable abiotic water conditions.

In short, many of the most negatively impacting species have arrived in ballast tanks which triggered the interest to develop globally applicable organisms transfer preventing measures, i.e., the *International Convention for the Control and Management of Ships' Ballast Water and Sediments* (BWM Convention).

Ballast Water Management Policy

Due to the lack of implemented, internationally agreed ballast water management standards, national BWM requirements arose. As shipping is a truly global business, regionally or nationally different standards are a disadvantage and globally uniform rules are essential to harmonise political, institutional and geographical heterogeneity regarding BWM. This aspect triggered the International Maritime Organization (IMO) mandate to address the ballast water problem originally. Subsequently IMO worked on the preparation of the BWM Convention, which was adopted in 2004, however it is not yet in force.

In light of this different national and regional BWM requirements continued to be implemented to protect the coastal waters from introductions of HAOP as the countries along these regions saw a need to implement such (sometimes voluntary) BWM requirements even before the BWM Convention enters into force. In most occasions, these regional initiatives follow exactly the requirements as set in the BWM Convention, but they just apply earlier. However, to our knowledge only the USA adopted BWM requirements which include D-2 standard related requirements and more stringent numerical standards are also considered. Upon entry into force of the BWM Convention many of these national and regional requirements are in most cases expected to be replaced by the BWM Convention requirements.

Ballast Water Management Convention

Agreements reached on a global level usually represent a combination of significant compromises coupled with action in the face of limited knowledge – and the BWM Convention is not an exception. During the BWM Convention negotiations, many issues were subject of controversial discussions and in certain cases it was extremely hard to reach a consensus, but when dealing with shipping we believe that solutions to an environmental problem should be sought at a global scale.

Although the movement of non-indigenous species usually receive predominant attention, the BWM Convention addresses all species, i.e. cryptogenic species and harmful native species are also included as IMO uses the term "Harmful Aquatic Organisms and Pathogens" (HAOP).

All IMO Conventions, Codes, Protocols etc., are written for ships involved in international voyages through international waters and may be adopted by states for domestic implementation. This Convention protects the coastal environments, mainly up to 50 NM with port State and flag state requirements relating to HAOP being discharged via ballast water into the receiving ports/areas. However, ballast water discharge can also affect international waters especially when ballast water is exchanged "on the high seas" according to the D-1 standard. The D-2 standard however relates to any discharge of ballast water from a vessel regardless of its location. The move to a discharge standard provides protection to high seas as well as coastal regions of the world's oceans and seas.

A country considering to become a Party to the BWM Convention must make resources available to ensure that the obligations resting on the country are ensured and not underestimated. The implementation of this Convention may involve significant costs for the shipping industry, e.g., to install and operate BWMS. However, we believe that an appropriate cost/benefit analysis would reveal that funds used to achieve the aims of the BWM Convention would be well spent, assuming that new biological invasions showing economic impacts are considerably reduced, and especially when considering the essentially important environment and human health protection.

The BWM Convention will enter into force 12 months after the date on which more than 30 states, with combined merchant fleets not less than 35 % of the gross tonnage of the world's merchant shipping, have signed this Convention. As of December 2013, 38 states ratified the BWM Convention, representing 30.38 % of the world merchant shipping gross tonnage (for an update visit Status of Conventions at www.imo.org). Several expert fora assume that the entry into force of the BWM Convention may occur in 2015 or 2016.

Ballast Water Management Systems

The development of ballast water management system (BWMS) and especially their efficiency is very important for an effective prevention of the transfer of harmful aquatic organisms and pathogens across natural barriers. The BWMS review conducted has shown that there are very good perspectives to equip vessels with BWMS as certified BWMS are available. However the BWM Convention requiring their installation is not yet into force, and there are no other binding regional or national requirements like the D-2 standard applying today that would force vessels to install BWMS. However, in the USA BWM standards start to become into effect according to the Vessel General Permit (VGP) requirements starting in December 2013. This includes avoidance areas for ballast water uptake, cleaning of ballast tanks regularly to remove sediments in mid-ocean or under controlled arrangements in a port, or at a dry dock and minimizing the discharge of ballast water essential for vessel operations while in the waters subject to the VGP. The implementation schedule for the first US numerical interim BWM standards starts in 2016.

More than 100 BWMS were identified and they use different treatment technologies mostly in combination to achieve required efficiency over a large variety of ballast water flow rates. BWMS are in different development stages, but more than 30 of them were already type approved by responsible authorities. This makes certified systems available for sales to the shipping industry, however some uncertainty remains if the BWMS production capacities will be able to accommodate the installation needs of the shipping industry over certain short periods after the BWM Convention entry into force. Furthermore, shipyards installation capacities may become a bottleneck to meet the demand. This is a fast developing field as the interest is triggered by a worldwide market of close to 70,000 vessels that will need to be equipped with such systems which may result in a peak demand of 45 BWMS to be installed per day.

We believe that it would be very important for the industry to grab the impetus of this moment and be involved in the development of the BWMS, as the economic perspectives of the global shipping market are very attractive. Furthermore, the involvement of administrations in the certification processes is also important to support a fast development and to ensure the performance quality and reliability of certified BWMS, and hence also better protect the world's oceans and seas, human health, property and resources from the transfer of harmful aquatic organisms and pathogens.

To meet the D-2 standard it may also be considered necessary to combine BWE and ballast water treatment until BWMS become more efficient. By doing so, the efficacy of existing BWMS may be enhanced when the ballast water taken onboard is treated during the exchange.

Risk Assessment

There are two fundamentally different implementation approaches of the BWM Convention, the selective and the blanket approach. The selective approach means that appropriate BWM measures are required depending on different risk levels posed by the intended ballast water discharge. The level of risk is a result of a risk assessment (RA), and the BWM measures are then adapted to the RA result and the acceptance of certain risks. Base on low level risk, an acceptable risk, under G7 Guidelines conditions vessels may be also exempted from BWM requirements up to 5 years, subject to renewal. On the other side, when unacceptable or even extreme risks are identified, BWM is required and some additional measures may need to be implemented.

RA may also support port State control actions. When high risk ballast water is being planned for discharge, a port State authority (PSA) may be interested to ascertain if all necessary BWM measures were undertaken properly, and that there was no failure in the BWM process. On the other side, when a vessel may not be able to comply with basic BWM requirements or was found non-compliant by port State control (PSC), but RA results in low risk level, in such a case PSA may have grounds to allow a vessel to discharge unmanaged ballast water, as this would be understood that such ballast water is not posing a threat to harm the environment, human health, property and resources. This may be a very important point in regards of the Articles 9 and 10 of the BWM Convention, which otherwise require PSA not to let the vessel that was found non-compliant to discharge ballast water which presents a threat of harm to the environment, human health, property or resources.

Reliability of environmental and biological data needed to conduct RA for BWM purposes was found to be crucial, what is in line with the precautionary approach when RA relates to environmental and human health protection. If there is no recent data available about the possible presence of HAOP in ports or areas where ballast water is being loaded or discharged, no species-specific and species' biogeographical RA can be conducted. To ensure biological data reliability, port baseline surveys should be undertaken, and as additional species may be introduced through time, regular monitoring programmes need to be established. When undertaking port baselines surveys, a harmonized approach for the sampling standards and protocols is needed so that all studies generate reliable and comparable results. In this process the frequency of studies, the habitats to be included, i.e., plankton, benthos, fouling, the number of sampling stations, and the availability of taxonomic expertise would need to be considered. If environmental matching RA results in acceptable low risk, no biological data is needed.

Ballast Water Sampling and Sample Processing

Many different ballast water sampling (BWS) methods and equipment have been used for different BWS purposes. Shipboard sampling is also conducted for BWMS testing for type approval. Hence, BWS methods for testing BWMS actually exist, and these have been approved by different national responsible authorities. However, studies have shown that BWS results may be biased by different sampling processes because of, e.g., patchy distribution of organisms in tanks, die-off of organisms during sampling etc. As there is still no commonly agreed BWS methodology or approach, this may impact representative sampling, and certain vessels may be found in compliance with BWM requirements in one port, but not in another due to different sampling methods and approaches chosen.

BWS studies have shown that different methods and sampling equipment may be used for different sampling goals, e.g., sampling for D-1 or D-2 standards, indicative or detailed sampling. Sampling methods and equipment also depend on ballast water access points, i.e., in-tank via manholes, sounding pipes or air vents, or in-line installed sampling points, and on the target groups of organisms, i.e., organisms greater than or equal to 50 μm in minimum dimension, organisms less than 50 μm and greater than or equal to 10 μm in minimum dimension, and indicator microbes.

Sampling inaccuracy remains a significant issue and it may therefore be easier to prove non-compliance rather than compliance to the D-2 standard. From a legal and biological perspective, proving non-compliance is easier and more defensible.

It is of prime importance to consider the appropriate BWS approach for compliance monitoring and enforcement (CME) according to the BWM Convention. The BWS methods described in the chapter "Ballast Water Sampling and Sample Analysis for Compliance Control" were extensively used on board vessels to test BWMS to proof compliance especially with the D-2 standard, and these methods were scientifically validated by additional tests and studies. These BWS methods have also shown to be relatively simple, cost effective and they are generally applicable on all vessel types and in all geographic regions. With this these BWS methods and recommendations may result in a workable, equitable and pragmatic solution to ease port State CME efforts, and to support the entry into force and efficient implementation of the BWM Convention. However, it is also believed that the developed sampling methods and approaches can be improved further, which highlights the need of future work on this subject.

There are two approaches to analyse ballast water samples to proof compliance with BWM requirements, i.e., the samples may be analysed indicatively or in detail. A comprehensive review of sample processing technology, conducted by the authors, revealed that organism detection technologies that enable both an indicative and detailed inspection of ballast water samples are already available today. This conclusion was also supported by our tests conducted on board of commercial vessels to evaluate the suitability of such technologies for practical work by PSC. In summary, for an indicative sample analysis, it is recommended to use Pulse-Amplitude Modulated (PAM) fluorometry to check for viable phytoplankton, use enzyme-chemistry for bacteria analysis and a stereomicroscope for the analysis of the zooplankton organisms above 50 μm in minimum dimension. It should be noted that the PAM method does not deliver organism counts, but it gives a semi-quantitative measurement so that the higher the reading of the instruments is, the higher is the viable biological content. Enzyme-chemistry for bacteria gives a presence/absence indication, but cannot evaluate colony forming units as required by the D-2 standard. However, the presence or absence of the indicator microbes are to be taken as an indication that the BWM method used was successful or not.

The instruments for indicative analysis referred to above are portable and, with the exception of the microscope, of hand-held design and deliver results possibly in less than 10 min so that PSC could check for compliance already on board of the inspected vessels. However, a certain training level is needed to use these organism detection tools that a PSC officer can operate the tools.

For a detailed sample analysis, the recommended methods are more cumbersome and include flow-cytometry and epifluorescence microscopy for the analysis of phytoplankton, with a viability test using stains. Zooplankton should be analysed by a microscope either using gentle poking or a stain to check the organism viability. For bacteria analysis it is recommended to use selective media and it seems that an incubation time of at least 48 h is needed to proof compliance with the D-2 standard so that these results may only become available when the vessel has already left the port. In these cases PSC may keep record of such a vessel for a future inspection of the vessel should she call for this port again or notify the next port of call. The sample processing methods for a detailed analysis are not portable and require a high experience level of a trained biologist so that the samples either need to be brought to a laboratory for subsequent analysis or a van may be equipped with these methods and driven to the port for a sample analysis on the pier.

Final Conclusions

Noting the problems caused by unmanaged ballast water movements naval architects considered to design vessels which would not require the use of water as ballast. Other attempts to solve the problem included a vessel design with continuous flow through of ballast water. However, all alternative ballast concepts so far did not reach a commercially viable level so that the use of ballast water in segregated ballast tanks and/or in cargo holds seems to be the only practicable ballast method today.

In the absence of the globally applicable BWM requirements of the BWM Convention, some countries and regions require BWM already today. Most of these initiatives are based upon BWE as BWMS are largely not installed on vessels. Although more than 30 BWMS are type approved already and annually this number increases, only few vessel owners started to install BWMS on their vessels. One of the reasons for this may be the (substantial) costs involved and the unclarity when the BWM Convention will enter into force.

Countries that wish to protect their seas from the introduction of HAOP via ballast water are confronted with the challenge of balancing the efficiency of BWM measures and the safety and higher costs in the shipping industry as the result of management efforts. For these reasons, the 'blanket approach' of requiring all vessels to undertake BWM is unreasonable in many cases. Alternatively, the 'selective approach' allows for the adjustment of the intensity of BWM measures to each vessel and voyage-specific RA, thus both reducing safety risks and costs to the shipping industry, while simultaneously allowing for improved environmental, human health, property and resources protection. However, a selective approach requires more extensive data gathering for port States, more data and reporting requirements for vessels, and may require higher skills and knowledge from port State personnel. All this can be overcome with an appropriate BWM decision support system (DSS).

A DSS is a supporting tool enhancing the decision-making process that uses a combination of models, analytical techniques, and information retrieval to help develop and evaluate appropriate decision alternatives. DSSs today are widely supporting decision-making processes in business, social programs, medicine, policy, games, information technologies, transport, and are major building blocks in environmental management and science. Decision-makers are frequently faced with taking decisions on very complex issues requiring a large data input, and forced to do so rapidly. This is also the case with the BWM issue. DSS helps decision makers to reduce uncertainties, as well as ease and speed-up the decision process.

The BWM DSS model presented in this book was developed in line with the BWM Convention and related guidelines, and further tested using real condition data from the Port of Koper (Slovenia). The geographical, hydrological, meteorological, important resources, shipping patterns, shipping safety and regulatory regimes were considered in the DSS model and analysed in relation with the effectiveness of the BWM. The results show some important advantages and effectiveness of the selective approach supported by the presented BWM DSS model, especially regarding problems that arise from proximity to the shore and limited water depths on existing vessel routes, as well as the length of voyages, demonstrated to be the main limiting factors for effective BWE. In such cases, implementing the blanket approach would practically mean that vessels would need to 'do nothing' to be compliant with the BWM Convention, until the D-2 standard enters into force and BWMSs are installed on vessels. The blanket approach, supported with a designated BWEA with requiring all vessels to use it as an additional measure, shows some potential, especially because it is relatively simple to implement. However, different vessels would be unnecessarily exposed to additional BWM measures. BWM DSS shows also different advantages when the D-2 standard will be in place, especially to support compliance monitoring and enforcement, and in cases when a BWM was not conducted satisfactory.

The BWM DSS model was designed to be transparent, adaptable and reviewable, if necessary. This yields the potential to be used in different parts of the world for more effective prevention of HAOP transfers via ballast waters, and concurrently to the sustainable development of the shipping industry.

Although some BWM related facts are unquestionable, issues to be clarified/solved remain. These may include:

- Our experience resulted in a sampling approach which we believe is representative of the ballast water discharged. However, future work on this subject may result in changes to this suggested sampling approach, which would need to be validated.
- Sample processing methods are available for both an indicative and a detailed analysis. Organism detection tool manufacturers have recognised the special needs to proof compliance with BWM approaches and currently new organism detection tools are under development. A testing and validation phase of these systems is required.
- Appropriate training of PSC officers is needed to address all implementation needs of the BWM Convention.
- Do the current BWM Convention requirements substantially reduce the number of new HAOP introductions or are stricter standards needed? However, this may be very difficult to document as other organism transport vectors may overlap with ballast water so that a clear identification of the responsible vector is impossible.
- Can BWMS systems be cost-efficiently enhanced in their performance to even achieve better protection, e.g., USA ballast water performance standards? Is a zero detectable organism discharge standard achievable?
- Sufficiently developed RA-based exemptions from BWM requirements are needed to address all requirements of the G7 Guidelines and the precautionary principle not to undermine the BWM Convention purpose.
- Self-funding mechanisms, such as fees and penalties, may be developed to support the implementation of all BWM Convention needs.
- The applied CME measures should be harmonised in minimum on a regional level to avoid that vessels are compliant in one port, but not in another, because different methods and approaches are implemented to proof compliance.
- As agreed by IMO, the BWM Convention and its guidelines may have to be reviewed as new knowledge developed and experience was gained. However, such a review process may only be initiated after its entry into force.

By summarizing BWM related aspects from many disciplines and by providing insights of latest research results and regulatory aspects we hope that this book clarified many ballast water issues. We also believe that the proposed RA and DSS approaches will reduce the BWM burden of ships by providing at the same time an adequate protection from HAOP introductions by ballast water.

Although some issues raised above are critical, our view is that the BWM Convention should enter into force soon to reduce the risks of future ballast water mediated species introductions.

Index

M. David, S. Gollasch (eds.), *Global Maritime Transport and Ballast Water Management*, Invading Nature - Springer Series in Invasion Ecology 8,
DOI 10.1007/978-94-017-9367-4

MIX
Papier aus verantwortungsvollen Quellen
Paper from responsible sources
FSC® C105338
FSC
www.fsc.org

If you have any concerns about our products,
you can contact us on
ProductSafety@springernature.com

In case Publisher is established outside the EU,
the EU authorized representative is:
Springer Nature Customer Service Center GmbH
Europaplatz 3, 69115 Heidelberg, Germany

Printed by Libri Plureos GmbH
in Hamburg, Germany